Elimination Methods in Polynomial Computer Algebra

Mathematics and Its Applications

Volume 448

Elimination Methods in Polynomial Computer Algebra

by

Valery Bykov

Computer Center,
Siberian Branch of the Russian Academy of Sciences,
Krasnoyarsk, Russia

Alexander Kytmanov

Krasnoyarsk State University,
Krasnoyarsk, Russia

and

Mark Lazman

Institute of Catalysis,
Siberian Branch of the Russian Academy of Sciences,
Krasnoyarsk, Russia

with

Mikael Passare (ed.)

Department of Mathematics,
Stockholm University,
Stockholm, Sweden

SPRINGER SCIENCE+BUSINESS MEDIA, B.V.

ISBN 978-94-010-6230-5 ISBN 978-94-011-5302-7 (eBook)
DOI 10.1007/978-94-011-5302-7

This is a completely revised and updated translation
of the original Russian work *Methods of Elimination in
Computer Algebra of Polynomials* (in Russian)
Novosibirsk, Nauka ©1991
Translation by A. Kytmanov.

Printed on acid-free paper

Table of Contents

INTRODUCTION

The subject of this book is connected with a new direction in mathematics, which has been actively developed over the last few years, namely the field of polynomial computer algebra, which lies at the intersection point of algebra, mathematical analysis and programming.

There were several incentives to write the book. First of all, there has lately been a considerable interest in applied nonlinear problems characterized by multiple stationary states. Practical needs have then in their turn led to the appearance of new theoretical results in the analysis of systems of nonlinear algebraic equations. And finally, the introduction of various computer packages for analytic manipulations has made it possible to use complicated elimination-theoretical algorithms in practical research. The structure of the book is accordingly represented by three main parts: Mathematical results driven to constructive algorithms, computer algebra realizations of these algorithms, and applications.

Nonlinear systems of algebraic equations arise in diverse fields of science. In particular, for processes described by systems of differential equations with a polynomial right hand side one is faced with the problem of determining the number (and location) of the stationary states in certain sets. This then leads to problems in polynomial computer algebra: The construction of algorithms for determining the number of roots of the given equation system in different sets; the determination of the roots themselves; the elimination of some of the unknowns from the system; the construction of a "triangular" system (Gröbner basis) equivalent to the given system. To deal with such questions one naturally has to develop methods for analytic computer manipulations.

The first chapter of this monograph is of an introductory character, and it basically consists of a presentation of well-known material.

The main new mathematical results are concentrated to the second chapter. Here we describe a modified elimination method proposed by Aizenberg. The idea of this method is to use a multidimensional residue formula to express the sum of the values of an arbitrary polynomial in the roots of a given algebraic system (without having to first determine these roots). We provide formulas for finding such sums for various systems of equations, and we show how to construct intermediate polynomials (analogues of resultants) for this method. All these methods and formulas can easily be turned into algorithms. In particular, for systems of three equations the calculations are carried out completely. A formula for finding the logarithmic derivative of the resultant of a system of algebraic equations is also obtained, and it allows one to compute the resultant itself as the sum of local residues (and hence to express the resultant in closed form). As a corollary we get a formula expressing the logarithmic derivative of the resultant through a sum of logarithmic derivatives of subsystems of the given system. These formulas allows us to write down the coefficients of the resultant in final form. Multidimensional analogues of the Newton recursion relations are given. It turns out that there are many such formulas, and that they are connected with different types of systems of equations. Moreover, some of the coefficients of these systems cannot be determined in terms of power sums of roots, so it is necessary to consider also "pseudopower" sums. Various other problems related to the modified elimination method are also discussed: The determination of resultants with respect to certain groups of variables, and the determination of the number of real roots in domains with real analytic boundary.

Computer realizations of the proposed methods are given in the fourth chapter. As our basic software for analytic manipulations we have chosen the system REDUCE. Two important problems of polynomial computer algebra are emphasized: The construction of resultants and computer realizations of elimination methods. In the fairly voluminous applications provided at the end of the book we have included the texts of the programs that we have developed together with some examples of results obtained by letting them run.

The applications of the theory (the third chapter and numerous examples) are within the field of mathematical kinetics, where one typically has to deal with systems of nonlinear algebraic equations. These systems provide a quite broad and practically significant class of examples. It should be mentioned that it was this type of applied problems that originated the team–work of the authors. To a considerable extent our new theoretical results and their computer realizations were obtained as answers to questions arising in the analysis of chemical kinetic equations. As a result of our subsequent work there appeared algorithms and programs operating on systems with symbolic coefficients of the most general form. Our experience with the use of these mathematical and software tools has made us confident that the methods of polynomial computer algebra will gain further development and be of a wide practical use in diverse fields of science and technology.

Since the methods of computer algebra are currently being rapidly developed, it is hardly possible to give a complete account of the subject, or even of the branch related to systems of algebraic equations. However, the authors do hope that their experience will prove to be of some use for the interested reader. We shall be grateful for any comments and requests regarding the subject of this monograph.

The material presented in this book was used by the authors in a special courses of lectures given at the Krasnoyarsk State University.

The authors consider it a pleasant duty to express their gratitude toward Lev Aizenberg for his support during this research and also toward Grigory Yablonskii for a long and fruitful cooperation. On the basis of theoretical results obtained by Aizenberg we were able to considerably modify the elimination methods for systems of nonlinear algebraic equations, and Yablonskii was the one who initiated the use of new mathematical methods in chemical kinetics, and the authors owe to him the physical and chemical interpretation of a number of joint results.

The authors are grateful to Leonid Bystrov, Andrey Levitskii, Alexander Semenov, Vassily Stelletskii, Leonid Provorov and Vladimir Topunov for their invaluable help with the computer realization of this work, and also to Alexander Gorban, Vyacheslav Novikov, August Tsikh and Alexander Yuzhakov for helpful discussions in the course of writing the book.

We use a double numbering of formulas, the first number denotes the number of the section, and the second indicates the number of the formula within the section. The applications at the end are numbered without reference to particular sections.

PREFACE TO THE ENGLISH EDITION

Five years have passed since the publication of our book in Russian. In this period some new results have been obtained, and a portion of the material has become obsolete. Therefore, apart from the translation work it was also necessary to revise some sections of the book. Let us briefly describe these alterations.

In Chapter 2, Section 9, the Theorem 9.1 and its corollaries are given in the more general situation of projective space and under milder restrictions on the system of equations. In Section 11 we have added some results (Theorem 11.5 and its corollaries) concerning nonlinear and non-algebraic systems of equations.

In Chapter 3, Section 14, Theorem 14.7 is proved completely. In Section 15 we give a general survey of the applications of our elimination method to systems of algebraic and non-algebraic equations arising in chemical kinetics.

The last chapter has been subject to a thorough revision. This is connected with the appearance of new computer algebra systems. In the russian edition we used the system REDUCE, but for the english version we have chosen the `Maple` system. All software is designed for this system, except subsections 18.3 and 19.3, where the previous variant is maintained.

We express our thanks to Tatyana Osetrova for software realization of a number of algorithms and to Sergey Znamenskii for consultations regarding the `Maple` system.

Special gratitude is due to our colleague, Professor Mikael Passare, who agreed to undertake the editing of our preliminary translation. This has helped to substantially improve the text. The responsibility for any remaining errors, mathematical and linguistic, is of course entirely ours.

The MAPLE software that is needed to follow the examples described in the fourth chapter is available upon request from the authors at kytmanov@math.kgr.krasnoyarsk.su.

During preparation of the manuscript the first author was partially supported by RFFI, grant 93-12-00596, and the second author was partially supported by RFFI, grant 96-01-00080.

PREFACE BY THE ENGLISH EDITOR

The marvellous discovery, that insight into problems on the real line can be gained by exploring the complex plane, is as old as the complex numbers themselves. Methods from complex analysis are classical tools in countless applications. It is however only fairly recently that ideas from several complex variables have begun to find their way into applied mathematics. So have pluricomplex methods lately been introduced in such diverse fields as theoretical physics, signal processing, medical tomography and economics.

The present monograph, written by a well composed troika of one pure mathematician, one computer scientist and one chemist, provides not only a thorough account of the use of multidimensional residues in polynomial algebra, together with computer implemenations — but also applications of these methods to equilibrium problems in chemical kinetics.

Classical elimination theory serves well enough for solving small systems of polynomial equations, but for slightly larger systems it is only of theoretical use. A more applicable method, which has become standard in modern computer algebra, is that of Gröbner bases. These allow one to avoid superfluous solutions and to bring the system into triangular form.

In this book the central approach is based on logarithmic residues in several variables, the theory of which to a substantial extent has been developed by the Krasnoyarsk pluricomplex school, see for instance the monograph by Aizenberg and Yuzhakov (American Mathematical Society 1983) and Tsikh (American Mathematical Society 1992).

Just as in one variable the method of residues combines elegance with efficiency. To be more precise, using a multidimensional analogue of the argument principle one obtains explicit integral formulas for the number of common roots of the given system, and also for the power sums of these roots. This substantially simplifies the computation of the corresponding resultants. In conclusion it is maybe worth remarking that similar methods occurred already in Poincaré's proof of the Weierstrass preparation theorem, and hence can be said to lie at very heart of local analytic geometry.

LIST OF SYMBOLS

$\mathbf{A}$ is the matrix of a linear transformation
a_{jk} is an entry of the matrix $\mathbf{A}$
$b_{\pm i}$ are kinetic coefficients
$cf(g, r)$ is the coefficient of the monomial r in the polynomial g
$\mathbf{C}$ is the field of complex numbers
$\mathbf{C}^n$ is the n-dimensional complex space
$\mathbf{CP}^n$ is the n-dimensional complex projective space
D is a domain in $\mathbf{C}^n$
∂D is the boundary of D
$\det \mathbf{A}$ is the determinant of the matrix $\mathbf{A}$
$\mathbf{J_f}(\mathbf{a})$ is the Jacobian of the mapping $\mathbf{f}$ at the point $\mathbf{a}$
$E_{\mathbf{f}}$ is the set of roots (or zeros) of the system $\mathbf{f} = \mathbf{0}$
i is the imaginary unit
$P(\mathbf{z})$, $Q(\mathbf{z})$ are polynomials
$\operatorname{res}_a f$ is the residue of the function f at the point a
$\mathbf{R}$ is the field of real numbers
$\mathbf{R}^n$ is the n-dimensional real space
$R(P, Q)$ is the resultant of the polynomials P, Q
$R(w)$ is the resultant in w of a system of stationary state equations
S_k are power sums
$U_r(\mathbf{a})$ is a ball of radius r with center at $\mathbf{a} \in \mathbf{C}^n$
w is a stationary rate of reaction
Z_i is an intermediate substance
$z = x + iy$ is a complex number, $x, y \in \mathbf{R}$
$||\boldsymbol{\alpha}|| = \alpha_1 + \ldots + \alpha_n$
$\boldsymbol{\alpha}! = \alpha_1! \ldots \alpha_n!$
$\boldsymbol{\alpha}^m = \alpha_1^m \ldots \alpha_n^m$
α_{ij}, β_{ij} are stoichiometric coefficients
γ is a smooth curve with given orientation
ν_i is the stoichiometric number at the i-th step
$\mathfrak{M}[Q(\mathbf{z})]$ is the functional computing the coefficient of a given monomial in the numerator of $Q(\mathbf{z})$
$<_T$ is a symbol for the ordering of the variables
$\underset{F}{\rightarrow}$ denotes reduction modulo the ideal F
□ is the end of a proof
Algebra 0.5, ANALITIK, ARAP, REFAL, SIRIUS, SRM, ALTRAN, FORMAC, MACSYMA, muMATH, Maple, SACH and Mathematica are systems for analytic manipulations on the computer
GB is a Gröbner basis
GCD is the greatest common divisor
s.s. is a steady state
SP is a S-polynomial

CHAPTER 1

BASIC MATHEMATICAL FACTS

1. The logarithmic residue

We recall some facts from complex analysis. All assertions and definitions can be found in any textbook on function theory in one complex variable (see, for example, [**66, 127, 134**]).

Let **C** be the field of complex numbers. The elements $z \in \mathbf{C}$ have the form $z = x + iy$, where x, y are real numbers ($x, y \in \mathbf{R}$) and i is the imaginary unit ($i^2 = -1$);

$$|z| = (x^2 + y^2)^{1/2}, \ \bar{z} = z - iy, \ \operatorname{Re} z = x, \ \operatorname{Im} z = y.$$

We can also represent a complex number in trigonometrical form: $z = r(\cos\varphi + i\sin\varphi)$ where $r = |z|$ and $\varphi = \operatorname{Arg} z$. The argument of a complex number is defined up to a multiple of 2π. By means of the Euler formula a complex number z can be represented in exponential form: $z = re^{i\varphi}$.

The topology in the field **C** is given by the metric $d(z, w) = |z - w|$. As usual a domain $D \subset \mathbf{C}$ is a non-empty open connected set and a compact set $K \subset \mathbf{C}$ is a closed bounded set. We denote by $C(F)$ the class of complex-valued continuous functions given on a set F.

Recall that a complex-valued function f is holomorphic in a domain D if for every point $z_0 \in D$ there exists a power series of the form

$$\sum_{n=0}^{\infty} c_n (z - z_0)^n$$

which converges to f in some disk of radius $r > 0$ with centre at z_0. Later on we shall denote such a disk by $U_r(z_0)$.

For example, any polynomial

$$P(z) = a_0 + a_1 z + \ldots + a_n z^n$$

($a_j \in \mathbf{C}$, $j = 0, 1, \ldots, n$) is a holomorphic function in **C**, and any rational function $R(z) = P(z)/Q(z)$ (P and Q are polynomials) is holomorphic in each domain D in which the denominator $Q \neq 0$. The functions $f(z) = |z|$ and $\varphi(z) = \bar{z}$ are not holomorphic in any domain D. The elementary functions defined for real values of the argument can be extended to holomorphic functions (in some domains) by means of their Taylor series expansion. For example, the functions:

$$e^z = \sum_{n=0}^{\infty} \frac{z^n}{n!}, \quad \sin z = \sum_{n=0}^{\infty} \frac{(-1)^n z^{2n+1}}{(2n+1)!}, \quad \cos z = \sum_{n=0}^{\infty} \frac{(-1)^n z^{2n}}{(2n)!}.$$

Moreover, these series converge on all of **C**.

A fundamental role in the theory of functions is played by the Cauchy theorem. Let a function f be holomorphic in a bounded domain D and continuous on its closure $\overline{D}$. Suppose the boundary ∂D of the domain D consists of a finite number of smooth curves (every curve γ contained in ∂D is oriented such that by passing along the curve γ the domain D remains on the left). If

$$f(z) = u(x, y) + iv(x, y)$$

where u, v are real-valued functions ($z = x + iy$) then

$$f(z)dz = (u + iv)(dx + idy) = udx - vdy + i(udy + vdx).$$

Hence we can define

$$\int_\gamma f(z)dz$$

(γ is a smooth curve with given orientation) as

$$\int_\gamma f(z)dz = \int_\gamma (udx - vdy) + i\int_\gamma (udy + vdx)$$

(a line integral of the second type).

THEOREM 1.1 (Cauchy). *Under the above assumptions the integral*

$$\int_{\partial D} f(z)dz = 0.$$

The Cauchy theorem is actually a corollary to the following fact: The holomorphic function f satisfies the Cauchy–Riemann equations

$$\frac{\partial u}{\partial x} = \frac{\partial v}{\partial y}, \quad \frac{\partial u}{\partial y} = -\frac{\partial v}{\partial x} \qquad (f = u + iv).$$

If we define the formal derivatives of the function f by

$$\frac{\partial f}{\partial z} = \frac{1}{2}\left(\frac{\partial f}{\partial x} - i\frac{\partial f}{\partial y}\right) = \frac{1}{2}\left(\frac{\partial u}{\partial x} + \frac{\partial v}{\partial y}\right) + \frac{1}{2}i\left(-\frac{\partial u}{\partial y} + \frac{\partial v}{\partial x}\right),$$

$$\frac{\partial f}{\partial \bar{z}} = \frac{1}{2}\left(\frac{\partial f}{\partial x} + i\frac{\partial f}{\partial y}\right) = \frac{1}{2}\left(\frac{\partial u}{\partial x} - \frac{\partial v}{\partial y}\right) + \frac{1}{2}i\left(\frac{\partial u}{\partial y} + \frac{\partial v}{\partial x}\right),$$

then the Cauchy–Riemann equations can be written more concisely:

$$\frac{\partial f}{\partial \bar{z}} = 0.$$

Let a function f be holomorphic in the disk $U_R(a)$ with center at the point a except (maybe) the point a, and let γ_r be the boundary of $U_r(a)$ where $0 < r < R$, $\gamma_r \subset U_R(a)$. The orientation of γ_r is taken to be counter-clockwise.

The (local) residue

$$\operatorname*{res}_a f$$

of the function f at the point a is the expression

$$\operatorname*{res}_a f = \frac{1}{2\pi i}\int_{\gamma_r} f(z)dz.$$

The Cauchy theorem shows that $\operatorname*{res}_a f$ does not depend on the radius r.

EXAMPLE 1.1. Let us find

$$\operatorname*{res}_a (z-a)^n,$$

where n is an integer ($n \in \mathbf{Z}$). Consider the following parameterization of $\gamma_r : z = a + re^{i\varphi}$, $0 \leq \varphi \leq 2\pi$. Then $(z-a)^n = r^n e^{in\varphi}$ and $dz = ire^{i\varphi} d\varphi$ on γ_r (we use the exponential form of a complex number z). Therefore

$$\int_{\gamma_r} (z-a)^n dz = ir^{n+1} \int_0^{2\pi} e^{i(n+1)\varphi} d\varphi =$$

$$= ir^{(n+1)} \int_0^{2\pi} (\cos(n+1)\varphi + i \sin(n+1)\varphi) d\varphi = 0,$$

if $n+1 \neq 0$. For $n = -1$ we have

$$\int_{\gamma_r} \frac{dz}{z-a} = i \int_0^{2\pi} d\varphi = 2\pi i.$$

Hence $\operatorname*{res}_a (z-a)^n = 0$ for $n \neq -1$ and $\operatorname*{res}_a (z-a)^{-1} = 1$. The appearance of the factor $1/(2\pi i)$ in the definition of the residue stems from the fact that

$$\int_\gamma \frac{dz}{z-a} = 2\pi i.$$

Let now $f(z)$ have the Laurent series expansion

$$f(z) = \sum_{n=-\infty}^{+\infty} c_n (z-a)^n$$

in $U_r(a)$. Using Example 1.1 we then get

$$\operatorname*{res}_a f = \frac{1}{2\pi i} \int_{\gamma_r} f(z) dz = \frac{1}{2\pi i} \sum_{n=-\infty}^{+\infty} c_n \int_{\gamma_r} (z-a)^n dz = c_{-1}.$$

It follows that the residue

$$\operatorname*{res}_a f$$

is equal to the coefficient c_{-1} of z^{-1} in the Laurent expansion of f centered at the point a. This property is a basic one for the computation of residues.

The logarithmic residue is a special case of the residue. It is closely connected with the number of zeros of functions. Let f be a holomorphic function in a disk $U_R(a)$, and let a be a root of the equation $f(z) = 0$ (i.e., let a be a zero of the function f). If $f(z) = (z-a)^n \varphi(z)$, where φ is holomorphic in $U_R(a)$ with $\varphi(a) \neq 0$, then the number n is called the multiplicity of the zero a of the function f. If $n = 1$ then the zero a is called simple (it is characterized by the condition $f'(a) \neq 0$). Let the function h be holomorphic in $U_R(a)$. The logarithmic residue of the function h at the point a is given by the integral

$$\frac{1}{2\pi i} \int_{\gamma_r} \frac{h df}{f} = \frac{1}{2\pi i} \int_{\gamma_r} \frac{h f' dz}{f} = \operatorname*{res}_a \frac{h f'}{f},$$

where, as before, γ_r is a circle of radius r and with center at the point a lying in $U_R(a)$. The name "logarithmic" reflects the fact that

$$\frac{f'}{f}$$

is the logarithmic derivative $(\ln f)'$ of the function f.

If a is a zero of multiplicity n of a function f then $f(z) = (z-a)^n\varphi(z)$ and $\varphi(a) \neq 0$. Therefore

$$\frac{f'(z)}{f(z)} = \frac{n(z-a)^{n-1}\varphi(z) + (z-a)^n\varphi'(z)}{(z-a)^n\varphi(z)} = \frac{n}{z-a} + \frac{\varphi'(z)}{\varphi(z)};$$

$$\operatorname*{res}_a \frac{hf'}{f} = \frac{n}{2\pi i}\int_{\gamma_r}\frac{h(z)dz}{(z-a)} + \frac{1}{2\pi i}\int_{\gamma_r}\frac{h(z)\varphi'(z)dz}{\varphi(z)}.$$

Since $\varphi(a) \neq 0$ we can assume $\varphi \neq 0$ in $U_R(a)$. Hence the function $h\varphi'/\varphi$ is holomorphic in $U_R(a)$, so by the Cauchy theorem

$$\int_{\gamma_r}\frac{h\varphi'}{\varphi}dz = 0.$$

Expanding h in a Taylor series in $U_R(a)$ we obtain

$$h(z) = \sum_{n=0}^{\infty} c_n(z-a)^n \quad (c_0 = h(a)).$$

Therefore

$$\frac{h(z)}{z-a} = \sum_{n=0}^{\infty} c_n(z-a)^{n-1} = \frac{h(a)}{z-a} + \sum_{n=0}^{\infty} c_{n+1}(z-a)^n$$

and hence (see example 1.1)

$$\int_{\gamma_r}\frac{h(z)dz}{z-a} = h(a)2\pi i.$$

Thus

$$\operatorname*{res}_a \frac{hf'}{f} = nh(a).$$

In particular,

$$\operatorname*{res}_a \frac{f'}{f} = n$$

for $h \equiv 1$.

From here, using the Cauchy theorem, we get the theorem on logarithmic residue. Let D be a domain bounded by a finite set of smooth curves (oriented in the same way as in the Cauchy theorem). Let the function h be holomorphic in D and continuous on $\overline{D}$, and let the function f be holomorphic in a neighborhood of $\overline{D}$. Assume moreover that $f \neq 0$ on ∂D. Then f has a finite set of zeros in D. We denote that set by E_f. If a is a zero of multiplicity n then we assume that n copies of a are contained in E_f.

THEOREM 1.2 (on logarithmic residue). *The formula*

$$\frac{1}{2\pi i}\int_{\partial D} h\frac{df}{f} = \sum_{a\in E_f} \operatorname{res}_a \frac{hf'}{f} = \sum_{a\in E_f} h(a) \tag{1.1}$$

holds.

COROLLARY 1.1. *If N is the number of zeros of the function f in the domain D (every zero is counted according to its multiplicity), then*

$$N = \frac{1}{2\pi i}\int_{\partial D} \frac{df}{f}. \tag{1.2}$$

As an application of this corollary it is not difficult to derive the Rouché theorem. Let the domain D be the same as in the Cauchy theorem.

THEOREM 1.3 (Rouché). *If f and g are holomorphic functions in a neighborhood of $\overline{D}$, such that for all $z \in \partial D$*

$$|f(z)| > |g(z)|,$$

then f and $f+g$ have the same number of zeros in D.

Proof. Consider the function $f + tg$ where $t \in [0,1]$. Under the condition in the theorem

$$|f+tg| \geq |f| - t|g| > 0$$

on ∂D. Hence the function $f+tg \neq 0$ on ∂D and therefore this function has a finite number of zeros in D. By corollary 1.1 the integral

$$\frac{1}{2\pi i}\int_{\partial D} \frac{d(f+tg)}{f+tg}$$

is an integer. On the other hand, this integral is a continuous function of $t \in [0,1]$. Hence

$$\int_{\partial D} \frac{d(f+tg)}{f+tg} \equiv \text{const}$$

and

$$\int_{\partial D} \frac{df}{f} = \int_{\partial D} \frac{d(f+g)}{f+g}. \quad \square$$

EXAMPLE 1.2. From the Rouché theorem it is easy to obtain the "Fundamental Theorem of Algebra": Every polynomial $P(z)$ of degree n has exactly n complex zeros. Indeed, let

$$P(z) = a_0 z^n + \ldots + a_n, \quad f(z) = a_0 z^n$$

and

$$g(z) = a_1 z^{n-1} + \ldots + a_n.$$

Then $P(z) = f(z) + g(z)$. By the lemma on the modulus of the leading term we have $|f(z)| > |g(z)|$ if $|z| > R$ (here R is a sufficiently large number), and hence functions f and $f+g$ have the same number of zeros in D.

It follows that the inequality $|f(z)| > |g(z)|$ holds on γ_R. Hence by the Rouché theorem the functions f and P have the same number of zeros in the disk $U_R(0)$. But the function $f(z) = a_0 z^n$ has exactly n zeros in $U_R(0)$. (The point $z = 0$ is a root of order n of the equation $f(z) = 0$).

2. The Newton recursion formulas

The Newton recursion formulas connect the coefficients of a polynomial to the power sums of its zeros. These formulas can be found, for example, in [**26**, **97**]. The proof presented here is new. We also consider a generalization of the Newton formulas to a wider class of functions.

Let a polynomial $P(z)$ have the form

$$P(z) = z^n + b_1 z^{n-1} + \ldots + b_{n-1}z + b_n.$$

Denote by $z_1, z_2, \ldots, z_n$ its zeros (some of them can be repeated). The power sums S_k are defined as follows:

$$S_k = z_1^k + \cdots + z_n^k.$$

THEOREM 2.1 (Newton recursion formulas). *The power sums S_k and the coefficients b_j are related in the following way:*

$$\begin{aligned} S_j + S_{j-1}b_1 + S_{j-2}b_2 + \ldots + S_1 b_{j-1} + jb_j = 0, \quad &\text{if } 1 \le j \le n, \\ \text{and} \quad S_j + S_{j-1}b_1 + \ldots + S_{j-n}b_n = 0, \quad &\text{if } j > n. \end{aligned} \tag{2.1}$$

Proof. We choose $R > 0$ so that all the zeros z_k of $P(z)$ lie in the disk $U_R(0)$, and we consider the integral

$$J = \int_{\gamma_R} \frac{P'(z)dz}{z^j}.$$

On the one hand

$$J = \sum_{k=1}^{n} \int_{\gamma_R} \frac{b_{n-k} k z^{k-1} dz}{z^j} = \sum_{k=1}^{n} k b_{n-k} \int_{\gamma_R} z^{k-j-1} dz.$$

Therefore, if $j > n$ or $j \le 0$ then $J = 0$, and if $1 \le j \le n$ then $J = 2\pi i b_{n-j}$ (see Example 1.1). On the other hand

$$J = \int_{\gamma_R} \frac{PdP}{z^j P} = \sum_{k=0}^{n} b_{n-k} \int_{\gamma_R} z^{k-j} \frac{dP}{P}.$$

If $k - j > 0$, then by Theorem 1.2 on logarithmic residues

$$\int_{\gamma_R} z^{k-j} \frac{dP}{P} = 2\pi i S_{k-j}.$$

If $k - j = 0$, then

$$\int_{\gamma_R} \frac{dP}{P} = (2\pi i)n.$$

For the case $k - j < 0$ we make the change of variables $z = 1/w$ in the integral

$$\int_{\gamma_R} z^{k-j} \frac{dP}{P}.$$

The circle γ_R is then transformed to $\gamma_{1/R}$ with opposite orientation,

$$P(z) \longrightarrow P(1/w) = \frac{1}{w^n} \left(\sum_{j=0}^{n} b_{n-j} w^{n-j} \right) \quad (b_0 = 1),$$

$$P'(z) \longrightarrow P'\left(\frac{1}{w}\right) = \frac{1}{w^{n-1}}\left(\sum_{j=0}^{n} jb_{n-j}w^{n-j}\right),$$

$$dz \to -\frac{1}{w^2}dw.$$

Hence

$$\int_{\gamma_R} z^{k-j}\frac{dP}{P} = \int_{\gamma_{1/R}} \frac{w^{j-k-1}\sum\limits_{j=1}^{n} jb_{n-j}w^{n-j}}{\sum\limits_{j=0}^{n} b_{n-j}w^{n-j}}dw.$$

Since all zeros of the polynomial $P(z)$ lie in the disk $U_R(0)$, the polynomial

$$w^nP\left(\frac{1}{w}\right) \neq 0$$

in the disk $U_{1/R}(0)$. Since $j-k-1 \geq 0$, the integrand in the last integral is holomorphic in $U_{1/R}(0)$, so by the Cauchy theorem this integral vanishes. Thus we get (for $1 \leq j \leq n$)

$$jb_{n-j} = nb_{n-j} + \sum_{k=j+1}^{n} b_{n-k}S_{k-j},$$

i.e.

$$(n-j)b_{n-j} + b_{n-j-1}S_1 + b_{n-j-2}S_2 + \ldots + b_1S_{n-j-1} + S_{n-j} = 0$$

(we recall that $b_0 = 1$). If $j \leq 0$, then

$$\sum_{k=0}^{n} b_{n-k}S_{k-j} = 0,$$

i.e.

$$b_nS_{-j} + b_{n-1}S_{1-j} + \ldots + b_1S_{n-1-j} + S_{n-j} = 0. \quad \square$$

The formulas (2.1) allow one to find the power sums of the zeros with the help of the coefficients of the polynomial P. And vice versa, they also permit one to find the coefficients of $P(z)$ by means of the power sums S_k.

EXAMPLE 2.1. Let $P(z) = z^2 + b_1z + b_2$, and let the power sums S_1 and S_2 be given. Then from (2.1) we obtain $S_1 + b_1 = 0$ for $j = 1$, i.e. $b_1 = -S_1$; and $S_2 + S_1b_1 + 2b_2 = 0$ for $j = 2$, i.e.

$$2b_2 = -S_2 - S_1b_1 = -S_2 + S_1^2, \quad b_2 = (-S_2 + S_1^2)/2.$$

These formulas actually show that for every set of numbers $S_1, \ldots, S_n$ there exists a polynomial P with given power sums $S_1, \ldots, S_n$.

The following formulas of Waring are often useful.

THEOREM 2.2 (Waring formulas). *The formulas*

$$S_k = (-1)^k \begin{vmatrix} b_1 & 1 & 0 & \ldots & 0 \\ 2b_2 & b_1 & 1 & \ldots & 0 \\ 3b_3 & b_2 & b_1 & \ldots & 0 \\ \ldots & \ldots & \ldots & \ldots & \ldots \\ kb_k & b_{k-1} & b_{k-2} & \ldots & 0 \end{vmatrix}, \tag{2.2}$$

$$b_k = \frac{(-1)^k}{k!} \begin{vmatrix} S_1 & 1 & 0 & \dots & 0 \\ S_2 & S_1 & 2 & \dots & 0 \\ S_3 & S_2 & S_1 & \dots & 0 \\ \dots & \dots & \dots & \dots & \dots \\ S_k & S_{k-1} & S_{k-2} & \dots & S_1 \end{vmatrix}, \tag{2.3}$$

hold for $1 \le k \le n$.

Proof. We multiply the second column of the determinant in (2.3) by b_1, the third by b_2 etc., and add them all to the first column. By the Newton formulas (2.1) we then get the determinant

$$\begin{vmatrix} 0 & 1 & 0 & \dots & 0 \\ 0 & S_1 & 2 & \dots & 0 \\ 0 & S_2 & S_1 & \dots & 0 \\ \dots & \dots & \dots & \dots & \dots \\ -kb_k & S_{k-1} & S_{k-2} & \dots & S_1 \end{vmatrix} = (-1)^k k! b_k.$$

In order to obtain (2.2) we must multiply the second column in (2.2) by S_1, the third by S_2 etc., and add them to the first column. Using formula (2.1) we now find

$$\begin{vmatrix} 0 & 1 & 0 & \dots & 0 \\ 0 & b_1 & 1 & \dots & 0 \\ 0 & b_2 & b_1 & \dots & 0 \\ \dots & \dots & \dots & \dots & \dots \\ -S_k & b_{k-1} & b_{k-2} & \dots & b_1 \end{vmatrix} = (-1)^k S_k.$$

In what follows we will need the recursion formulas for a wider class of functions than just polynomials, and to deal with such functions we have to consider power sums S_k with $k < 0$.

The proof of these generalized formulas could be carried out as for the formulas (2.1). But we prefer to take another approach.

Let the function $f(z)$ be holomorphic in a neighborhood $U_R(0)$ and $f(z) \neq 0$ in $U_R(0)$. Then we can consider the function $g(z) = \ln f(z)$ by choosing the principal branch of the logarithm (i.e. $\ln 1 = 0$). The function $g(z)$ is also holomorphic in $U_R(0)$. Suppose that the Taylor expansions for the functions f and g have the form

$$f(z) = b_0 + b_1 z + \ldots + b_k z^k + \ldots,$$

$$g(z) = \ln f(z) = a_0 + a_1 z + \ldots + a_k z^k + \ldots,$$

$$b_0 \neq 0 \quad \text{and} \quad \ln b_0 = a_0.$$

LEMMA 2.1. *The recursion relations*

$$\sum_{j=0}^{k-1} (k-j) b_j a_{k-j} = k b_k, \quad k \ge 1, \tag{2.4}$$

hold. Moreover,

$$a_k = \frac{(-1)^{k-1}}{kb_0^k} \begin{vmatrix} b_1 & b_0 & 0 & \dots & 0 \\ 2b_2 & b_1 & b_0 & \dots & 0 \\ 3b_3 & b_2 & b_1 & \dots & 0 \\ \dots & \dots & \dots & \dots & \dots \\ kb_k & b_{k-1} & b_{k-2} & \dots & b_1 \end{vmatrix}, \tag{2.5}$$

$$b_k = \frac{b_0}{k!} \begin{vmatrix} a_1 & -1 & 0 & \dots & 0 \\ 2a_2 & a_1 & -2 & \dots & 0 \\ \dots & \dots & \dots & \dots & \dots \\ ka_k & (k-1)a_{k-1} & (k-2)a_{k-2} & \dots & a_1 \end{vmatrix}. \tag{2.6}$$

Proof. Since

$$f(z) = e^{g(z)} = b_0 + b_1 z + \ldots + b_k z^k + \ldots,$$

it follows that

$$f(z)g'(z) = b_1 + 2b_2 z + \ldots + kb_k z^{k-1} + \ldots.$$

From this we get

$$\sum_{j=0}^{\infty} b_j z^j \sum_{s=1}^{\infty} s a_s w^{s-1} = \sum_{k=1}^{\infty} kb_k z^{k-1}.$$

Multiplying the series and equating coefficients at equal degrees in z we get (2.4). The formulas (2.5) and (2.6) are obtained as the Waring formulas, but instead of (2.1) we must use (2.4). □

Now suppose that the function $f(z)$ has the form

$$f(z) = \frac{h(z)}{u(z)},$$

where $h(z)$, $u(z)$ are entire functions represented as infinite products

$$h(z) = b_0 \prod_{j=1}^{\infty} \left(1 - \frac{z}{w_j}\right), \quad u(z) = \prod_{j=1}^{\infty} \left(1 - \frac{z}{c_j}\right),$$

with $w_j \neq 0$, $c_j \neq 0$, $j = 1, 2, \ldots$, the series

$$\sum_{j=1}^{\infty} \frac{1}{|c_j|}$$

and

$$\sum_{j=1}^{\infty} \frac{1}{|w_j|}$$

being convergent. (The convergence of these series ensures the convergence of the infinite products.) Denote by σ_k the power sums of the form

$$\sigma_k = \sum_{j=1}^{\infty} \frac{1}{w_j^k} - \sum_{j=1}^{\infty} \frac{1}{c_j^k}, \quad k \geq 1.$$

These series also converge, and we have

$$\frac{f'}{f} = \sum_{j=1}^{\infty} \frac{1}{z - w_j} - \sum_{j=1}^{\infty} \frac{1}{z - c_j}.$$

From here we see that

$$k! a_k = \frac{d^k \ln f}{dz^k}(0) = -(k-1)! \sigma_k,$$

i.e. $ka_k = -\sigma_k$, and (using Lemma 2.1) we arrive at the following theorem:

THEOREM 2.3. (NEWTON AND WARING RECURSION FORMULAS FOR MEROMORPHIC FUNCTIONS) *The formulas*

$$\sum_{j=0}^{k-1} b_j \sigma_{k-j} + k b_k = 0, \quad k \geq 1, \tag{2.7}$$

hold. Moreover,

$$\sigma_k = \frac{(-1)^k}{b_0^k} \begin{vmatrix} b_1 & b_0 & 0 & \dots & 0 \\ 2b_2 & b_1 & b_0 & \dots & 0 \\ 3b_3 & b_2 & b_1 & \dots & 0 \\ \dots & \dots & \dots & \dots & \dots \\ kb_k & b_{k-1} & b_{k-2} & \dots & b_1 \end{vmatrix},$$

$$b_k = \frac{(-1)^k}{k!} b_0 \begin{vmatrix} \sigma_1 & 1 & 0 & \dots & 0 \\ \sigma_2 & \sigma_1 & 2 & \dots & 0 \\ \dots & \dots & \dots & \dots & \dots \\ \sigma_k & \sigma_{k-1} & \sigma_{k-2} & \dots & \sigma_1 \end{vmatrix}.$$

In particular these formulas are valid for rational functions and for polynomials. If f is a polynomial then formula (2.7) is actually (2.1), but in (2.7) $\sigma_k = S_{-k}$ and the orders of the coefficients of f in (2.1) and (2.7) are opposite.

EXAMPLE 2.2. Consider the function

$$f(z) = \frac{\sin\sqrt{z}}{\sqrt{z}} = \sum_{k=0}^{\infty} \frac{(-1)^k z^k}{(2k+1)!} = \prod_{k=1}^{\infty} \left(1 - \frac{z}{\pi^2 k^2}\right).$$

Then

$$b_k = \frac{(-1)^k}{(2k+1)!},$$

$$\sigma_k = \frac{1}{\pi^{2k}} \sum_{n=1}^{\infty} \frac{1}{n^{2k}} = \frac{1}{\pi^{2k}} \zeta(2k),$$

where $\zeta(w)$ is the Riemann ζ-function.

The formulas (2.7) thus provide a relation between b_k and $\zeta(2k)$, in particular, they permit us to calculate $\zeta(2k)$. Taking into account that $\zeta(2k)$ are also connected to the Bernoulli numbers

$$B_k = \frac{2(2k)!}{(2\pi)^{2k}} \zeta(2k) = \frac{(2k)!}{2^{2k-1}} \sigma_k,$$

we find the formula

$$B_k = \frac{(-1)^k(2k)!}{2^{2k-1}} \begin{vmatrix} -\frac{1}{3!} & 1 & 0 & \dots & 0 \\ \frac{2}{5!} & -\frac{1}{3!} & 1 & \dots & 0 \\ \dots & \dots & \dots & \dots & \dots \\ \frac{k(-1)^k}{(2k+1)!} & \frac{(-1)^{k-1}}{(2k-1)!} & \dots & \dots & -\frac{1}{3!} \end{vmatrix}.$$

3. Localization theorems for the real zeros of a polynomial

We describe some methods of counting the real zeros of a real polynomial on a real interval: The Hermite method, the Sturm method, the Descartes sign rule, and the Budan–Fourier theorem.

These methods can be found in [**75, 97**], and they are further developed in [**92, 94**].

At first we consider the Hermite method, or the method of quadratic forms, for determining the distinct real roots of a polynomial P with real coefficients. We begin by recalling some concepts from the theory of quadratic forms.

Let

$$T(x) = \sum_{j,k=1}^{n} a_{jk} x_j x_k$$

be a quadratic form. Its matrix

$$\mathbf{A} = ||a_{jk}||_{j,k=1}^{n}$$

is symmetric (i.e. $a_{jk} = a_{kj}$ for all k, j), and type entries a_{jk} are real. The rank r of the quadratic form $T(x)$ is equal to the rank of the matrix $\mathbf{A}$. The quadratic form $T(x)$ can be reduced to a sum of independent squares by various methods, i.e. we can represent $T(x)$ in the form

$$T(x) = \sum_{j=1}^{r} a_j y_j^2, \quad \text{where } a_j \in \mathbf{R} \quad \text{and } y_j = \sum_{k=1}^{n} c_{kj} x_k,$$

the linear forms $y_1, \dots, y_r$ being linearly independent. Denote by μ the number of positive coefficients a_j and by ν the number of negative a_j. Then the principle of inertia for quadratic forms asserts that μ and ν do not depend on the particular representation of $T(x)$ as a sum of independent squares. Finally, we have $\mu + \nu = r$, and the number $\mu - \nu$ is called the signature of the quadratic form T.

Denote by

$$D_k = \det ||a_{js}||_{j,s=1}^{k}$$

the principal minors of a matrix $\mathbf{A}$.

THEOREM 3.1 (Jacobi). *Let $T(x)$ be a quadratic form with non-vanishing minors D_k, $k = 1, \dots, r$. Then the number μ of positive squares and the number ν of negative squares of $T(x)$ coincide respectively with the number P of sign repetitions and with the number V of sign changes in the sequence of numbers $1, D_1, \dots, D_r$. Moreover, the signature is given by $\sigma = r - 2V$.*

If $b_1, \ldots, b_k$ are real numbers different from zero then the number of sign repetitions in this sequence is usually denoted $P(b_1, \ldots, b_k)$, while the number of sign changes is written $V(b_1, \ldots, b_k)$. It is clear that $P + V = k - 1$. The Jacobi theorem asserts that the signature satisfies $\sigma = r - 2V(1, D_1, \ldots, D_r)$.

Suppose now that

$$P(x) = a_0 x^n + a_1 x^{n-1} + \ldots + a_n$$

is a polynomial with real coefficients, and that we want to determine the number of distinct (real) roots of P. We could immediately get rid of any multiple zeros by dividing P with the greatest common divisor of P and its derivative P' (using the Euclid algorithm), but here we choose not to perform this operation. Assuming instead that $P(z)$ has distinct zeros $\alpha_1, \ldots, \alpha_q$ of multiplicities $n_1, \ldots, n_q$, respectively, we have the factorization

$$P(z) = a_0(z - \alpha_1)^{n_1} \ldots (z - \alpha_q)^{n_q},$$

and the power sums

$$S_k = \sum_{j=1}^{q} n_j \alpha_j^k.$$

Consider now the quadratic form

$$T_m(x) = \sum_{j,k=0}^{m-1} S_{j+k} x_j x_k,$$

where m is any number such that $m \geq q$ (for example $m = n$).

THEOREM 3.2 (Hermite). *The number of all distinct zeros of the polynomial $P(z)$ is equal to the rank of the form $T_m(x)$, and the number of distinct real zeros is equal to the signature of the form $T_m(x)$.*

Proof. From the definition of the power sums S_k and the form T_m we get

$$T_m(x) = \sum_{j,k=0}^{m-1} \sum_{p=1}^{q} n_p \alpha_p^{j+k} x_j x_k =$$

$$= \sum_{p=1}^{q} n_p (x_0 + \alpha_p x_1 + \alpha_p^2 x_2 + \ldots + \alpha_p^{m-1} x_{m-1})^2 = \sum_{p=1}^{q} n_p y_p^2, \tag{3.1}$$

where

$$y_p = x_0 + \alpha_p x_1 + \alpha_p^2 x_2 + \ldots + \alpha_p^{m-1} x_{m-1}.$$

These linear forms $y_1, \ldots, y_q$ are linearly independent since the determinant formed from their coefficients is the Vandermonde determinant, which it is not equal to zero (here we use the fact that numbers $\alpha_1, \ldots, \alpha_q$ are distinct). Therefore the rank of $T_m(x)$ is equal to q. But in general the coefficients in y_p can be complex, and then the representation of $T_m(x)$ as a sum of squares needs to be modified. In the representation (3.1) positive squares $n_p y_p^2$ correspond to real zeros α_p. Complex zeros occur in pairs α_p and $\bar{\alpha}_p$, and hence we can write $y_p = u_p + iv_p$, where u_p and v_p are real, and obtain

$$y_p^2 = u_p^2 - v_p^2 + 2iu_pv_p.$$

For the conjugate zero $\bar{\alpha}_p$ we have

$$\bar{y}_p = u_p - iv_p, \quad \bar{y}_p^2 = u_p^2 - v_p^2 - 2iu_pv_p,$$

and hence $n_py_p^2 + n_p\bar{y}_p^2 = 2n_pu_p^2 - 2n_pv_p^2$.

This shows that every pair of conjugate zeros corresponds to one positive and one negative square. □

Using the Jacobi theorem we now get

COROLLARY 3.1. *Let* $D_0 = 1,\ D_1 = S_0 = n,$

$$D_2 = \begin{vmatrix} n & S_1 \\ S_1 & S_2 \end{vmatrix}, \dots, \quad D_r = \begin{vmatrix} n & S_1 & \dots & S_{r-1} \\ S_1 & S_2 & \dots & S_r \\ \dots & \dots & \dots & \dots \\ S_{r-1} & S_r & \dots & S_{2r-2} \end{vmatrix}.$$

If $D_j \neq 0,\ j = 0, \dots, r$, *then the number of distinct real zeros of* $P(z)$ *is equal to*

$$r - 2V(D_0, D_1, \dots, D_r).$$

The power sums S_k can be computed by means of the Newton formulas (2.1).

EXAMPLE 3.1. Let $P(x) = x^2 + px + q,\ n = 2$. We then have the quadratic form

$$T_2(x) = \sum_{j,k=0}^{1} S_{j+k}x_jx_k.$$

First we compute S_0, S_1, S_2. By formulas (2.1) we have $S_0 = 2,\ S_1 + p = 0,\ S_1 = -p,\ S_2 + S_1p + 2q = 0,\ S_2 = p^2 - 2q$. We thus obtain the matrix

$$\mathbf{A} = ||S_{j+k}||_{j,k=0}^{1} = \begin{Vmatrix} 2 & -p \\ -p & p^2 - 2q \end{Vmatrix}$$

with minors

$$D_0 = 1,\ D_1 = 2,\ D_2 = p^2 - 4q,$$

and hence

$$V(1, 2, p^2 - 4q) = 0, \qquad \text{if } p^2 - 4q > 0$$
$$V(1, 2, p^2 - 4q) = 1, \qquad \text{if } p^2 - 4q < 0$$

We arrive at a well-known conclusion: If $p^2 - 4q > 0$ then $P(x)$ has two distinct real zeros, if $p^2 - 4q < 0$ then $P(x)$ has no real zeros.

The form $T_m(x)$ is a Hankel form, and hence the Jacobi theorem can be generalized to the case where some principal minors vanish. Corollary 3.1 is then generalized accordingly (see [**75**, ch. 10, sec. 10; ch. 16, sec. 10]). The Hermite method can be applied also to other problems connected with the localization of zeros (see [**94**]).

Now we turn to the question of determination the number of real zeros of a real polynomial P on an interval (a, b). This number can be calculated with the help of the theorem on logarithmic residues. In fact, if γ is a curve surrounding the segment

$[a, b]$ and not containing any non-real zeros of the polynomial P, then by formula (1.2) we have

$$\frac{1}{2\pi i}\int_\gamma \frac{dP}{P} = N, \tag{3.2}$$

where N is the number of zeros on $[a, b]$. But here arises the problem of calculating the given integral in terms of the coefficients of our polynomial P. First we shall compute this integral by means of the so-called Cauchy index.

Let $R(x)$ be a real rational function. The Cauchy index

$$J_a^b R(x)$$

is the number of times that $R(x)$ jumps from $-\infty$ to $+\infty$ minus the number of times that $R(x)$ jumps from $+\infty$ to $-\infty$ as x goes from a to b. We assume here that a and b are points of continuity of $R(x)$.

Thus if

$$R(x) = \sum_{j=1}^{p} \frac{A_j}{y - \alpha_j} + R_1(x),$$

where α_j are real and $R_1(x)$ has no real singular points then

$$J_a^b R(x) = \sum_{a<\alpha_j<b} \operatorname{sign} A_j.$$

In particular if

$$R(x) = \frac{P'(x)}{P(x)}$$

then

$$R(x) = \sum_{j=1}^{q} \frac{n_j}{x - \alpha_j},$$

where α_j are distinct zeros of P and n_j are multiplicities of α_j. If $\alpha_1, \ldots, \alpha_p$ are real zeros and $\alpha_{p+1}, \ldots, \alpha_q$ are complex zeros then

$$R(x) = \sum_{j=1}^{p} \frac{n_j}{x - \alpha_j} + R_1(x),$$

where R_1 has no real singular points. Hence the index $J_a^b R(x)$ is equal to the number of distinct real zeros of $P(x)$ on (a, b). If all $n_j = 1$, then

$$J_a^b \frac{P'}{P}(x) = \frac{1}{2\pi i}\int_\gamma \frac{dP}{P}$$

by formula (3.2).

The Cauchy index of a rational function can be computed using so-called Sturm series. A system of real polynomials $P_1(x), \ldots, P_m(x)$ is called a Sturm series on the interval (a, b) if:

1. For all $x \in (a, b)$ such that $P_k(x) = 0$ for $k \geq 2$, we have that the neighboring functions $P_{k-1}(x)$ and $P_{k+1}(x)$ do not vanish at x, and

$$P_{k-1}(x)P_{k+1}(x) < 0;$$

2. The last function $P_m(x)$ does not vanish on (a,b);
3. If $P_1(x) = 0$ then $P_2(x) \neq 0$.

Denote by $V(x)$ the number of sign changes in the sequence $P_1(x), \dots, P_m(x)$, i.e. $V(x) = V(P_1(x), \dots, P_m(x))$. If $P_1(x) \neq 0$ then $V(x)$ is well-defined even at points where some function $P_k(x), k \geq 2$, happens to vanish. Indeed, if $P_k(x) = 0$ then by virtue of condition 1) $P_{k-1}(x)P_{k+1}(x) < 0$, so we can delete $P_k(x)$ from our series $P_1, \dots, P_m$. $V(a)$ equals $V(a+\varepsilon)$, where $\varepsilon > 0$ is sufficiently small and chosen so that the functions $P_j(x)$ do not vanish on $(a, a+\varepsilon]$. If all the functions $P_j(x)$ are non-zero for $x = a$, then $V(a) = V(P_1(a), \dots, P_m(a))$. $V(b)$ is equal to $V(b-\varepsilon)$.

As x varies from a to b the value of $V(x)$ can change only when passing a zero of some function of the Sturm series. But by virtue of condition 1) the value $V(x)$ remains unchanged when passing a zero of P_k, $k \geq 2$, while passing a zero of $P_1(x)$ means that exactly one sign change is lost or added.

We have the following theorem.

THEOREM 3.3 (Sturm). *If $P_1, \dots, P_m$ is a Sturm series on (a,b) and if $V(x)$ is the number of sign changes in this series, then*

$$J_a^b \frac{P_2(x)}{P_1(x)} = V(a) - V(b). \tag{3.3}$$

If we multiply all the terms of the Sturm series by one polynomial $d(x)$ then nothing changes in equality (3.3). It follows that the Sturm theorem remains true also for generalized Sturm series.

We can construct generalized Sturm series with the help of the Euclid algorithm. Let P and Q be two polynomials with $\deg P \geq \deg Q$. Define $P_1 = P$ and $P_2 = Q$. Let further $-P_3$ be the remainder after division of P_1 by P_2, let $-P_4$ be the remainder after division of P_2 by P_3 etc. Then we get

$$P_1 = q_1 P_2 - P_3, \dots, P_{k-1} = q_{k-1} P_k - P_{k+1}, \dots, P_{m-1} = q_{m-1} P_m,$$

where P_m is the greatest common divisor of P and Q, and also a divisor of all functions P_j. If $P_m(x) \neq 0$ on (a,b) then the obtained series is a Sturm series. If $P_1(x) = 0$ then $P_2(x) \neq 0$, for otherwise $P_3(x) = 0, \dots, P_m(x) = 0$. In exactly the same way we see that if $P_k(x) = 0$, then $P_{k-1}(x) \neq 0$ and $P_{k+1} \neq 0$ and moreover, $P_{k-1}(x) = -P_{k+1}(x)$. If $P_m(x)$ vanishes at some points of (a,b) then we have obtained a generalized Sturm series.

As a corollary we get the Sturm theorem on the number of distinct zeros of a polynomial $P(x)$ on (a,b).

THEOREM 3.4 (Sturm). *Let $P(x)$ be a real polynomial, $P(a) \neq 0$, $P(b) \neq 0$. Construct for $P(x)$ the generalized Sturm series starting with $P_1(x) = P(x)$ and $P_2(x) = P'(x)$ using the Euclid algorithm. Then the number of distinct real zeros of P on (a,b) is equal to $V(a) - V(b)$.*

EXAMPLE 3.2. Suppose $P_1 = P(x) = x^2 + px + q$. Then we have $P_2 = P'(x) = 2x + p$, and dividing P by P' we get $P_3 = p^2/4 - q$. Assume $p^2/4 - q \neq 0$. Then the number of distinct real zeros of $P(x)$ on (a,b) is equal to

$$V(a^2 + ap + q, 2a + p, p^2 - 4q) - V(b^2 + bp + q, 2b + p, p^2 - 4q).$$

The Sturm method is rather inconvenient. Therefore one usually applies the Hermite theorem in combination with the Descartes sign rule or the Budan–Fourier theorem (see, for example, [**97**]).

THEOREM 3.5 (Descartes). *Let $P(x)$ be a real polynomial of the form*

$$P(x) = a_0x^n + a_1x^{n-1} + \ldots + a_n.$$

The number of positive zeros of $P(x)$ (counted according to multiplicities) is either equal to the number of sign changes in the sequence of coefficients of P (i.e. $V(a_0, a_1, \ldots, a_n)$) (zero coefficients are to be neglected), or less than this value by an even integer. If all the zeros of $P(x)$ are real then the number of positive zeros is always equal to $V(a_0, a_1, \ldots, a_n)$.

THEOREM 3.6 (Budan–Fourier). *Let $P(x)$ be a real polynomial of degree n. The number of real zeros of P on the segment $[a, b]$ is equal either to the difference*

$$V(P(a), P'(a), \ldots, P^{(n)}(a)) - V(P(b), P'(b), \ldots, P^{(n)}(b)),$$

or less than this value by an even integer. (Here it is assumed that $P(a) \neq 0$, $P(b) \neq 0$. If some derivative vanishes at a or b then $V(P(a), \ldots, P^{(n)}(a))$ and $V(P(b), \ldots, P^{(n)}(b))$ are defined as in the Sturm theorem.)

The proofs of these theorems are based on the Rolle theorem.

Here we have not mentioned the subresultant method (see [**92**] and section 19), since we shall have no need for it.

We have thus seen that there exist several quite effective methods for determining the number of real or complex zeros of a polynomial on an interval. Some of these methods are also applicable to wider classes of functions (see [**125**]). Turning to systems of equations we shall see that the situation is completely different.

4. The local residue (of Grothendieck)

We consider a generalization of the concept of a local residue for systems of holomorphic functions of several complex variables (see [**80**]). It is the so-called Grothendieck residue. Consider the space $\mathbf{C}^n$ consisting of vectors $\mathbf{z} = (z_1, \ldots, z_n)$, where $z_j = x_j + iy_j$, $x_j, y_j \in \mathbf{R}$, $j = 1, \ldots, n$. We use the notation

$$\operatorname{Re}\mathbf{z} = (\operatorname{Re} z_1, \ldots, \operatorname{Re} z_n) = (x_1, \ldots, x_n),$$

$$\operatorname{Im}\mathbf{z} = (\operatorname{Im} z_1, \ldots, \operatorname{Im} z_n) = (y_1, \ldots, y_n),$$

$$\bar{\mathbf{z}} = (\bar{z}_1, \ldots, \bar{z}_n),$$

$$|\mathbf{z}| = \sqrt{|z_1|^2 + \ldots + |z_n|^2} = \sqrt{|x_1|^2 + \ldots + |x_n|^2 + |y_1|^2 + \ldots + |y_n|^2}.$$

The topology in $\mathbf{C}^n$ is given by the metric $d(\mathbf{z}, \mathbf{w}) = |\mathbf{z} - \mathbf{w}|$, $\mathbf{z}, \mathbf{w} \in \mathbf{C}^n$. As usual a domain D is an open connected set and a compact K is a closed bounded set.

The power series in $\mathbf{C}^n$ (centered at the origin) has the form

$$\sum_{||\alpha|| \geq 0} c_\alpha \mathbf{z}^\alpha,$$

where $\boldsymbol{\alpha} = (\alpha_1, \dots, \alpha_n)$ is a multi-index, i.e. α_j are non-negative integers, the monomials $\mathbf{z}^{\boldsymbol{\alpha}} = z_1^{\alpha_1} \cdots z_n^{\alpha_n}$ are of total degree $||\boldsymbol{\alpha}|| = \alpha_1 + \dots + \alpha_n$, and $c_{\boldsymbol{\alpha}}$ are complex numbers (the coefficients of series). Denote by $U_R(\mathbf{a}) = \{\mathbf{z} : |\mathbf{z} - \mathbf{a}| < R\}$ the open ball of radius R centered at the point $\mathbf{a}$.

A complex-valued function $f(\mathbf{z})$ is said to be holomorphic in the domain $D \subset \mathbf{C}^n$ if for every point $\mathbf{a} \in D$ there exists a power series

$$\sum_{||\boldsymbol{\alpha}|| \geq 0} c_{\boldsymbol{\alpha}} (\mathbf{z} - \mathbf{a})^{\boldsymbol{\alpha}}$$

which converges to $f(\mathbf{z})$ in some neighborhood $U_R(\mathbf{a})$ of the point $\mathbf{a}$. For example, every polynomial

$$P(\mathbf{z}) = \sum_{0 \leq ||\boldsymbol{\alpha}|| \leq m} c_{\boldsymbol{\alpha}} \mathbf{z}^{\boldsymbol{\alpha}}$$

is a holomorphic function in $\mathbf{C}^n$. If at least one of the leading coefficients $c_{\boldsymbol{\alpha}}$, $||\boldsymbol{\alpha}|| = m$, is non-zero then P is said to be of degree m ($\deg P = m$).

A rational function

$$R(\mathbf{z}) = \frac{P(\mathbf{z})}{Q(\mathbf{z})}$$

is holomorphic wherever $Q(\mathbf{z}) \neq 0$ (here $P(\mathbf{z})$ and $Q(\mathbf{z})$ are polynomials).

We now define the local (Grothendieck) residue in $\mathbf{C}^n$. Suppose $h, f_1, \dots, f_n$ are holomorphic functions in some neighborhood $U_R(\mathbf{a})$ of the point $\mathbf{a}$. Suppose also that

$$f_j(\mathbf{a}) = 0, \ j = 1, \dots, n,$$

and that $\mathbf{a}$ is an isolated zero of the mapping

$$\mathbf{f} = (f_1, \dots, f_n).$$

Denote by ω the meromorphic differential form

$$\omega = \frac{h dz_1 \wedge \dots \wedge dz_n}{f_1 \cdots f_n} = \frac{h \mathbf{dz}}{f_1 \cdots f_n}$$

$$(\mathbf{dz} = dz_1 \wedge dz_2 \wedge \dots \wedge dz_n),$$

and consider the set

$$\Gamma_{\mathbf{f}} = \{\mathbf{z} \in U_R(\mathbf{a}) : |f_1(\mathbf{z})| = \varepsilon_1, \dots, |f_n(\mathbf{z})| = \varepsilon_n\},$$

where $\varepsilon_1 > 0, \dots, \varepsilon_n > 0$ are sufficiently small. By Sard's theorem the set $\Gamma_{\mathbf{f}}$ is a smooth compact surface of dimension n, i.e. a cycle, for almost all sufficiently small ε_j. The orientation of $\Gamma_{\mathbf{f}}$ is given by the condition

$$d \arg f_1 \wedge \dots \wedge d \arg f_n \geq 0.$$

The integral

$$\operatorname*{res}_{\mathbf{a}} \omega = \frac{1}{(2\pi i)^n} \int_{\Gamma_{\mathbf{f}}} \frac{h \mathbf{dz}}{f_1 \cdots f_n}$$

is called the local (Grothendieck) residue

$$\operatorname*{res}_{\mathbf{a}} \omega = \operatorname*{res}_{\mathbf{a}\ \mathbf{f}} h$$

of the form ω at the point $\mathbf{a}$.

In just the same way as in the case of one complex variable it can be shown that $\operatorname{res}_{\mathbf{a}} \omega$ does not depend on ε_j.

EXAMPLE 4.1. Let $f_j = z_j^{\alpha_j}$, $\mathbf{a} = 0$ and

$$h = \mathbf{z}^{\beta} = z_1^{\beta_1} \dots z_n^{\beta_n}.$$

Then

$$\Gamma_{\mathbf{f}} = \{\mathbf{z} : \ |z_j| = \varepsilon_j, \ j = 1, \dots, n\}$$

is a standard torus and

$$\operatorname{res}_{0} \omega = \frac{1}{(2\pi i)^n} \int_{\Gamma_{\mathbf{f}}} \frac{\mathbf{z}^{\beta} \mathbf{dz}}{\mathbf{z}^{\alpha}} = \frac{1}{(2\pi i)^n} \prod_{j=1}^{n} \int_{|z_j| = \varepsilon_j} z_j^{\beta_j - \alpha_j} dz_j =$$

$$= \begin{cases} 0, & \text{if } \alpha_j - \beta_j \neq 1 \text{ at least for one } j, \\ 1, & \text{if } \alpha_j - \beta_j = 1 \text{ for all } j \end{cases}$$

(see example 1.1).

We shall say that a mapping $\mathbf{f} = (f_1, \dots, f_n)$ is nonsingular at the point $\mathbf{a}$ if its Jacobian

$$J_{\mathbf{f}}(\mathbf{a}) = \frac{\partial(f_1, \dots, f_n)}{\partial(z_1, \dots, z_n)}(\mathbf{a}) = \begin{vmatrix} \frac{\partial f_1}{\partial z_1} & \dots & \frac{\partial f_1}{\partial z_n} \\ \dots & \dots & \dots \\ \frac{\partial f_n}{\partial z_1} & \dots & \frac{\partial f_n}{\partial z_n} \end{vmatrix} (\mathbf{a}) \neq 0.$$

We can then assume $J_{\mathbf{f}}(\mathbf{z}) \neq 0$ in $U_R(\mathbf{a})$, so by the implicit function theorem the mapping $\mathbf{w} = \mathbf{f}(\mathbf{z})$ is one-to-one in $U_R(\mathbf{a})$. Moreover, it is biholomorphic, i.e. the inverse mapping $\mathbf{z} = \mathbf{f}^{-1}(\mathbf{w})$ is also holomorphic in a neighborhood of the origin.

Consider now the integral

$$\operatorname{res}_{\mathbf{a}} \omega = \frac{1}{(2\pi i)^n} \int_{\Gamma_{\mathbf{f}}} \frac{h \mathbf{dz}}{f_1 \dots f_n},$$

and make the change of variables $w_j = f_j(\mathbf{z})$, $j = 1, \dots, n$. We then have

$$\mathbf{dz} = J_{\mathbf{f}^{-1}} \mathbf{dw} = \frac{\mathbf{dw}}{J_{\mathbf{f}}\left(\mathbf{f}^{-1}(\mathbf{w})\right)},$$

while $\Gamma_{\mathbf{f}}$ transforms to

$$\Gamma_{\mathbf{w}} = \{\mathbf{w} : \ |w_j| = \varepsilon_j, \ j = 1, \dots, n\}.$$

Therefore we can write

$$\operatorname{res}_{\mathbf{a}} \omega = \frac{1}{(2\pi i)^n} \int_{\Gamma_{\mathbf{w}}} \frac{h\left(\mathbf{f}^{-1}(\mathbf{w})\right) \mathbf{dw}}{J_{\mathbf{f}}\left(\mathbf{f}^{-1}(\mathbf{w})\right) w_1 \dots w_n}.$$

Now, the function $h/J_{\mathbf{f}}$ is holomorphic in a neighborhood of the point 0 (as long as $J_{\mathbf{f}} \neq 0$) and hence it can be expanded in a power series

$$\sum_{||\alpha|| \geq 0} c_{\alpha} \mathbf{w}^{\alpha},$$

with

$$c_{(0,\dots,0)} = \frac{h}{J_{\mathbf{f}}}(\mathbf{f}^{-1}(0)) = \frac{h(\mathbf{a})}{J_{\mathbf{f}}(\mathbf{a})}.$$

In view of Example 4.1 we thus get

$$\operatorname*{res}_{\mathbf{a}} \omega = \frac{h(\mathbf{a})}{J_{\mathbf{f}}(\mathbf{a})}.$$

So in the case of a nonsingular mapping $\mathbf{f}$ the local residue is easily computed. It is easy to compute it also if h has the form $h_1 f_1 + \ldots + h_n f_n$, where h_j are holomorphic in $U_R(\mathbf{a})$. Indeed,

$$\operatorname*{res}_{\mathbf{a}} \omega = \frac{1}{(2\pi i)^n} \int_{\Gamma_{\mathbf{f}}} \frac{h d\mathbf{z}}{f_1 \ldots f_n} = \sum_{j=1}^{n} \frac{1}{(2\pi i)^n} \int_{\Gamma_{\mathbf{f}}} \frac{h_j d\mathbf{z}}{f_1 \ldots f_{j-1} f_{j+1} \ldots f_n},$$

so it suffices to consider, say, the integral

$$\int_{\Gamma_{\mathbf{f}}} \frac{h_1 d\mathbf{z}}{f_2 \ldots f_n}.$$

The integrand $h_1/(f_2 \ldots f_n)$ is holomorphic in the open set

$$U_R(\mathbf{a}) - \cup_{j=2}^{n} \{\mathbf{z} : f_j(\mathbf{z}) = 0\},$$

which contains the chain

$$U_1 = \{\mathbf{z} : |f_1| \leq \varepsilon_1, |f_2| = \varepsilon_2, \ldots, |f_n| = \varepsilon_n\}.$$

On the other hand we have $\partial U_1 = \Gamma_f$, so by the Stokes formula we get

$$\int_{\Gamma_{\mathbf{f}}} \frac{h_1 d\mathbf{z}}{f_2 \ldots f_n} = \int_{\partial U_1} \frac{h_1 d\mathbf{z}}{f_2 \ldots f_n} = \int_{U_1} d\left(\frac{h_1 d\mathbf{z}}{f_2 \ldots f_n}\right).$$

But

$$d\left(\frac{h_1}{f_2 \ldots f_n} d\mathbf{z}\right) = d\frac{h_1}{f_2 \ldots f_n} \wedge d\mathbf{z} = 0,$$

since $h_1, f_2 \ldots, f_n$ are holomorphic, and it follows that

$$\operatorname*{res}_{\mathbf{a}} \omega = 0, \quad \text{if } h = h_1 f_1 + \ldots + h_n f_n.$$

Next we consider a transformation formula for local residues, which will play a fundamental role in the sequel. Let $\mathbf{f} = (f_1, \ldots, f_n)$ and $\mathbf{g} = (g_1, \ldots, g_n)$ be holomorphic mappings defined in $U_R(\mathbf{a})$. The point $\mathbf{a}$ is assumed to be an isolated zero of the mappings $\mathbf{f}$ and $\mathbf{g}$, and in addition

$$g_k(\mathbf{z}) = \sum_{j=1}^{n} a_{kj}(\mathbf{z}) f_j(\mathbf{z}), \quad k = 1, \ldots, n, \tag{4.1}$$

with a_{kj} being holomorphic in $U_R(\mathbf{a})$, $k, j = 1, \ldots, n$. In matrix notation $\mathbf{g} = \mathbf{A}\mathbf{f}$, where

$$\mathbf{A} = ||a_{kj}(\mathbf{z})||_{k,j=1}^{n}.$$

THEOREM 4.1 (The transformation law for local residues). *If h is holomorphic in $U_R(\mathbf{a})$ then*

$$\operatorname*{res}_{\mathbf{a}\,\mathbf{f}} h = \operatorname*{res}_{\mathbf{a}\,\mathbf{g}} h \det \mathbf{A}. \tag{4.2}$$

In other words

$$\int_{\Gamma_{\mathbf{f}}} \frac{h d\mathbf{z}}{f_1 \dots f_n} = \int_{\Gamma_{\mathbf{g}}} \frac{h \det \mathbf{A}\, d\mathbf{z}}{g_1 \dots g_n}.$$

We prove this formula for nonsingular mappings $\mathbf{f}$ and $\mathbf{g}$, the general case can be found in [**80**, ch. 5].

So let $\mathbf{f}$ and $\mathbf{g}$ be nonsingular mappings, i.e. $J_{\mathbf{f}}(\mathbf{a}) \neq 0$ and $J_{\mathbf{g}}(\mathbf{a}) \neq 0$. It follows that $\det \mathbf{A}(\mathbf{a}) \neq 0$. We can assume that these determinants do not vanish in $U_R(\mathbf{a})$. Then

$$\operatorname*{res}_{\mathbf{a}\ \mathbf{f}} h = \frac{h(\mathbf{a})}{J_{\mathbf{f}}(\mathbf{a})}.$$

$$\operatorname*{res}_{\mathbf{a}\ \mathbf{g}} \det \mathbf{A} = \frac{h(\mathbf{a}) \det \mathbf{A}(\mathbf{a})}{J_{\mathbf{g}}(\mathbf{a})}.$$

But by virtue of (4.1), and since $\mathbf{f}(\mathbf{a}) = \mathbf{g}(\mathbf{a}) = 0$, we have

$$J_{\mathbf{g}}(\mathbf{a}) = J_{\mathbf{f}}(\mathbf{a}) \det \mathbf{A}(\mathbf{a}).$$

Therefore

$$\operatorname*{res}_{\mathbf{a}\ \mathbf{f}} h = \operatorname*{res}_{\mathbf{a}\ \mathbf{g}} h \det \mathbf{A}.$$

If $\mathbf{f}$ and $\mathbf{g}$ are singular then one considers a slight perturbation $\mathbf{f}_\varepsilon$ of $\mathbf{f}$, having a finite set of zeros $\mathbf{z}^k$ in $U_R(\mathbf{a})$, with $J_{\mathbf{f}_\varepsilon}(\mathbf{z}^k) \neq 0$. For these zeros formula (4.2) is applied and then letting $\varepsilon \to 0$ one gets the general case. □

The transformation formula makes it possible to compute local residues quite explicitly. Consider an arbitrary residue

$$\operatorname*{res}_{\mathbf{a}} \omega = \operatorname*{res}_{\mathbf{a}\ \mathbf{f}} h.$$

Since the functions $z_j - a_j$ vanish at the point $\mathbf{a}$ it follows from Hilbert's Nullstellensatz that there exist $\alpha_j \geq 0$ such that

$$g_j(\mathbf{z}) = (z_j - a_j)^{\alpha_j + 1} = \sum_{k=1}^{n} a_{jk}(\mathbf{z}) f_k(\mathbf{z}).$$

By formula (4.2) we then have

$$\operatorname*{res}_{\mathbf{a}\ \mathbf{f}} h = \operatorname*{res}_{\mathbf{a}\ \mathbf{g}} h \det \mathbf{A} = \frac{1}{(2\pi i^n)} \int_{\Gamma_{\mathbf{g}}} \frac{h \det \mathbf{A} d\mathbf{z}}{(z_1 - a_1)^{\alpha_1 + 1} \dots (z_n - a_n)^{\alpha_n + 1}}.$$

Expanding $h \det \mathbf{A}$ in a power series

$$h \det \mathbf{A} = \sum_{||\beta|| \geq 0} c_\beta (\mathbf{z} - \mathbf{a})^\beta$$

we get

$$\operatorname*{res}_{\mathbf{a}\ \mathbf{f}} h = c_{\alpha_1, \dots, \alpha_n} = \frac{1}{\alpha_1! \dots \alpha_n!} \cdot \frac{\partial^{||\alpha||}(h \det \mathbf{A})}{\partial z_1^{\alpha_1} \dots \partial z_n^{\alpha_n}}(\mathbf{a}).$$

5. The multidimensional logarithmic residue

Just as in the case of one complex variable the logarithmic residue is a particular case of the local residue. The results of this section are to be found in [**11**].

Later on we shall need an analog of the Cauchy theorem in $\mathbf{C}^n$, sometimes referred to as the Cauchy–Poincaré theorem.

THEOREM 5.1 (Cauchy–Poincaré). *Let $f(\mathbf{z})$ be a holomorphic function in the domain $D \subset \mathbf{C}^n$ and σ let be a smooth relatively compact oriented surface of (real) dimension $n+1$ having a smooth boundary $\partial\sigma$. Assume that $\sigma \cup \partial\sigma \subset D$, and let the orientation of $\partial\sigma$ be induced by the orientation of σ. Then*

$$\int_{\partial\sigma} f(\mathbf{z})d\mathbf{z} = \int_{\partial\sigma} f(\mathbf{z})dz_1 \wedge \ldots \wedge dz_n = 0.$$

This theorem is a direct corollary of the Stokes formula and the Cauchy–Riemann equations:

$$\frac{\partial f}{\partial \bar{z}_j} = 0, \quad j = 1, \ldots, n.$$

Now we proceed to define the multiplicity of a zero of a holomorphic mapping. Suppose that we are given a holomorphic mapping

$$\mathbf{w} = \mathbf{f}(\mathbf{z}) = (f_1(\mathbf{z}), \ldots, f_n(\mathbf{z}))$$

in a neighborhood $U_R(\mathbf{a})$ of the point $\mathbf{a}$, and suppose moreover that $\mathbf{a}$ is an isolated zero of the mapping $\mathbf{f}$. If $\mathbf{f}$ is nonsingular at the point $\mathbf{a}$, i.e. $J_{\mathbf{f}}(\mathbf{a}) \neq 0$, then $\mathbf{a}$ is called a simple zero of the mapping $\mathbf{f}$ and its multiplicity is equal to 1.

Suppose now instead of this that $J_{\mathbf{f}}(\mathbf{a}) = 0$. Consider a perturbation of the mapping $\mathbf{f}$, e.g. the mapping

$$\mathbf{f}_{\boldsymbol{\xi}}(\mathbf{z}) = \mathbf{f}(\mathbf{z}) - \boldsymbol{\xi} = (f_1(\mathbf{z}) - \xi_1, \ldots, f_n(\mathbf{z}) - \xi_n),$$

where $|\boldsymbol{\xi}|$ is sufficiently small. Since $|\mathbf{f}(\mathbf{z})| \neq 0$ for $\mathbf{z} \neq \mathbf{a}$ we have that $|\mathbf{f}| \geq \varepsilon > 0$ on the boundary $\partial U_R(\mathbf{a})$. Taking $\boldsymbol{\xi}$ so that $|\boldsymbol{\xi}| < \varepsilon$ we get that $|\mathbf{f}_{\boldsymbol{\xi}}| > 0$ on $\partial U_R(\mathbf{a})$, i.e. $\mathbf{f}_{\boldsymbol{\xi}} \neq 0$ on $\partial U_R(\mathbf{a})$, and hence the set

$$E_{\mathbf{f}_{\boldsymbol{\xi}}} = \{\mathbf{z} \in U_R(\mathbf{a}) : \mathbf{f}_{\boldsymbol{\xi}}(\mathbf{z}) = 0\}$$

is compact. By a well-known theorem of complex analysis the set $E_{\mathbf{f}_{\boldsymbol{\xi}}}$ must then be finite. It can be shown that for almost all $\boldsymbol{\xi}$ this set consists of the same number of points. The number of elements of $E_{\mathbf{f}_{\boldsymbol{\xi}}}$ is called the multiplicity $\mu_{\mathbf{a}}(\mathbf{f})$ of the zero $\mathbf{a}$ of the mapping $\mathbf{f}$ at the point $\mathbf{a}$. We notice that for almost all $\boldsymbol{\xi}$ the elements of $E_{\mathbf{f}_{\boldsymbol{\xi}}}$ are simple zeros of the mapping $\mathbf{f}_{\boldsymbol{\xi}}$. This is the "dynamic" definition of the multiplicity of zeros (see [**11**]).

EXAMPLE 5.1. Consider the mapping $w_1 = z_1, w_2 = z_2^2 + az_1^2$. The point $(0,0)$ is a zero of this mapping of multiplicity 2. Indeed, making a perturbation we find that the zeros of the mapping $(z_1 - \xi_1, z_2^2 + az_1^2 - \xi_2)$ are given by

$$z_1 = \xi_1, \ z_2 = \pm\sqrt{a\xi_1^2 - \xi_2}.$$

Thus, if $a\xi_1^2 - \xi_2 \neq 0$, i.e. for almost all (ξ_1, ξ_2), the perturbed mapping has two zeros.

Let the holomorphic mapping $\mathbf{f}(\mathbf{z}) = (f_1, \dots, f_n)$ be given in a domain $D \subset \mathbf{C}^n$ and that it has a finite zero set $E_\mathbf{f}$ in D. Consider the special analytic polyhedron

$$D_\mathbf{f} = \{\mathbf{z} \in D : |f_1(\mathbf{z})| < \varepsilon_1, \dots, |f_n(\mathbf{z})| < \varepsilon_n\}.$$

If $\varepsilon_1, \dots, \varepsilon_n$ are sufficiently small positive numbers then D_f consists of a finite number of relatively compact domains in D and every component of D_f contains only one zero of a mapping $\mathbf{f}$. The distinguished boundary

$$\Gamma_\mathbf{f} = \{\mathbf{z} \in D : \ |f_j(\mathbf{z})| = \varepsilon_j, \ j = 1, \dots, n\}$$

of this polyhedron is accordingly divided into a finite number of connected surfaces (of the kind that we already considered for local residues). Moreover, by the Sard theorem Γ_f consists of smooth surfaces. Recall that orientation of $\Gamma_\mathbf{f}$ is given by the relation

$$d \arg f_1 \wedge \dots \wedge d \arg f_n \geq 0$$

on Γ_f. In other words, it is induced by the orientation of D_f.

Suppose h is holomorphic in D. An integral of the form

$$\frac{1}{(2\pi i)^n} \int_{\Gamma_\mathbf{f}} h(\mathbf{z}) \frac{df_1 \wedge \dots \wedge df_n}{f_1 \dots f_n} = \frac{1}{(2\pi i)^n} \int_{\Gamma_\mathbf{f}} h \frac{J_\mathbf{f} d\mathbf{z}}{f_1 \dots f_n} \tag{5.1}$$

is called the logarithmic residue of a function h with respect to $\mathbf{f}$ in D. If $\mathbf{f}$ has only one zero $\mathbf{a}$ in D then the logarithmic residue is a local residue

$$\operatorname*{res}_{\mathbf{a}\,\mathbf{f}} hJ_\mathbf{f}.$$

From the Cauchy–Poincaré theorem it is not difficult to deduce that the logarithmic residue does not depend on ε_j.

THEOREM 5.2 (Cacciopolli–Martinelli–Sorani). *For every holomorphic function h in the domain D the formula*

$$\frac{1}{(2\pi i)^n} \int_{\Gamma_\mathbf{f}} h(\mathbf{z}) \frac{J_\mathbf{f} d\mathbf{z}}{f_1 \dots f_n} = \sum_{\mathbf{a} \in E_\mathbf{f}} h(\mathbf{a}) \tag{5.2}$$

holds.

As for $n = 1$ every zero $\mathbf{a}$ is counted according to its multiplicity.

Theorem 5.2 is the multidimensional logarithmic residue theorem.

Proof. Let $U_R(\mathbf{a})$ be mutually disjoint neighborhoods of the different points $\mathbf{a}$. Take ε_j sufficiently small to ensure that for all $\mathbf{a}$ there is precisely one connected component of D_f contained in each $U_R(\mathbf{a})$. Denote this component by

$$\Gamma_{\mathbf{f},\mathbf{a}} = \Gamma_\mathbf{f} \cap U_R(\mathbf{a}).$$

By the Cauchy–Poincaré theorem we have

$$\int_{\Gamma_\mathbf{f}} h \frac{J_\mathbf{f} d\mathbf{z}}{f_1 \dots f_n} = \sum \int_{\Gamma_{\mathbf{f},\mathbf{a}}} h \frac{J_\mathbf{f} d\mathbf{z}}{f_1 \dots f_n},$$

so it remains only to be shown that

$$\frac{1}{(2\pi i)^n} \int_{\Gamma_{\mathbf{f},\mathbf{a}}} h \frac{J_\mathbf{f} d\mathbf{z}}{f_1 \dots f_n} = \operatorname*{res}_{\mathbf{a}\,\mathbf{f}} J_\mathbf{f} h = \mu_\mathbf{a}(\mathbf{f}) h(\mathbf{a}).$$

If $\mathbf{f}$ is nonsingular at the point $\mathbf{a}$ then we already know (see Section 4) that

$$\operatorname*{res}_{\mathbf{a}\ \mathbf{f}} hJ_{\mathbf{f}} = \frac{h(\mathbf{a})J_{\mathbf{f}}(\mathbf{a})}{J_{\mathbf{f}}(\mathbf{a})} = h(\mathbf{a}).$$

If $\mathbf{a}$ is not simple zero, then considering a perturbation $\mathbf{f}_\xi$ with only simple zeros, we get by the Cauchy–Poincaré theorem

$$\frac{1}{(2\pi i)^n}\int_{\Gamma_{\mathbf{f}}} \frac{h(\mathbf{a})J_{\mathbf{f}_\xi}\mathbf{dz}}{f_{1\xi}\cdots f_{n\xi}} = \sum_{\mathbf{b}_\xi\in E_{f_\xi}} h(\mathbf{b}_\xi).$$

Letting ξ tend to 0 (so that $\mathbf{b}_\xi \to \mathbf{a}$) we reach the required assertion. □

COROLLARY 5.1. *If the holomorphic mapping* $\mathbf{f}$ *has a finite number* N *of zeros in* D, *then*

$$\frac{1}{(2\pi i)^n}\int_{\Gamma_{\mathbf{f}}} \frac{J_{\mathbf{f}}\mathbf{dz}}{f_1\cdots f_n} = N$$

(every zero is counted according to its multiplicity).

In conclusion we formulate (without proof) a version of the Rouché theorem which is due to A.P.Yuzhakov (see [**11**, sec. 4]).

THEOREM 5.3 (Yuzhakov). *Suppose that* $D, \mathbf{f}, h$ *are as in Theorem 5.2 and that the holomorphic mapping* $\mathbf{g}$ *satisfies the condition*

$$|g_j| < |f_j|$$

on $\Gamma_{\mathbf{f}}$, $j = 1,\ldots,n$. *Then: a) the mappings* $\mathbf{f}$ *and* $\mathbf{f}+\mathbf{g}$ *have the same number of zeros in* D; *b) the formula*

$$\frac{1}{(2\pi i)^n}\int_{\Gamma_{\mathbf{f}}} h(\mathbf{z})\frac{d(f_1+g_1)\wedge\ldots d(f_n+g_n)}{(f_1+g_1)\ldots(f_n+g_n)} = \sum_{\mathbf{a}\in E_{\mathbf{f}+\mathbf{g}}} h(\mathbf{a})$$

holds.

Notice that the integral in this theorem is taken over $\Gamma_{\mathbf{f}}$ and not over $\Gamma_{\mathbf{f}+\mathbf{g}}$.

EXAMPLE 5.2. Find the multiplicity of the zero (0,0) of the mapping

$$w_1 = z_1 - az_2^2, \quad w_2 = bz_1^2 - z_2^3.$$

Since on the distinguished boundary

$$\Gamma = \{\mathbf{z} : |z_1| = \varepsilon^7,\ |z_2| = \varepsilon^4\}$$

we have the inequalities $|a||z_2|^2 < |z_1|$ and $|z_2|^3 > |b||z_1|^2$ for sufficiently small positive ε, it follows by Theorem 5.3 that the multiplicity of the point (0,0) is equal to the integral

$$J = \frac{1}{(2\pi i)^2}\int_\Gamma \frac{d(z_1 - az_2^3)\wedge d(bz_1^2 - z_2^2)}{(z_1 - az_2^2)(bz_1^2 - z_2^2)} =$$

$$= \frac{1}{(2\pi i)^2}\int_\Gamma \frac{(-3z_2^2 + 4abz_1z_2)dz_1\wedge dz_2}{(z_1 - az_2^2)(b_1z_1^2 - z_2^3)}.$$

We integrate first with respect to z_1, keeping z_2 fixed. Since $|z_1| > |a||z_2|^2$ and $|z_2|^3 > |b||z_1|^2$ it follows that in the disk $|z_1| < \varepsilon^7$ there is only one singular point

$z_1 = az_2^2$ of the integrand. Hence by the one-dimensional theorem on logarithmic residues (see Sec. 1, Theorem 1.2)

$$J = \frac{1}{2\pi i}\int_{|z_2|=\varepsilon^4} \frac{-3z_2^2 + 4abz_2^3}{a^2bz_2^4 - z_2^3}dz_2 = \frac{1}{2\pi i}\int_{|z_2|=\varepsilon^4} \frac{-3 + 4abz_2}{(a^2bz_2 - 1)z_2}dz_2 = 3,$$

because the root $z_2 = 1/(a^2b)$ does not lie in the disk $|z_2| < \varepsilon^4$ for sufficiently small $\varepsilon > 0$.

6. The classical scheme for elimination of unknowns

The results presented in this section one can find in [**97, 142**]. First we consider a classical result of Sylvester. Take two polynomials:

$$P(z) = a_0z^n + a_1z^{n-1} + \ldots + a_n,$$

$$Q(z) = b_0z^s + b_1z^{s-1} + \ldots + b_s,$$

$z \in \mathbf{C}$, $a_j, b_j \in \mathbf{C}$ and $a_0 \neq 0,\ b_0 \neq 0$.

How can we determine whether or not these polynomials have any common zeros? Suppose $\alpha_1, \ldots, \alpha_n$ are the zeros of $P(z)$ and $\beta_1, \ldots, \beta_s$ the zeros of $Q(z)$. Form the expression

$$R(P,Q) = a_0^s b_0^n \prod_{k=1}^{n}\prod_{j=1}^{s}(\alpha_k - \beta_j), \tag{6.1}$$

which is known as the resultant of the polynomials P and Q. It is clear that $R(P,Q) = 0$ if the polynomials P and Q have at least one zero in common.

The resultant (6.1) can also be calculated directly from the coefficients of P and Q without knowing their zeros. To see this we start with the observation that

$$R(P,Q) = (-1)^{ns}R(Q,P).$$

Since

$$Q(z) = b_0\prod_{j=1}^{s}(z - \beta_j),$$

it follows that

$$Q(\alpha_k) = b_0\prod_{j=1}^{s}(\alpha_k - \beta_j),$$

and hence

$$R(P,Q) = a_0^s\prod_{k=1}^{n}Q(\alpha_k).$$

In exactly the same way we see that

$$R(Q,P) = b_0^n\prod_{j=1}^{s}P(\beta_j).$$

THEOREM 6.1 (Sylvester). *The resultant of the polynomials P and Q is equal to the determinant*

$$D = \begin{vmatrix} a_0 & a_1 & \dots & a_n & 0 & \dots & 0 \\ 0 & a_0 & \dots & a_{n-1} & a_n & \dots & 0 \\ \dots & \dots & \dots & \dots & \dots & \dots & \dots \\ 0 & 0 & \dots & a_0 & a_1 & \dots & a_n \\ b_0 & b_1 & \dots & b_s & 0 & \dots & 0 \\ 0 & b_0 & \dots & b_{s-1} & b_s & \dots & 0 \\ \dots & \dots & \dots & \dots & \dots & \dots & \dots \\ 0 & 0 & \dots & b_0 & b_1 & \dots & b_s \end{vmatrix} \begin{matrix} \left.\vphantom{\begin{matrix}a\\a\\a\\a\end{matrix}}\right\} s \text{ lines} \\ \left.\vphantom{\begin{matrix}a\\a\\a\\a\end{matrix}}\right\} n \text{ lines} \end{matrix}.$$

Proof. Consider the auxiliary determinant

$$M = \begin{vmatrix} \beta_1^{n+s-1} & \beta_2^{n+s-1} & \dots & \beta_s^{n+s-1} & \alpha_1^{n+s-1} & \alpha_2^{n+s-1} & \dots & \alpha_n^{n+s-1} \\ \beta_1^{n+s-2} & \beta_2^{n+s-2} & \dots & \beta_s^{n+s-2} & \alpha_1^{n+s-2} & \alpha_2^{n+s-2} & \dots & \alpha_n^{n+s-2} \\ \dots & \dots & \dots & \dots & \dots & \dots & \dots & \dots \\ \beta_1^2 & \beta_2^2 & \dots & \beta_s^2 & \alpha_1^2 & \alpha_2^2 & \dots & \alpha_n^2 \\ \beta_1 & \beta_2 & \dots & \beta_s & \alpha_1 & \alpha_2 & \dots & \alpha_n \\ 1 & 1 & \dots & 1 & 1 & 1 & \dots & 1 \end{vmatrix}.$$

Since M is a Vandermonde determinant we have the formula

$$M = \prod_{1\le k<j\le s} (\beta_k - \beta_j) \prod_{j=1}^{s}\prod_{k=1}^{n} (\beta_j - \alpha_k) \prod_{1\le k<j\le n} (\alpha_k - \alpha_j).$$

Hence we obtain, by the definition of $R(P,Q)$, that

$$a_0^s b_0^n DM = DR(Q,P) \prod_{1\le k<j\le s} (\beta_k - \beta_j) \prod_{1\le k<j\le n} (\alpha_k - \alpha_j). \tag{6.2}$$

On the other hand we can calculate the product DM of the determinants D and M by using the formula for the determinant of a product of matrices. Multiplying the matrices corresponding to our determinants, and taking into account that the α_j are zeros of P and the β_k are zeros of Q, we get

$$DM = \begin{vmatrix} \beta_1^{s-1}P(\beta_1) & \beta_2^{s-1}P(\beta_2) & \dots & \beta_s^{s-1}P(\beta_s) & 0 & \dots & 0 \\ \dots & \dots & \dots & \dots & \dots & \dots & \dots \\ \beta_1 P(\beta_1) & \beta_2 P(\beta_2) & \dots & \beta_s P(\beta_s) & 0 & \dots & 0 \\ P(\beta_1) & P(\beta_2) & \dots & P(\beta_s) & 0 & \dots & 0 \\ 0 & 0 & \dots & 0 & \alpha_1^{n-1}Q(\alpha_1) & \dots & \alpha_n^{n-1}Q(\alpha_n) \\ 0 & 0 & \dots & 0 & \alpha_1^{n-2}Q(\alpha_1) & \dots & \alpha_n^{n-2}Q(\alpha_n) \\ \dots & \dots & \dots & \dots & \dots & \dots & \dots \\ 0 & 0 & \dots & 0 & Q(\alpha_1) & \dots & Q(\alpha_n) \end{vmatrix} =$$

$$= \prod_{j=1}^{s} P(\beta_j) \prod_{k=1}^{n} Q(\alpha_k) \begin{vmatrix} \beta_1^{s-1} & \beta_2^{s-1} & \dots & \beta_s^{s-1} \\ \dots & \dots & \dots & \dots \\ \beta_1 & \beta_2 & \dots & \beta_s \\ 1 & 1 & \dots & 1 \end{vmatrix} \begin{vmatrix} \alpha_1^{n-1} & \alpha_2^{n-1} & \dots & \alpha_n^{n-1} \\ \dots & \dots & \dots & \dots \\ \alpha_1 & \alpha_2 & \dots & \alpha_n \\ 1 & 1 & \dots & 1 \end{vmatrix} =$$

$$= \prod_{j=1}^{s} P(\beta_j) \prod_{k=1}^{n} Q(\alpha_k) \prod_{k<j} (\beta_k - \beta_j) \prod_{k<j} (\alpha_k - \alpha_j).$$

This means that

$$a_0^s b_0^n DM = R(P,Q)R(Q,P) \prod_{k<j} (\beta_k - \beta_j) \prod_{k<j} (\alpha_k - \alpha_j). \tag{6.3}$$

Comparing (6.2) and (6.3), and dividing out the common factors that are not identically zero, we find that $R(P,Q) = D$. □

In practice one frequently encounters situations when the coefficients of P and Q depend on some parameters. In this case one can no longer assert that $a_0 \neq 0$ and $b_0 \neq 0$. Hence it is in general not to be expected that the vanishing of the resultant should be equivalent to the existence of a common zero of P and Q. Indeed, if $a_0 = b_0 = 0$ then $R(P,Q) = 0$, but P and Q do not necessarily have any common zeros. It turns out that this is the only case when the vanishing of the resultant does not imply the existence of a common zero.

COROLLARY 6.1. *For given polynomials P and Q with arbitrary leading coefficients the resultant $R(P,Q)$ equals zero if and only if either P and Q have a common zero or both leading coefficients vanish ($a_0 = b_0 = 0$). (In the case $a_0 = 0$ or $b_0 = 0$ we use the identity $R(P,Q) = D$).*

EXAMPLE 6.1. Let $P = a_0 z^2 + a_1 z + a_2$, and $Q = b_0 z^2 + b_1 z + b_2$. Then

$$R(P,Q) = \begin{vmatrix} a_0 & a_1 & a_2 & 0 \\ 0 & a_0 & a_1 & a_2 \\ b_0 & b_1 & b_2 & 0 \\ 0 & b_0 & b_1 & b_2 \end{vmatrix} = (a_0 b_2 - a_2 b_0)^2 - (a_0 b_1 - a_1 b_0)(a_1 b_2 - a_2 b_1).$$

In particular, if $P = z^2 - 4z - 5$ and $Q = z^2 - 7z + 10$ we get $R(P,Q) = 0$, and a common root of P and Q is $z = 5$.

Let us now derive another important property of the resultant $R(P,Q)$. We write down the system

$$\begin{aligned} z^{s-1}P(z) &= a_0 z^{n+s-1} + a_1 z^{n+s-2} + \ldots + a_n z^{s-1}, \\ z^{s-2}P(z) &= a_0 z^{n+s-2} + a_1 z^{n+s-3} + \ldots + a_n z^{s-2}, \\ \ldots & \quad \ldots\ldots\ldots\ldots \\ P(z) &= a_0 z^n + a_1 z^{n-1} + \ldots + a_n, \\ z^{n-1}Q(z) &= b_0 z^{n+s-1} + b_1 z^{n+s-2} + \ldots + b_s z^{n-1}, \\ z^{n-2}Q(z) &= b_0 z^{n+s-2} + b_1 z^{n+s-3} + \ldots + b_s z^{n-2}, \\ \ldots & \quad \ldots\ldots\ldots\ldots \\ Q(z) &= b_0 z^s + b_1 z^{s-1} + \ldots + b_s. \end{aligned} \tag{6.4}$$

Considering each right hand side as a polynomial of degree $n + s - 1$, we see that the determinant formed by their coefficients is exactly equal to $R(P,Q)$. So if we multiply each of the equations by the minor associated with the corresponding entry in the last column of $R(P,Q)$ and add them together, i.e. we develop the determinant along the last column, then we get the identity

$$AP + BQ \equiv R(P,Q),$$

where A is a polynomial of degree $s-1$ and B is a polynomial of degree $n-1$.

Theorem 6.1 and Corollary 6.1 can be applied to eliminate one of the unknowns in a system of two algebraic equations in two variables. Let $P(z,w)$ and $Q(z,w)$ be polynomials in two complex variables z and w. Grouping their terms according to degrees of z we have

$$\begin{aligned} P(z,w) &= a_0(w)z^k + a_1(w)z^{k-1} + \ldots + a_n(w), \\ Q(z,w) &= b_0(w)z^s + b_1(w)z^{s-1} + \ldots + b_s(w), \end{aligned}$$

where the coefficients $a_j(w)$ and $b_j(w)$ are polynomials in w. We assume that $a_0 \not\equiv 0$ and $b_0 \not\equiv 0$.

Consider the resultant $R(P,Q)$ of these polynomials (in the variable z). It will be some polynomial $R(w)$. If the vector (α,β) is a root of the system

$$P(z,w) = 0, \quad Q(z,w) = 0, \tag{6.5}$$

then substituting the value $w = \beta$ we get that the system of polynomials $P(z,\beta)$ and $Q(z,\beta)$ has the common zero $z = \alpha$, hence by Theorem 6.1 $R(\beta) = 0$. Conversely, if β is a zero of the polynomial $R(w)$ then, according to Corollary 6.1, either the polynomials $P(z,\beta)$ and $Q(z,\beta)$ have a common zero at α (i.e. the system (6.5) has the solution (α,β)) or β is a zero of the polynomials $a_0(w)$ and $b_0(w)$. In particular, if these polynomials have no common zeros (which can be checked by finding the resultant of a_0 and b_0), then every zero of $R(w)$ is the second coordinate of a root of the system (6.5).

EXAMPLE 6.2. Consider the system

$$\begin{aligned} P(z,w) &= wz^2 + 3wz + 2w + 3 = 0, \\ Q(z,w) &= 2zw - 2z + 2w + 3 = 0. \end{aligned}$$

First we re-write P and Q in the form

$$P = w \cdot z^2 + 3w \cdot z + (2w+3), \quad Q = 2(w-1) \cdot z + (2w+3).$$

Then we get

$$R(P,Q) = \begin{vmatrix} w & 3w & 2w+3 \\ 2(w-1) & 2w+3 & 0 \\ 0 & 2(w-1) & 2w+3 \end{vmatrix} =$$

$$= (2w+3)\left(\begin{vmatrix} 2(w-1) & 2w+3 \\ 0 & 2(w-1) \end{vmatrix} + \begin{vmatrix} w & 3w \\ 2(w-1) & 2w+3 \end{vmatrix}\right) =$$

$$= (2w+3)(4w^2 - 8w + 4 + 2w^2 + 3w - 6w^2 + 6w) =$$

$$= (2w+3)(w+4).$$

The zeros of $R(w)$ are $w_1 = -\frac{3}{2}$ and $w_2 = -4$. They are not zeros of the leading coefficients of P and Q, i.e. the polynomials w and $2(w-1)$. For $w = -4$ we obtain the system

$$-4z^2 - 12z - 5 = 0, \quad -10z - 5 = 0,$$

from which we get $z = -5$. For $w = -\frac{3}{2}$ we have

$$-\frac{3}{2}z^2 - \frac{9}{2} = 0, \quad -5z = 0,$$

and hence $z = 0$. Thus, the initial system has the two roots $\left(0, -\frac{3}{2}\right)$ and $(-5, -4)$.

Before moving on to more general systems of equations we consider the question: When does a polynomial $P(z)$ have multiple zeros? This problem has already been touched upon earlier (see Sec. 3). Here we analyze it using the concept of a discriminant of the polynomial P. Suppose

$$P(z) = a_0 z^n + a_1 z^{n-1} + \ldots + a_n,$$

and $\alpha_1, \ldots, \alpha_n$ are its zeros. Denote by $D(P)$ the expression

$$D(P) = a_0^{2n-2} \prod_{n \geq k > j \geq 1} (a_k - a_j)^2 = a_0^{2n-2}(-1)^{\frac{n(n-1)}{2}} \prod_{j \neq k} (a_k - a_j).$$

It is clear that $D(P) = 0$ if and only if the polynomial P has multiple zeros.

The discriminant $D(P)$ is easily expressed in terms of the resultant $R(P, P')$. In fact, we know that

$$R(P, P') = a_0^{n-1} \prod_{k=1}^{n} P'(\alpha_k).$$

Differentiating the equality

$$P(z) = a_0 \prod_{k=1}^{n} (z - \alpha_k),$$

we obtain

$$P'(z) = a_0 \sum_{k=1}^{n} \prod_{j \neq k} (z - \alpha_j),$$

from which it follows that

$$P'(\alpha_k) = a_0 \prod_{j \neq k} (\alpha_k - \alpha_j).$$

Hence we have

$$R(P, P') = a_0^{2n-1} \prod_{j \neq k} (\alpha_k - \alpha_j) = a_0^{2n-1}(-1)^{\frac{n(n-1)}{2}} \prod_{n \geq k > j \geq 1} (\alpha_k - \alpha_j)^2 =$$

$$= a_0(-1)^{\frac{n(n-1)}{2}} D(P),$$

and therefore

$$R(P, P') = a_0(-1)^{\frac{n(n-1)}{2}} D(P).$$

There are also other ways of representing the discriminant $D(P)$.

Theorem 6.2. *The discriminant $D(P)$ satisfies*

$$D(P) = a_0^{2n-2} \begin{vmatrix} n & S_1 & S_2 & \ldots & S_{n-1} \\ S_1 & S_2 & S_3 & \ldots & S_n \\ \ldots & \ldots & \ldots & \ldots & \ldots \\ S_{n-1} & S_n & S_{n+1} & \ldots & S_{2n-2} \end{vmatrix}, \tag{6.6}$$

where S_j are the power sums of the zeros of $P(z)$.

Proof. We construct the Vandermonde determinant

$$d = \begin{vmatrix} 1 & 1 & \dots & 1 \\ \alpha_1 & \alpha_2 & \dots & \alpha_n \\ \dots & \dots & \dots & \dots \\ \alpha_1^{n-1} & \alpha_2^{n-1} & \dots & \alpha_n^{n-1} \end{vmatrix} = \prod_{n \geq k > j \geq 1} (\alpha_k - \alpha_j),$$

and observe that $D(P) = a_0^{2n-2} d^2$. Then we calculate d^2 in a different way by multiplying the matrix of d by its transposed matrix. □

Notice that the determinant in this theorem coincides with the determinant of the matrix of a quadratic form that arose in the proof of the Hermite theorem (see Sec. 3). It should also be noted that the size of the determinant (6.6) is equal to n, whereas the size of the determinant $R(P, P')$ is $(2n - 1)$.

EXAMPLE 6.3. If $P = z^2 + bz + c$ then

$$D(P) = \begin{vmatrix} 2 & S_1 \\ S_1 & S_2 \end{vmatrix}, \quad \text{where } S_1 = -b,\ S_2 = b^2 - 2c,$$

and therefore $D(P) = b^2 - 4c$.

Now we turn to systems consisting of several equations.

THEOREM 6.3. *Given r polynomials $P_1(z), \dots, P_r(z)$ in one variable $z \in \mathbf{C}$, there exists a system $R_1, \dots, R_k$ of polynomials in the coefficients of $P_1, \dots, P_r$, such that the conditions $R_1 = 0, \dots, R_k = 0$ are necessary and sufficient for the system of equations*

$$P_1 = 0, \dots, P_r = 0 \tag{6.7}$$

to have a common root, or the leading coefficients of all the polynomials $P_1, \dots, P_r$ to vanish simultaneously.

Proof. We begin by transforming the polynomials $P_1, \dots, P_r$ so that they have equal degree. If n is the greatest of the degrees of $P_1, \dots, P_r$ then we multiply every polynomial P_m of degree n_m by z^{n-n_m} and by $(z-1)^{n-n_m}$. In this way we get two polynomials of degree n, whose leading coefficients are the same as in P_m. Denote by $Q_1, \dots, Q_s$ all the new polynomials thus obtained from the system $P_1, \dots, P_r$. The system (6.7) and the system $Q_1 = 0, \dots, Q_s = 0$ have the same roots.

We consider now the polynomials

$$Q_u = u_1 Q_1 + \dots + u_s Q_s, \quad Q_v = v_1 Q_1 + \dots + v_s Q_s,$$

where u_j and v_j are additional variables. If Q_u and Q_v have a common factor (for arbitrary u_j and v_j) then this factor can not depend on u_j or v_j, and hence it must be a factor of all the Q_j. Conversely: A common factor of all the Q_j is a factor of both Q_u and Q_v. Since the degrees of all Q_j coincide, the vanishing of the leading coefficients of Q_u and Q_v is equivalent to the vanishing of the leading coefficients of all Q_j.

For the polynomials Q_u and Q_v we form the resultant $R(Q_u, Q_v)$. By Theorem 6.1 and Corollary 6.1 $R(Q_u, Q_v) = 0$ if and only if either Q_u and Q_v have a common zero (i.e. a common factor), or the leading coefficients of Q_u and Q_v vanish simultaneously. Since $R(Q_u, Q_v)$ is a polynomial in u_j and v_j we have $R = 0$ if and only if its coefficients $R_1, \dots, R_k$ all vanish.

The system $R_1, \dots, R_k$ is called the system of resultants for $P_1, \dots, P_r$.

Consider now an algebraic system of the form

$$P_1(\mathbf{z}) = 0, \dots, P_r(\mathbf{z}) = 0, \tag{6.8}$$

where $\mathbf{z} = (z_1, \dots, z_n) \in \mathbf{C}^n$. We can approach it as follows. Suppose the polynomial P_1 contains a monomial of greatest degree in z_1 of the form z_1^α and that its coefficient is a non-zero constant. (We can always achieve this by making a linear change of variables.) Then, applying Theorem 6.3 with respect to z_1 to the system (6.8), we are led to the new system of equations

$$d_1 = 0, \dots, d_l = 0, \tag{6.9}$$

which now only involves $z_2, \dots, z_n$, and which is equivalent to the system (6.8). To this new system we apply the same method and so on. After iterating this procedure s times we are either left with the zero system, or a system of constants which are not all equal to zero. In the second case the initial system (6.8) had no common roots. In the first case we have a triangular system of the form

$$\begin{aligned} P_1 = 0, \dots, P_r &= 0, \\ d_1 = 0, \dots, d_l &= 0, \\ \dots\dots\dots&\dots\dots \\ R_1 = 0, \dots, R_k &= 0, \end{aligned} \tag{6.10}$$

where $R_1, \dots, R_k$ are polynomials in $z_s, \dots, z_n$ and the following step (with respect to z_s) produces the zero system. We can then assign to the variables $z_{s+1}, \dots, z_n$ arbitrary values $\xi_{s+1}, \dots, \xi_n$. Finding the common solutions ξ_s to the system $R_1 = 0, \dots, R_k = 0$, we substitute them into the previous system and so on, up to the original system (6.8).

Of course the system (6.10) may contain superfluous equations. A further development of the elimination theory is realized through the method of Gröbner bases (see [**1, 28**]), which permits a reduction of the system (6.8) to triangle form without unnecessary equations. We see that the classical method of elimination is quite awkward and practically useless when the number of equations is greater than two.

In the following chapter we shall describe a new method of elimination based on the multidimensional logarithmic residue formula.

CHAPTER 2

A MODIFIED ELIMINATION METHOD

7. A generalized transformation formula for local residues

In Sec. 4 we have seen that the transformation law for local residues allows one to explicitly calculate them. Often however, more complicated integrals appear. Therefore we need a transformation formula also for such integrals (see [98]).

Let $h, f_1, \ldots, f_n, g_1, \ldots, g_n$ be holomorphic functions in a neighborhood $U_R(\mathbf{a})$ of a point $\mathbf{a} \in \mathbf{C}^n$. The point $\mathbf{a}$ is assumed to be an isolated zero of the mappings $\mathbf{f} = (f_1, \ldots, f_n)$ and $\mathbf{g} = (g_1, \ldots, g_n)$. Moreover, $\mathbf{f}$ and $\mathbf{g}$ are connected via relations

$$g_j(\mathbf{z}) = \sum_{k=1}^{n} a_{jk}(\mathbf{z}) f_k(\mathbf{z}), \tag{7.1}$$

the functions $a_{jk}(\mathbf{z})$ being holomorphic in $U_R(\mathbf{a})$. Introducing the matrix

$$\mathbf{A} = ||a_{jk}(\mathbf{z})||_{j,k=1}^{n},$$

we can re-write the formula (7.1) in the form

$$\mathbf{g} = \mathbf{A}\mathbf{f}. \tag{7.2}$$

Connected with the mapping $\mathbf{f}$ is the distinguished boundary

$$\Gamma_{\mathbf{f}} = \{\mathbf{z} \in U_R(\mathbf{a}) : |f_j(\mathbf{z})| = \varepsilon_j, \ \ j = 1, \ldots, n\}$$

of the associated analytic polyhedron. For almost every sufficiently small $\varepsilon_j > 0$ the set $\Gamma_{\mathbf{f}}$ is a smooth compact closed orientable surface of dimension n, i.e. an n-dimensional cycle. The orientation of Γ_f is determined by the condition

$$d(\arg f_1) \wedge \ldots \wedge d(\arg f_n) > 0.$$

In Sec. 4 we derived the transformation formula

$$\operatorname*{res}_{\mathbf{a}\,\mathbf{f}} h = \operatorname*{res}_{\mathbf{a}\,\mathbf{g}} h \det \mathbf{A}$$

for the local residue, see (4.2). We now consider a transformation formula for integrals of the form

$$\int_{\Gamma_{\mathbf{f}}} \frac{h d\mathbf{z}}{f_{i_1} \ldots f_{i_m} f_1 \ldots f_n} = \int_{\Gamma_{\mathbf{f}}} \frac{h d\mathbf{z}}{f_1^{\alpha_1+1} \ldots f_n^{\alpha_n+1}}, \quad i_1, \ldots, i_m = 1, \ldots, n.$$

THEOREM 7.1 (Kytmanov). *If the holomorphic mappings* $\mathbf{f}$ *and* $\mathbf{g}$ *are connected by the relation (7.2), then*

$$c_{i_1 \ldots i_m} \int_{\Gamma_{\mathbf{f}}} \frac{h d\mathbf{z}}{f_{i_1} \ldots f_{i_m} f_1 \ldots f_n} =$$

$$= \sum_{j_1,\dots,j_m=1}^{n} c_{j_1\dots j_m} \int_{\Gamma_\mathbf{g}} \frac{h a_{j_1 i_1}\dots a_{j_m i_m} \det \mathbf{A} d\mathbf{z}}{g_{j_1}\dots g_{j_m} g_1 \dots g_n}. \tag{7.3}$$

Here the constants $c_{i_1\dots i_m}$ (and $c_{j_1\dots j_m}$) are defined as follows: If in the set $(i_1,\dots,i_m)$ the number of ones is α_1, the number of twos is α_2 and so on, then $c_{i_1\dots i_m} = \alpha_1!\dots\alpha_n!$.

Proof. We reason as in the proof of formula (4.2). First we consider the case when the mappings $\mathbf{f}$ and $\mathbf{g}$ are nonsingular, i.e.

$$J_\mathbf{f}(\mathbf{a}) \neq 0, \quad J_\mathbf{g}(\mathbf{a}) \neq 0$$

and $\det \mathbf{A}(\mathbf{a}) \neq 0$. We shall assume that these functions are non-zero in all of $U_R(\mathbf{a})$, and that the mappings $\mathbf{f}$ and $\mathbf{g}$ are biholomorphic. Therefore, both $(f_1,\dots,f_n)$ and $(g_1,\dots,g_n)$ can be taken as new coordinates in $U_R(\mathbf{a})$.

Assume to begin with that the function h satisfies

$$\frac{\partial^\beta h}{\partial \mathbf{z}^\beta}(\mathbf{a}) = \frac{\partial^{||\beta||} h}{\partial z_1^{\beta_1}\dots\partial z_n^{\beta_n}} = 0$$

for all multiindeces β, with $||\boldsymbol{\beta}|| = \beta_1 + \dots + \beta_n < m$. Then, by the rule for differentiation of composite functions, we have

$$\frac{\partial^\beta h}{\partial g^\beta}(\mathbf{a}) = 0$$

and

$$\frac{\partial^\beta h}{\partial f^\beta}(\mathbf{a}) = 0$$

for all $\boldsymbol{\beta}$, with $||\boldsymbol{\beta}|| < m$.

Making the change of variables $\mathbf{w} = \mathbf{f}(\mathbf{z})$ in $U_R(\mathbf{a})$ we get

$$c_{i_1\dots i_m} \int_{\Gamma_\mathbf{f}} \frac{h d\mathbf{z}}{f_{i_1}\dots f_{i_m} f_1\dots f_n} = c_{i_1\dots i_m} \int_{\Gamma_{mathbfw}} \frac{h(\mathbf{f}^{-1})}{J_\mathbf{f}(\mathbf{f}^{-1})} \cdot \frac{d\mathbf{w}}{w_{i_1}\dots w_{i_m} w_1 \dots w_n} =$$

$$= (2\pi i)^n \frac{\partial^m}{\partial w_{i_1}\dots\partial w_{i_m}} \left[\frac{h}{J_\mathbf{f}}(\mathbf{f}^{-1}(0))\right] = \frac{(2\pi i)^n}{J_\mathbf{f}(\mathbf{a})} \cdot \frac{\partial^m h}{\partial f_{i_1}\dots\partial f_{i_m}}(\mathbf{a}).$$

In exactly the same way

$$c_{j_1\dots j_m} \int_{\Gamma_\mathbf{g}} \frac{h a_{j_1 i_1}\dots a_{j_m i_m} \det \mathbf{A}}{g_{j_1}\dots g_{j_m} g_1\dots g_n} d\mathbf{z} =$$

$$= (2\pi i)^n \frac{\partial^m (h a_{j_1 i_1}\dots a_{j_m i_m} \det \mathbf{A}/J_\mathbf{g})}{\partial g_{j_1}\dots\partial g_{j_m}}(\mathbf{a}) =$$

$$= (2\pi i)^n \frac{a_{j_1 i_1}(\mathbf{a})\dots a_{j_m i_m}(\mathbf{a}) \det \mathbf{A}(\mathbf{a})}{J_\mathbf{g}(\mathbf{a})} \cdot \frac{\partial^m h}{\partial g_{j_1}\dots\partial g_{j_m}}(\mathbf{a}).$$

Since

$$J_\mathbf{g}(\mathbf{a}) = \det \mathbf{A}(\mathbf{a}) J_\mathbf{f}(\mathbf{a})$$

and

$$\frac{\partial^m h}{\partial f_{i_1}\dots\partial f_{i_m}}(\mathbf{a}) = \sum_{j_1,\dots,j_m=1}^{n} a_{j_1 i_1}(\mathbf{a})\dots a_{j_m i_m}(\mathbf{a}) \frac{\partial^m h}{\partial g_{j_1}\dots\partial g_{j_m}}(\mathbf{a}),$$

so the equality (7.3) follows in this case.

We continue the proof by induction on m. Formula (7.3) has been proved for $m = 1$ and for a function h, such that $h(\mathbf{a}) = 0$. We must prove it also for a function h, which is not equal to zero at the point $\mathbf{a}$. Letting $h = J_{\mathbf{f}}$, and $\mathbf{w} = \mathbf{f}(\mathbf{z})$, we have

$$\int_{\Gamma_{\mathbf{f}}} \frac{J_{\mathbf{f}} d\mathbf{z}}{f_p f_1 \ldots f_n} = \int_{\Gamma_{\mathbf{w}}} \frac{d\mathbf{w}}{w_p w_1 \ldots w_n} = 0.$$

(See Sec. 4, Example 4.1.)

Since $\mathbf{g} = \mathbf{A}\mathbf{f}$ it follows that

$$\mathbf{E} = \mathbf{A}\frac{\partial \mathbf{f}}{\partial \mathbf{g}} + \sum_{k=1}^{n} f_k \mathbf{A}_k,$$

where $\mathbf{E}$ is the unit matrix, $\partial \mathbf{f}/\partial \mathbf{g}$ is the matrix

$$\left\| \frac{\partial f_i}{\partial g_k} \right\|_{j,k=1}^{n}$$

consisting of partial derivatives, and A_k is the matrix

$$\left\| \frac{\partial a_{jk}}{\partial g_s} \right\|_{s,j=1}^{n}$$

We can therefore write

$$\det \mathbf{A} \cdot \det \frac{\partial \mathbf{f}}{\partial \mathbf{g}} = 1 - \sum_{k=1}^{n} f_k B_k + \sum_{k,j=1}^{n} f_k f_j B_{kj} + \ldots, \tag{7.4}$$

with

$$B_k = \sum_{r=1}^{n} \frac{\partial a_{rk}}{\partial g_r}.$$

From here we see that

$$\sum_{j=1}^{n} \int_{\Gamma_{\mathbf{g}}} \frac{a_{ip} \det \mathbf{A} \det \frac{\partial \mathbf{f}}{\partial \mathbf{g}} d\mathbf{g}}{g_j g_1 \ldots g_n} =$$

$$= \sum_{j=1}^{n} \left(\int_{\Gamma_{\mathbf{g}}} \frac{a_{jp} d\mathbf{g}}{g_j g_1 \ldots g_n} - \sum_{k=1}^{n} \int_{\Gamma_{\mathbf{g}}} \frac{a_{jp} B_k f_k d\mathbf{g}}{g_j g_1 \ldots g_n} \right).$$

(Here we used the equality

$$\det \frac{\partial \mathbf{f}}{\partial \mathbf{g}} dg_1 \wedge \ldots \wedge dg_n = J_{\mathbf{g}} d\mathbf{z}$$

and the condition that $f_j(\mathbf{a}) = 0$. Therefore all the integrals containing products of the functions f_j vanish).

From the already proved part of formula (7.3), i.e. when $\mathbf{f}(\mathbf{a}) = 0$, we have

$$\sum_{j,k=1}^{n} \int_{\Gamma_{\mathbf{g}}} \frac{a_{ip} B_k f_k d\mathbf{g}}{g_j g_1 \ldots g_n} = \sum_{k=1}^{n} \int_{\Gamma_{\mathbf{f}}} \frac{B_k f_k J_{\mathbf{g}} d\mathbf{z}}{\det \mathbf{A} \cdot f_p f_1 \ldots f_n} =$$

$$= (2\pi i)^n \frac{B_p(\mathbf{a}) J_{\mathbf{g}}(\mathbf{a})}{J_{\mathbf{f}}(\mathbf{a}) \det \mathbf{A}(\mathbf{a})} = (2\pi i)^n B_p(\mathbf{a}),$$

and

$$\sum_{j=1}^{n}\int_{\Gamma_{\mathbf{g}}}\frac{a_{ip}\mathbf{dg}}{g_j g_1\cdots g_n}=\sum_{j=1}^{n}(2\pi i)^n\frac{\partial a_{jp}}{\partial g_j}(\mathbf{a}).$$

Hence from (7.4) we get

$$\sum_{j=1}^{n}\int_{\Gamma_{\mathbf{g}}}\frac{a_{ip}\det\mathbf{A}J_{\mathbf{f}}\mathbf{dz}}{g_j g_1\cdots g_n}=0.$$

Let now m be arbitrary. We must prove the equality (7.3) for all monomials $\mathbf{z}^{\beta}$, such that $0\le||\beta||<m$. It is sufficient to prove (7.3) for monomials $\mathbf{g}^{\beta}$, $0<||\beta||<m$, and for the function $J_{\mathbf{f}}$, since $\mathbf{z}^{\beta}$ can be expanded in a series with respect to the functions $\mathbf{g}^{\gamma}$, $||\gamma||\ge||\beta||$. Consider first the case $h=\mathbf{g}^{\gamma}$, $||\gamma||>0$. It is not difficult to verify that formula the (7.3) transforms into a similar formula with $h=1$ and $m-||\gamma||<m$. It remains to consider the case $h=J_{\mathbf{f}}$. Then we have

$$\int_{\Gamma_{\mathbf{f}}}\frac{J_{\mathbf{f}}\mathbf{dz}}{f_{i_1}\cdots f_{i_m}f_1\cdots f_n}=0.$$

LEMMA 7.1. *If* $\mathbf{f}$ *and* $\mathbf{g}$ *are non-singular and connected by the relation (7.2), then*

$$\sum_{j_1,\dots,j_m=1}^{n}\frac{\partial^m}{\partial g_{j_1}\dots\partial g_{j_m}}\left(\det\mathbf{A}\det\frac{\partial\mathbf{f}}{\partial\mathbf{g}}a_{j_1i_1}\cdots a_{j_mi_m}\right)(\mathbf{a})=0.$$

Proof. The proof of this lemma consists in a direct calculation based on (7.4).
Using Lemma 7.1 we now get

$$\sum_{j_1,\dots,j_m=1}^{n}c_{j_1\dots j_m}\int_{\Gamma_{\mathbf{g}}}\frac{a_{j_1j_1}\cdots a_{j_mi_m}\det\mathbf{A}J_{\mathbf{f}}\mathbf{dz}}{g_{j_1}\cdots g_{j_m}g_1\cdots g_n}=$$

$$=(2\pi i)^n\sum_{j_1,\dots,j_m=1}^{n}\frac{\partial^m}{\partial g_{j_1}\dots\partial g_{j_m}}\left(a_{j_1i_1}\dots a_{j_mi_m}\det\mathbf{A}\det\frac{\partial\mathbf{f}}{\partial\mathbf{g}}\right)(\mathbf{a})=0.$$

In the case when $\mathbf{f}$ and $\mathbf{g}$ are singular at the point $\mathbf{a}$ the proof of (7.3) is carried out in just the same way as the proof of formula (4.2). □

Formula (7.3) can be written in the alternative form

$$\int_{\Gamma_{\mathbf{f}}}\frac{h\mathbf{dz}}{\mathbf{f}^{\alpha+\mathbf{I}}}=\int_{\Gamma_{\mathbf{f}}}\frac{h\,dz_1\wedge\ldots\wedge dz_n}{f_1^{\alpha_1+1}\cdots f_n^{\alpha_n+1}}=$$

$$=\sum_{k_{sj}\ge0}\frac{\prod_{s=1}^{n}\left(\sum_{j=1}^{n}k_{sj}\right)!}{\prod_{s,j=1}^{n}(k_{sj})!}\int_{\Gamma_{\mathbf{g}}}\frac{h\prod_{s,j=1}^{n}a_{sj}^{k_{sj}}\det\mathbf{A}d\mathbf{z}}{g_1^{\beta_1+1}\cdots g_n^{\beta_n+1}},\tag{7.5}$$

where

$$\beta_s=\sum_{j=1}^{n}k_{sj},$$

and the summation on the right hand side is performed over all matrices

$$||k_{sj}||_{s,j=1}^{n}$$

with nonnegative entries such that

$$\sum_{s=1}^{n} k_{sj} = \alpha_j.$$

Let now $\mathbf{f} = (f_1, \dots , f_n)$ and $\mathbf{g} = (g_1, \dots , g_n)$ be polynomial mappings and let their zero sets $E_{\mathbf{f}}$, $E_{\mathbf{g}}$ be finite (recall that each root is counted as many times as its multiplicity). The complete sum of residues, or the global residue, of the polynomial h with respect to $\mathbf{f}$ is the expression

$$\operatorname*{Res}_{\mathbf{f}} h = \sum_{\mathbf{a}\in E_{\mathbf{f}}} \operatorname*{res}_{\mathbf{a}\,\mathbf{f}} h.$$

COROLLARY 7.1 (Kytmanov). *If* $\mathbf{f}$ *and* $\mathbf{g}$ *are connected through the relation (7.1), and if the* a_{jk} *are also polynomials, then*

$$\sum_{a\in E_{\mathbf{f}}} c_{i_1\dots i_m} \int_{\Gamma_{\mathbf{f}}} \frac{h d\mathbf{z}}{f_{i_1}\cdots f_{i_m} f_1 \cdots f_n} =$$

$$= \sum_{\mathbf{a}\in E_{\mathbf{g}}} \sum_{j_1,\dots ,j_m=1} c_{j_1\dots j_m} \int_{\Gamma_{\mathbf{g}}} \frac{h \det \mathbf{A} a_{j_1 i_1} \dots a_{j_m i_m} d\mathbf{z}}{g_{j_1}\cdots g_{j_m} g_1 \cdots g_n}.$$

In particular, the following assertion holds ([**137, 158**]).

COROLLARY 7.2 (Yuzhakov). *Under the conditions of Corollary 7.1*

$$\operatorname*{Res}_{\mathbf{f}} h = \operatorname*{Res}_{\mathbf{g}} h \det \mathbf{A}, \tag{7.6}$$

i.e. the transformation formula (4.2) for local residues carries over to the case of a global residue.

8. A modified elimination method

Along with the space $\mathbf{C}^n$ of complex variables $\mathbf{z} = (z_1, \dots , z_n)$ we shall consider its compactification $\mathbf{CP}^n$, i.e. the complex projective space. We recall that $\mathbf{CP}^n$ consists of vectors $\boldsymbol{\xi} = (\xi_0, \xi_1, \dots , \xi_n)$, where $\boldsymbol{\xi}$ and $\boldsymbol{\eta}$ are considered as equivalent if $\boldsymbol{\xi} = \lambda\boldsymbol{\eta}$, where $\lambda \in \mathbf{C}$ and $\lambda \neq 0$. Thus the elements in $\mathbf{CP}^n$ may be thought of as complex one-dimensional subspaces (complex lines) in $\mathbf{C}^{n+1}$. To each point $\mathbf{z} \in \mathbf{C}^n$ we associate the point

$$\boldsymbol{\xi} \in \mathbf{CP}^n \quad \mathbf{z} \to \boldsymbol{\xi} = (1, z_1, \dots , z_n).$$

In this way we get $\mathbf{C}^n \subset \mathbf{CP}^n$. In other words, the image of $\mathbf{C}^n$ in $\mathbf{CP}^n$ consists of all points for which $\xi_0 \neq 0$. The points $\boldsymbol{\xi} = (0, \xi_1 \dots , \xi_n) \in \mathbf{CP}^n$ form the hyperplane "at infinity" Π. Thus $\mathbf{CP}^n = \mathbf{C}^n \cup \Pi$.

Let f be a polynomial in $\mathbf{C}^n$ of the form

$$f = \sum_{||\alpha||\le k} c_\alpha \mathbf{z}^\alpha, \tag{8.1}$$

where $\boldsymbol{\alpha} = (\alpha_1, \dots, \alpha_n)$ and $\mathbf{z}^{\boldsymbol{\alpha}} = z_1^{\alpha_1} \dots z_n^{\alpha_n}$. We assume that its degree ($\deg f$) is really equal to k, i.e. there exists a coefficient $c_{\alpha} \neq 0$ in (8.1) with $||\boldsymbol{\alpha}|| = k$. We shall often expand polynomials like f in homogeneous components

$$P_j = \sum_{||\alpha||=j} c_{\alpha} \mathbf{z}^{\alpha}.$$

Here P_j is a homogeneous polynomial of degree j and $f = P_k + P_{k-1} + \dots + P_0$, the polynomial P_k is the leading (highest) homogeneous part of f. The projectivization f^* of the polynomial f is given by the homogeneous polynomial

$$f^*(\boldsymbol{\xi}) = f^*(\xi_0, \xi_1, \dots, \xi_n) = P_k(\xi_1, \dots, \xi_n) +$$
$$+ \xi_0 P_{k-1}(\xi_1, \dots, \xi_n) + \dots + \xi_0^k P_0.$$

Notice that f^* is a polynomial in $\mathbf{C}^{n+1}$ and that

$$f^*(1, \xi_1, \dots, \xi_n) = f(\xi_1, \dots, \xi_n).$$

We shall consider the following system of algebraic equations in $\mathbf{C}^n$:

$$\begin{aligned} f_1 &= 0, \\ &\dots \\ f_n &= 0; \end{aligned} \tag{8.2}$$

where f_j is a polynomial of degree k_j. Let $f_j = P_j + Q_j$, where P_j is the highest homogeneous part of f_j, and Q_j is of degree less than k_j.

Along with the system (8.2) we shall also consider the system

$$\begin{aligned} P_1 &= 0, \\ &\dots \\ P_n &= 0. \end{aligned} \tag{8.3}$$

We shall say that the system (8.2) is nondegenerate if the system (8.3) has only one root, namely the point 0. Often we shall deal with polynomials f_j whose coefficients depend on some parameters. In this case we call the system (8.2) nondegenerate if the system (8.3) has one single root for almost every value on the parameters.

LEMMA 8.1. *Nondegenerate systems of the form (8.2) may be characterized as follows: They have a finite number of roots in* $\mathbf{C}^n$ *and they have no roots on the hyperplane at infinity* Π.

Proof. Consider the system consisting of projectivizations of f_j, i.e.

$$\begin{aligned} f_1^*(\boldsymbol{\xi}) &= 0, \\ &\dots \\ f_n^*(\boldsymbol{\xi}) &= 0. \end{aligned} \tag{8.4}$$

The system (8.2) have a root on Π, if the system (8.4) is equal to zero at some point $\boldsymbol{\xi}$ of the form $\boldsymbol{\xi} = (0, \xi_1, \dots, \xi_n)$. Further,

$$f^*(\boldsymbol{\xi}) = P_j(\xi_1, \dots, \xi_n) + \xi_0 \tilde{Q}_j(\xi_0, \xi_{,1}, \dots, \xi_n), \quad j = 1, \dots, n,$$

so it follows that the system (8.2) has a root on Π if and only if the system (8.3) has a nontrivial root $\boldsymbol{\xi}$.

Now we consider the unit sphere

$$S = \{\mathbf{z} : |\mathbf{z}|^2 = |z_1|^2 + \ldots + |z_n|^2 = 1\}$$

in $\mathbf{C}^n$. If $\mathbf{z}^0 \in S$ then, by virtue of the nondegeneracy of the system (8.2), there exists a number j such that $P_j(\mathbf{z}^0) \neq 0$. Since

$$P_j(\lambda \mathbf{z}^0) = \lambda^{k_j} P_j(\mathbf{z}^0)$$

for $\lambda \in \mathbf{C}$, we see that

$$f(\lambda \mathbf{z}^0) = \lambda^{k_j} P_j(\mathbf{z}^0) + Q_j(\lambda \mathbf{z}^0),$$

where the degree of Q_j (in λ) is less than k_j. By the lemma on the modulus of leading terms we necessarily have an inequality

$$|\lambda|^{k_j} |P_j(\mathbf{z}^0)| > |Q_j(\lambda \mathbf{z}^0)|$$

for sufficiently large $|\lambda|$. In other words, for sufficiently large $|\lambda|$ the system (8.2) has no roots. Using now the Borel–Lebesgue theorem we find that the system (8.2) has no roots outside some ball of sufficiently large radius. The set of roots of (8.2) is thus compact, and being an analytic set it must therefore be finite. □

If the system (8.2) has a finite set of roots in $\mathbf{CP}^n$, then by a linear-fractional mapping it can be made non-degenerate. This is because by means of such a mapping we can make a plane with no roots of the system (8.2) become the hyperplane at infinity.

Later on we will be dealing with non-degenerate systems of equations. We shall be interested in the problem of eliminating from (8.2) all variables except one, i.e. we want to determine certain resultants $R(z_j)$. We have seen that the classical method of elimination (see Sec. 6) is not convenient for practical purposes, if the number of equation is greater than 2. It may also change the root multiplicities.

We are now going to describe in detail an elimination method based on the multidimensional logarithmic residue formula (5.2), which first appeared in a paper of Aizenberg [**2**] (see also [**11**, sec. 21]). Aizenberg noticed that for non-degenerate systems the formula (5.2) can be written explicitly in terms of the coefficients of the polynomials f_j. By the same token one can calculate power sums of roots without actually finding the roots. Using this one can then express the elementary symmetric polynomials in the roots by the Newton recursion formulas (2.1), and write out the resultant in each variable of the system (8.2).

Our plan here is first to present an improvement of the Aizenberg formula; second, to give a detailed description of the complete algorithm; and third, to describe an algorithm permitting to calculate the coefficients of the resultant independently of each other.

So let us return to the non-degenerate system of equations (8.2). By the Hilbert Nullstellensatz (see [**142**, Sec. 130]) there exists a matrix

$$\mathbf{A} = ||a_{jk}||_{j,k=1}^{n}$$

consisting of homogeneous polynomials a_{jk}, such that

$$z_j^{N_j+1} = \sum_{k=1}^{n} a_{jk}(\mathbf{z})P_k(\mathbf{z}), \quad j = 1,\dots,n.$$

Moreover, we can choose the number N_j less than the number $k_1 + \ldots + k_n - n$ (thanks to the Macaulay theorem [**115, 136**]).

THEOREM 8.1 (Kytmanov). *For each polynomial $R(\mathbf{z})$ of degree M we have*

$$\sum_{\mathbf{a}\in E_{\mathfrak{f}}} R(\mathbf{a}) =$$

$$= \sum_{\substack{k_{s,j}\geq 0 \\ \sum_{s,j}^{n} k_{s,j}\leq M}} \frac{(-1)^{\sum k_{s,j}} \prod_{s=1}^{n}\left(\sum_{j=1}^{n} k_{s,j}\right)!}{\prod_{s,j=1}^{n}(k_{s,j})!} \mathfrak{M}\left[\frac{P\mathbf{Q}^{\alpha} \det \mathbf{A} J_{\mathfrak{f}} \prod_{s,j=1}^{n} a_{sj}^{k_{s,j}}}{\prod_{j=1}^{n} z_j^{\beta_j N_j+\beta_j+N_j}}\right],$$

where

$$\alpha_j = \sum_{s=1}^{n} k_{s,j}, \qquad \beta_s = \sum_{j=1}^{n} k_{s,j},$$

$E_{\mathfrak{f}}$ is the set of roots of system (8.2) and $\mathfrak{M}$ is the functional that assigns to a Laurent polynomial

$$\left[\frac{P}{z_1^{j_1}\ldots z_n^{j_n}}\right]$$

its constant term, $J_{\mathfrak{f}}$ is Jacobian of system (8.2).

Proof. Let

$$D_{\mathbf{P}} = \{\mathbf{z}\in\mathbf{C}^n : \ |P_j| < \rho_j, \ j = 1,\dots,n\}$$

be a special analytic polyhedron and let

$$\Gamma_{\mathbf{P}} = \{\mathbf{z} : \ |P_j| = \rho_j, \ j = 1,\dots,n\}$$

be its distinguished boundary. By virtue of the non-degeneracy of system (8.2) the domain $D_{\mathbf{P}}$ is bounded, and by Sard's theorem $\Gamma_{\mathbf{P}}$ is a smooth manifold for almost every ρ_j. We choose ρ_j sufficiently large so that the inequality $|P_j| > |Q_j|$ holds for all j on $\Gamma_{\mathbf{P}}$. Then by Theorem 5.3 we have the identity

$$\sum_{\mathbf{a}\in E_{\mathfrak{f}}} R(\mathbf{a}) = \frac{1}{(2\pi i)^n}\int_{\Gamma_{\mathbf{P}}} R(\mathbf{z})\frac{\mathbf{df}}{\mathbf{f}}, \tag{8.5}$$

where R is an arbitrary polynomial in $\mathbf{C}^n$, and

$$\mathbf{df} = df_1\wedge\ldots\wedge df_n; \quad \frac{\mathbf{df}}{\mathbf{f}} = \frac{f_1}{f_1}\wedge\ldots\wedge\frac{df_n}{f_n}.$$

Since $|P_j| > |Q_j|$ on $\Gamma_{\mathbf{P}}$ it follows that

$$\frac{1}{f_j} = \frac{1}{P_j+Q_j} = \frac{1}{P_j}\cdot\frac{1}{1+Q_j/P_j} = \sum_{s=0}^{\infty}(-1)^s\frac{Q_j^s}{P_j^{s+1}}.$$

By substituting these expressions into (8.5) we get

$$(2\pi i)^n \sum_{\mathbf{a}\in E_\mathfrak{f}} R(\mathbf{a}) = \sum_{||\alpha||\geq 0} (-1)^{||\alpha||} \int_{\Gamma_\mathbf{P}} \frac{R(\mathbf{z})\mathbf{Q}^\alpha J_\mathfrak{f} \mathbf{dz}}{\mathbf{P}^{\alpha+\mathbf{I}}}, \tag{8.6}$$

where $\alpha = (\alpha_1, \dots, \alpha_n)$,

$$\mathbf{Q}^\alpha = Q_1^{\alpha_1} \dots Q_n^{\alpha_n}; \quad \mathbf{P}^{\alpha+\mathbf{I}} = P_1^{\alpha_1+1} \dots P_n^{\alpha_n+1}.$$

We shall show that only a finite number of terms of the sum in (8.6) are different from zero. Indeed, in view of the homogeneity of P_j we must have

$$\int_{\Gamma_\mathbf{P}} \frac{R\mathbf{Q}^\alpha J_\mathfrak{f} \mathbf{dz}}{\mathbf{P}^{\alpha+\mathbf{I}}} = 0,$$

if the degree of the numerator (together with $\mathbf{dz}$) is less than the degree of the denominator. The degree of the numerator is equal to

$$\deg R + n + \sum_{j=1}^n (k_j - 1) + \sum_{j=1}^n \alpha_j(k_j - 1) = \deg R + ||\mathbf{k}|| - ||\alpha|| + \sum_{j=1}^n \alpha_j k_j,$$

while the degree of the denominator equals

$$\sum_{j=1}^n (\alpha_j + 1)k_j = \sum_{j=1}^n \alpha_j k_j + ||\mathbf{k}||.$$

From here we see that for $\deg R < ||\alpha||$ the integral is equal to zero, and so (8.6) reduces to

$$(2\pi i)^n \sum_{\mathbf{a}\in E_\mathfrak{f}} R(\mathbf{a}) = \sum_{||\alpha||\leq M} (-1)^{||\alpha||} \int_{\Gamma_\mathbf{P}} \frac{R\mathbf{Q}^\alpha J_\mathfrak{f} \mathbf{dz}}{\mathbf{P}^{\alpha+\mathbf{I}}}, \tag{8.7}$$

where $M = \deg R$.

In particular, if $R = 1$, i.e. $M = 0$, then the left hand of (8.7) expresses the number of roots of system (8.2) counted with multiplicities, and on the right hand of (8.7) we have the integral

$$\int_{\Gamma_\mathbf{P}} \frac{J_\mathfrak{f} \mathbf{dz}}{\mathbf{P}} = \int_{\Gamma_\mathbf{P}} \frac{\mathbf{dP}}{\mathbf{P}} = (2\pi i)^n k_1 \dots k_n,$$

since the system (8.3) has only one root, and its multiplicity is $k_1 \dots k_n$. We thus recover the Bezout theorem for the system (8.2). In order to calculate the integrals occurring in (8.7) we can use the generalized transformation formula (7.5) for the Grothendieck residue. Let the polynomials P_j and g_j be connected by the relations

$$g_j(\mathbf{z}) = \sum_{k=1}^n a_{jk}(\mathbf{z}) P_k(\mathbf{z}), \quad j = 1, \dots, n, \quad g_j = z_j^{N_j+1},$$

where the matrix

$$\mathbf{A} = ||a_{jk}(\mathbf{z})||_{j,k=1}^n$$

has polynomial entries. Then we get

$$\int_{\Gamma_{\mathbf{P}}} \frac{R\mathbf{dz}}{\mathbf{P}^{\alpha+\mathbf{I}}} = \sum_{\substack{k_{sj}\geq 0 \\ \sum_{s=1}^{n} k_{sj}=\alpha_j,\ \sum_{j=1}^{n} k_{sj}=\beta_s}} \frac{\boldsymbol{\beta}!}{\prod_{s,j=1}^{n}(k_{sj})!} \int_{\Gamma_{\mathbf{g}}} \frac{R\det\mathbf{A}\prod_{s,j=1}^{n} a_{sj}^{k_{sj}}\mathbf{dz}}{\mathbf{g}^{\beta+\mathbf{I}}}, \tag{8.8}$$

where $\boldsymbol{\beta}! = \beta_1!\ldots\beta_n!$, $\boldsymbol{\beta} = (\beta_1,\ldots,\beta_n)$, and the summation in formula (8.8) is taken over all non-negative integer matrices

$$||k_{sj}||_{s,j=1}^{n},$$

which satisfy

$$\sum_{s=1}^{n} k_{sj} = \alpha_j \quad \sum_{j=1}^{n} k_{sj} = \beta_s.$$

Substituting (8.8) in (8.7) and replacing g_j by $z_j^{N_j+1}$ we get

$$(2\pi i)^n \sum_{\mathbf{a}\in E_{\mathfrak{f}}} R(\mathbf{a}) = \sum_{k_{sj}\geq 0} \frac{(-1)^{||\alpha||}\boldsymbol{\beta}!}{\prod_{s,j}(k_{sj})!} \int_T \frac{R\det\mathbf{A}\mathbf{Q}^{\alpha}J_{\mathfrak{f}}\prod_{s,j=1}^{n} a_{sj}^{k_{sj}}\mathbf{dz}}{\prod_{j=1}^{n} z_j^{(\beta_j+1)(N_j+1)}},$$

where $T = \{\mathbf{z} : |z_j| = R_0\}$. □

As a corollary we obtain a result of Aizenberg (see [**11**, sec. 21]), corresponding to

$$P_j = z_j^{k_j}.$$

In this case $N_j + 1 = k_j$, and the matrix $\mathbf{A}$ is just the unit matrix, so in Theorem 8.1 we must take only diagonal matrices k_{sj}, with $k_{jj} = \beta_j = \alpha_j$. This gives us

COROLLARY 8.1 (Aizenberg). *If R is a polynomial of degree M, then*

$$\sum_{\mathbf{a}\in E_{\mathfrak{f}}} R(\mathbf{a}) = \sum_{||\alpha||\leq M} (-1)^{||\alpha||}\mathfrak{M}\left[RJ_{\mathfrak{f}}\frac{Q_1^{\alpha_1}\ldots Q_n^{\alpha_n}}{z_1^{\alpha_1 k_1+k_1-1}\ldots z_n^{\alpha_n k_n+k_n-1}}\right]. \tag{8.9}$$

Formula (8.9) is much simpler than the formula in Theorem 8.1. It is actually valid for a wider class of systems than the system in Corollary 8.1. Let us consider a system

$$f_j(\mathbf{z}) = z_j^{k_j} + \sum_{k=1}^{j-1} z_k\varphi_{jk}(\mathbf{z}) + Q_j(\mathbf{z}) = 0,$$

where φ_{jk} are homogeneous polynomials of degree $k_j - 1$, and where $\deg Q_j < k_j$. As before, $J_{\mathfrak{f}}$ denotes the Jacobian of the system.

THEOREM 8.2 (Aizenberg, Tsikh [**10, 136**]). *The equality*

$$\sum_{\mathbf{a}\in E_{\mathfrak{f}}} R(\mathbf{a}) = \mathfrak{M}\left[\frac{RJ_{\mathfrak{f}}z_1\ldots z_n}{z_1^{k_1}\ldots z_n^{k_n}}\sum_{||\alpha||\geq 0}(-1)^{||\alpha||}\left(\frac{R_1}{z_1^{k_1}}\right)^{\alpha_1}\ldots\left(\frac{R_n}{z_n^{k_n}}\right)^{\alpha_n}\right],$$

holds for any polynomial R of degree m_j in the variable z_j and of total degree M. Here $R_j = f_j - z_j^{k_j}$, and the summation in this formula is taken over all n-tuples $\boldsymbol{\alpha}$ with integer coordinates in the parallelepiped $\sigma = \{\boldsymbol{\alpha} = (\alpha_1, \dots, \alpha_n)$:

$$0 \le \alpha_1 \le M,\ 0 \le \alpha_2 \le k_1(||\mathbf{m}||+1) - m_1 - 2,\ \dots,$$

$$0 \le \alpha_j \le k_1 \dots k_{j-1}(||\mathbf{m}||+1) - k_2 \dots k_{j-1}(m_1+1) - k_3 \dots k_{j-1}(m_2+1) - \dots -$$

$$-k_{j-1}(m_{j-2}+1) - (m_{j-1}+1), \qquad j = s, \dots, n\},$$

where $||\mathbf{m}|| = m_1 + \dots + m_n$.

In [**53**] Cattani, Dickenstein and Sturmfels presented symbolic algorithms for evaluating global residues of systems of **w**–homogeneous polynomials.

Let $\mathbf{w} = (w_1, \dots, w_n) \in \mathbf{N}^n$ be a positive weight vector. The weighted degree of a monomial $\mathbf{z}^\alpha$ is

$$\deg_{\mathbf{w}}(\mathbf{z}^\alpha) = \sum_{j=1}^{n} w_j \alpha_j.$$

Write each polynomial f_j from the system (8.2) as

$$f_j(\mathbf{z}) = P_j(\mathbf{z}) + Q_j(\mathbf{z}),$$

where P_j is **w**–homogeneous, and

$$k_j = \deg_{\mathbf{w}}(P_j) = \deg_{\mathbf{w}}(f_j), \quad \deg_{\mathbf{w}}(Q_j) < \deg_{\mathbf{w}}(f_j), \quad j = 1, \dots, n.$$

In the paper [**53**] the following assumption is made:

$$P_1(\mathbf{z}) = \dots = P_n(\mathbf{z}) = 0 \quad \text{if and only if} \quad \mathbf{z} = 0.$$

Under this assumption the global residue may be expressed as a single residue integral with respect to the initial forms. When the input equations form a Gröbner basis with respect to a term order, this leads to an efficient algorithm for evaluating the global residue. Therefore these results generalize Theorems 8.1 and 8.2.

Yuzhakov also considered in [**157**] a more general class of systems for which there is no need to find the matrix **A**. We consider his result at the end of this section.

Let us consider an example of how formula (8.9) can be applied.

EXAMPLE 8.1. Suppose

$$f_1 = z_1^2 + a_1 z_1 + b_1 z_2 + c_1 = 0, \quad f_2 = z_2^2 + a_2 z_1 + b_2 z_2 + c_2 = 0. \tag{8.10}$$

For R we take the monomials z_1^M, $M \ge 1$. Then we have

$$Q_1 = a_1 z_1 + b_1 z_2 + c_1, \quad Q_2 = a_2 z_1 + b_2 z_2 + c_2,$$

$$J_{\mathbf{f}} = 4 z_1 z_2 + 2 z_2 a_1 + 2 z_1 b_2 + a_1 b_2 - b_1 a_2.$$

Hence

$$S_M = \sum_{\mathbf{a} \in E_{\mathbf{f}}} R(\mathbf{a}) = \sum_{\alpha_1 + \alpha_2 \le M} (-1)^{\alpha_1 + \alpha_2} \mathfrak{M} \left[\frac{J_{\mathbf{f}} Q_1^{\alpha_1} Q_2^{\alpha_2}}{z_1^{2\alpha_1 + 1 - M} z_2^{2\alpha_2 + 1}} \right].$$

For $M = 1$ we get (for $\boldsymbol{\alpha} = (0,0)$)

$$\mathfrak{M}\left[\frac{J_{\mathbf{f}}}{z_2}\right] = 2a_1,$$

for $\boldsymbol{\alpha} = (1,0)$

$$\mathfrak{M}\left[\frac{J_f Q_1}{z_1^2 z_2}\right] = 4a_1,$$

and for $\boldsymbol{\alpha} = (0,1)$

$$\mathfrak{M}\left[\frac{J_f Q_2}{z_2^3}\right] = 0,$$

i.e. $S_1 = 2a_1 - 4a_1 = -2a_1$. In exactly the same way we find $S_2,\ S_3,\ S_4$.

$$S_2 = 2b_1b_2 + 2a_1^2 - 4c_1,$$

$$S_3 = -2a_1^3 - 3a_2b_1^2 - 3a_1b_1b_2 + 6a_1c_1,$$

$$S_4 = 2a_1^4 + 4a_1^2b_1b_2 + 2b_1^2b_2^2 - 8a_1^2c_1 - 4b_1^2c_2 + 4c_1^2 + 8a_1a_2b_1^2 - 4b_1b_2c_1.$$

The system (8.10) has four roots. If we denote their first coordinates by

$$\delta_j, \quad j = 1,2,3,4,$$

then the expressions S_1, S_2, S_3, S_4 are power sums of δ_j, i.e.

$$S_k = \sum_{j=1}^{4} \delta_j^k.$$

In order to write down the resultant $R(z_1)$ of system (8.10) in the variable z_1

$$\left(R(z_1) = z_1^4 + \gamma_1 z_1^3 + \gamma_2 z_1^2 + \gamma_3 z_1 + \gamma_4\right)$$

we must use the Newton recursion formulas (2.1):

$$S_k + \gamma_1 S_{k-1} + \ldots + \gamma_{k-1}S_1 + k\gamma_k = 0, \quad k \leq \deg R(z_1).$$

This yields the following identities:

$$\gamma_1 = -S_1 = 2a_1,$$

$$S_2 + \gamma_1 S_1 + 2\gamma_2 = 0,$$

$$\gamma_2 = a_1^2 - b_1b_2 + 2c_1,$$

$$S_3 + \gamma_1 S_2 + \gamma_2 S_1 + 3\gamma_3 = 0,$$

$$\gamma_3 = a_2b_1^2 + 2a_1c_1 - a_1b_1b_2,$$

$$S_4 + \gamma_1 S_3 + \gamma_2 S_2 + \gamma_3 S_1 + 4\gamma_4 = 0,$$

$$\gamma_4 = b_1^2c_2 + c_1^2 - b_1b_2c_1.$$

It follows then that

$$R(z_1) = z_1^4 + 2a_1z_1^3 + (a_1^2 - b_1b_2 + 2c_1)z_1^2 +$$
$$+(a_2b_1^2 + 2a_1c_1 - a_1b_1b_2)z_1 + b_1^2c_2 + c_1^2 - b_1b_2c_1.$$

If we directly eliminate z_2 in (8.10), then we arrive at the same answer.

The general case can be handled in the same way. For a given system (8.2) we find the power sums

$$S_k, \quad k = 1,2\ldots, k_1\ldots k_n,$$

by using Theorem 8.1 with $R(\mathbf{z})$ equal to z_1^k. Then, by the Newton formulas (2.1) we find the coefficients γ_j of the polynomial $R(z_1)$. Its roots are the first coordinates of the roots of system (8.2). In this way we eliminate from (8.2) all variables except z_1.

The principal difficulty of this method is the determination of the matrix **A**. We now discuss some ways of calculating **A**.

1. The general method for finding **A** is the method of undetermined coefficients. Since we have an estimate of the degree N_j, we can write down $a_{jk}(\mathbf{z})$ with unknown coefficients, then perform arithmetic operations, equating the coefficients at equal degrees of $\mathbf{z}$, and hence get a system (in general under determined) of linear equations. By solving it we can find $a_{jk}(\mathbf{z})$.

EXAMPLE 8.2. Consider the system

$$\begin{aligned} P_1 &= \alpha_1 z_1^2 - \alpha_2 z_2^2, \\ P_2 &= \beta_1 z_2 - \beta_2 z_3, \\ P_3 &= \gamma_1 z_3 - \gamma_2 z_1. \end{aligned}$$

Here we have $N_j \le 1$, and hence (for homogeneity reasons) $\deg a_{j1} = 0$ and $\deg a_{j2} = \deg a_{j3} = 1$. This means that

$$a_{11} = a = \text{const}, \quad a_{12} = b_1 z_1 + b_2 z_2 + b_3 z_3, \quad a_{13} = c_1 z_1 + c_2 z_2 + c_3 z_3.$$

From the equation $z_1^2 = a_{11}P_1 + a_{12}P_2$ we get a system of equations:

$$\begin{aligned} &\alpha_1 a - \gamma_2 c_1 = 1, \\ &-\alpha_2 a + \beta_1 b_2 = 0, \\ &-\beta_2 b_3 + \gamma_1 c_3 = 0, \\ &\beta_1 b_1 - \gamma_2 c_2 = 0, \\ &-\beta_2 b_1 - \gamma_2 c_3 + \gamma_1 c_1 = 0, \\ &\beta_1 b_3 - \beta_2 b_2 + \gamma_1 c_2 = 0. \end{aligned}$$

Solving this system we find

$$\begin{aligned} a &= \frac{\gamma_1^2\beta_1^2}{\alpha_1\beta_1^2\gamma_1^2 - \alpha_2\beta_2^2\gamma_2^2}, \\ c_1 &= \frac{\alpha_2\beta_2^2\gamma_2}{\alpha_1\beta_1^2\gamma_1^2 - \alpha_2\beta_2^2\gamma_2^2}, \\ b_2 &= \frac{\alpha_2\beta_2\gamma_1^2}{\alpha_1\beta_1^2\gamma_1^2 - \alpha_2\beta_2^2\gamma_2^2}, \end{aligned}$$

$$b_3 = \frac{\gamma_1}{\beta_2}c_3, \qquad b_1 = \frac{\alpha_2\beta_1\gamma_1\gamma_2}{\alpha_1\beta_1^2\gamma_1^2 - \alpha_2\beta_2^2\gamma_2^2} - \frac{\gamma_2}{\beta_2}c_3,$$

$$c_2 = \frac{\alpha_2\beta_1\beta_2\gamma_1}{\alpha_1\beta_1^2\gamma_1^2 - \alpha_2\beta_2^2\gamma_2^2} - \frac{\beta_1}{\beta_2}c_3.$$

Setting $c_3 = 0$ we then get

$$a = \frac{\gamma_1^2\beta_1^2}{\alpha_1\beta_1^2\gamma_1^2 - \alpha_2\beta_2^2\gamma_2^2}, \quad b_1 = \frac{\alpha_2\beta_2\gamma_1\gamma_2}{\alpha_1\beta_1^2\gamma_1^2 - \alpha_2\beta_2^2\gamma_2^2},$$

$$b_2 = \frac{\alpha_2\beta_1\gamma_1^2}{\alpha_1\beta_1^2\gamma_1^2 - \alpha_2\beta_2^2\gamma_2^2}, \quad b_3 = 0, \quad c_3 = 0,$$

$$c_1 = \frac{\alpha_2\beta_2^2\gamma_2}{\alpha_1\beta_1^2\gamma_1^2 - \alpha_2\beta_2^2\gamma_2^2}, \quad c_2 = \frac{\alpha_2\beta_1\beta_2\gamma_1}{\alpha_1\beta_1^2\gamma_1^2 - \alpha_2\beta_2^2\gamma_2^2}.$$

In the same manner we can find polynomials a_{2k} and a_{3k}, $k = 1,2,3$. We notice that the number of equations in this method quickly increases, and already for $n > 3$ it becomes very unwieldy.

2. Suppose $n = 2$ and consider the polynomials

$$P_1 = a_0x^k + a_1x^{k-1}y + \ldots + a_ky^k,$$
$$P_2 = b_0x^m + b_1x^{m-1}y + \ldots + b_my^m,$$
$$z_1 = x,\ z_2 = y,\ k_1 = k,\ k_2 = m.$$

Consider also the square matrix of dimension $(m+k)$:

$$\mathrm{Res} = \begin{pmatrix} a_0 & a_1 & a_2 & \ldots & a_k & 0 & \ldots & 0 \\ 0 & a_0 & a_1 & \ldots & a_{k-1} & a_k & \ldots & 0 \\ \ldots & \ldots & \ldots & \ldots & \ldots & \ldots & \ldots & \ldots \\ 0 & 0 & 0 & a_0 & \ldots & \ldots & \ldots & a_k \\ b_0 & b_1 & b_2 & \ldots & b_m & 0 & \ldots & 0 \\ 0 & b_0 & b_1 & \ldots & b_{m-1} & b_m & \ldots & 0 \\ \ldots & \ldots & \ldots & \ldots & \ldots & \ldots & \ldots & \ldots \\ 0 & 0 & 0 & \ldots & b_0 & \ldots & \ldots & b_m \end{pmatrix}.$$

The entries a_j are all in the first m lines and the b_j are contained in the last k lines. The determinant of this matrix

$$\Delta = \det \mathrm{Res}$$

is the classical resultant for the system of functions P_1 and P_2 (see Sec. 6). Let Δ_j denote the cofactor to the j-th element of the last column, and let $\tilde{\Delta}_j$ denote the cofactor to the j-th element of the first column.

Consider then the system (compare with (6.4)):

$$\begin{aligned} x^{m-1}P_1 &= a_0x^{m+k-1} + a_1yx^{m+k-2} + \ldots + a_ky^kx^{m-1}, \\ yx^{m-2}P_1 &= a_0yx^{m+k-2} + a_1y^2x^{m+k-3} + \ldots + a_ky^{k+1}x^{m-2}, \\ \ldots & \quad \ldots\ldots\ldots\ldots\ldots\ldots \\ y^{m-1}P_1 &= a_0y^{m-1}x^k + a_1y^mx^{k-1} + \ldots + a_ky^{m+k-1}, \\ x^{k-1}P_2 &= b_0x^{m+k-1} + b_1yx^{m+k-2} + \ldots + b_my^mx^{k-1}, \\ yx^{k-2}P_2 &= b_0yx^{m+k-2} + b_1y^2x^{m+k-3} + \ldots + b_my^{m+1}x^{k-2}, \\ \ldots & \quad \ldots\ldots\ldots\ldots\ldots\ldots \\ y^{k-1}P_2 &= b_0y^{k-1}x^m + b_1y^kx^{m-1} + \ldots + b_my^{m+k-1}. \end{aligned}$$

Multiplying the first equation by Δ_1, the second by Δ_2 and so on, and adding them together, we get (equating the coefficients at equal degrees and taking into account the properties of the determinant):

$$(\Delta_1x^{m-1} + \Delta_2yx^{m-2} + \ldots + \Delta_my^{m-1})P_1 +$$
$$+(\Delta_{m+1}x^{k-1} + \Delta_{m+2}yx^{k-2} + \ldots + \Delta_{m+k}y^{k-1})P_2 = \Delta \cdot y^{m+k-1}.$$

Similarly, multiplying this system consecutively by $\tilde{\Delta}_j$ and summing up we get:

$$(\tilde{\Delta}_1x^{m-1} + \tilde{\Delta}_2yx^{m-2} + \ldots + \tilde{\Delta}_my^{m-1})P_1 +$$
$$+(\tilde{\Delta}_{m+1}x^{k-1} + \tilde{\Delta}_{m+2}yx^{k-2} + \ldots + \tilde{\Delta}_{m+k}y^{k-1})P_2 = \Delta \cdot x^{m+k-1}.$$

Since the system $P_1 = 0,\ P_2 = 0$ has only one common root, namely the point 0, we must have $\Delta \neq 0$ (see Sec. 6), and hence we can take the following polynomials as our a_{jk}:

$$\begin{aligned}
a_{11} &= \frac{1}{\Delta}\sum_{j=1}^{m} \tilde{\Delta}_j y^{j-1} x^{m-j},\\
a_{12} &= \frac{1}{\Delta}\sum_{j=1}^{k} \tilde{\Delta}_{m+j} y^{j-1} x^{k-j},\\
a_{21} &= \frac{1}{\Delta}\sum_{j=1}^{m} \Delta_j y^{j-1} x^{m-j},\\
a_{22} &= \frac{1}{\Delta}\sum_{j=1}^{k} \Delta_{m+j} y^{j-1} x^{k-j}.
\end{aligned}$$

A computation of det **A** now gives:

$$\det \mathbf{A} = \frac{1}{\Delta^2}\sum_{l=1}^{m+k-2} y^l x^{m+k-l-2} \sum_{\substack{j+s=l\\ 0\le j\le m-1\\ 0\le s\le k-1}} \begin{vmatrix} \tilde{\Delta}_{j+1} & \tilde{\Delta}_{m+s+1} \\ \Delta_{j+1} & \Delta_{m+s+1} \end{vmatrix}.$$

Observe that Δ_j/Δ are the entries in the last line of the inverse matrix of Res, whereas $\tilde{\Delta}_j/\Delta$ are the entries in the first line of this matrix. By the formula for the determinant of an inverse matrix (see [**75**, p.31]) we get

$$\begin{vmatrix} \Delta_{j+1} & \Delta_{s+m+1} \\ \tilde{\Delta}_{j+1} & \tilde{\Delta}_{s+m+1} \end{vmatrix} = \Delta\cdot\Delta(j+1,\ s+m+1),$$

where $\Delta(j+1,\ s+m+1)$ is the cofactor to the second order minor of the matrix Res, corresponding to the entries of the first and last columns and the $(j+1)$-th and $(k+m+1)$-th lines.

This gives us

$$\det \mathbf{A} = \frac{1}{\Delta}\sum_{l=0}^{m+k-2} y^l x^{m+k-l-2} \sum_{\substack{j+s=l\\ 0\le j\le m-1\\ 0\le s\le k-1}} \Delta(j+1,\ s+m+1).$$

EXAMPLE 8.3. Starting with

$$\begin{aligned}
P_1 &= a_0x^2 + a_1xy + a_2y^2,\\
P_2 &= b_0x^2 + b_1xy + b_2y^2,
\end{aligned}$$

we obtain the following identities:

$$\text{Res} = \begin{pmatrix} a_0 & a_1 & a_2 & 0 \\ 0 & a_0 & a_1 & a_2 \\ b_0 & b_1 & b_2 & 0 \\ 0 & b_0 & b_1 & b_2 \end{pmatrix},$$

$$\tilde{\Delta}_1 = \begin{vmatrix} a_0 & a_1 & a_2 \\ b_1 & b_2 & 0 \\ b_0 & b_1 & b_2 \end{vmatrix} = a_2 b_1^2 - a_2 b_0 b_2 + a_0 b_2^2 - a_1 b_1 b_2,$$

$$\tilde{\Delta}_2 = -\begin{vmatrix} a_1 & a_2 & 0 \\ b_1 & b_2 & 0 \\ b_0 & b_1 & b_2 \end{vmatrix} = a_2 b_1 b_2 - a_1 b_2^2,$$

$$\tilde{\Delta}_3 = \begin{vmatrix} a_1 & a_2 & 0 \\ a_0 & a_1 & a_2 \\ b_0 & b_1 & b_2 \end{vmatrix} = a_1^2 b_2 - a_1 a_2 b_1 + a_2^2 b_0 - a_0 a_2 b_2,$$

$$\tilde{\Delta}_4 = -\begin{vmatrix} a_1 & a_2 & 0 \\ a_0 & a_1 & a_2 \\ b_1 & b_2 & 0 \end{vmatrix} = a_1 a_2 b_2 - a_2^2 b_1,$$

$$\Delta_1 = -\begin{vmatrix} 0 & a_0 & a_1 \\ b_0 & b_1 & b_2 \\ 0 & b_0 & b_1 \end{vmatrix} = a_0 b_0 b_1 - a_1 b_0^2,$$

$$\Delta_2 = \begin{vmatrix} a_0 & a_1 & a_2 \\ b_0 & b_1 & b_2 \\ 0 & b_0 & b_1 \end{vmatrix} = a_0 b_1^2 - a_0 b_0 b_2 - a_1 b_0 b_1 + a_2 b_0^2,$$

$$\Delta_3 = -\begin{vmatrix} a_0 & a_1 & a_2 \\ 0 & a_0 & a_1 \\ 0 & b_0 & b_1 \end{vmatrix} = a_0 a_1 b_0 - a_0^2 b_1,$$

$$\Delta_4 = \begin{vmatrix} a_0 & a_1 & a_2 \\ 0 & a_0 & a_1 \\ b_0 & b_1 & b_2 \end{vmatrix} = a_0^2 b_2 - a_0 a_1 b_1 - a_0 a_1 b_1 + a_1^2 b_0,$$

$$\Delta = a_0 a_2 b_1^2 - 2 a_0 a_2 b_0 b_2 + a_0^2 b_2^2 - a_0 a_1 b_1 b_2 + a_1^2 b_0 b_2 - a_1 a_2 b_0 b_1 + a_2^2 b_0^2.$$

Therefore

$$\begin{aligned}
a_{11} &= \frac{1}{\Delta}(x(a_2 b_1^2 - a_2 b_0 b_2 + a_0 b_2^2 - a_1 b_1 b_2) + y(a_2 b_1 b_2 - a_1 b_2^2)), \\
a_{12} &= \frac{1}{\Delta}(x(a_1^2 b_2 - a_1 a_2 b_1 + a_2^2 b_0 - a_0 a_2 b_2) + y(a_1 a_2 b_2 - a_2^2 b_1)), \\
a_{21} &= \frac{1}{\Delta}(x(a_0 b_0 b_1 - a_1 b_0^2) + y(a_0 b_1^2 - a_0 b_0 b_2 - a_1 b_0 b_1 + a_2 b_0^2)), \\
a_{22} &= \frac{1}{\Delta}(x(a_0 a_1 b_0 - a_0^2 b_1) + y(a_0^2 b_2 - a_0 a_1 b_1 - a_0 a_2 b_0 + a_1^2 b_0)).
\end{aligned}$$

It is not difficult to verify that

$$x^3 = a_{11}P_1 + a_{12}P_2,$$
$$y^3 = a_{21}P_1 + a_{22}P_2,$$

$$\det \mathbf{A} = \frac{1}{\Delta}\left(-x^2\begin{vmatrix} a_0 & a_1 \\ b_1 & b_2 \end{vmatrix} - y^2\begin{vmatrix} a_1 & a_2 \\ b_1 & b_2 \end{vmatrix} + xy\left(\begin{vmatrix} a_0 & a_1 \\ b_1 & b_2 \end{vmatrix} + \begin{vmatrix} a_1 & a_2 \\ b_0 & b_1 \end{vmatrix}\right)\right) =$$
$$= \frac{1}{\Delta}(x^2(a_1b_0 - a_0b_1) + xy(a_0b_2 - b_0a_2) + y^2(a_2b_1 - a_1b_2)).$$

3. The method we just described for determining the matrix $\mathbf{A}$ carries over to $\mathbf{C}^n$, $n > 2$, in many cases. Let there be given a non-degenerate system (8.2) and let

$$P_1, \dots, P_n$$

be the highest homogeneous parts of this system. We consider the polynomials P_1 and P_j as polynomials in the variable z_1 and we make up the resultant of this pair. In this way we obtain a homogeneous polynomial R_j in the variables $z_2, \dots, z_n$, and the previous reasoning shows (see item 2) that

$$R_j = R_j' P_1 + R_j'' P_2,$$

where R_j', R_j'' are polynomials in $(z_1, \dots, z_n)$. Thus we arrive to a system of homogeneous polynomials R_j in the variables $z_2, \dots, z_n$. We also have

$$R_j \not\equiv 0$$

since the initial system (8.2) was non-degenerate.

Then, applying the same arguments to the pairs R_2, R_j and eliminating z_2, we get a system of polynomials

$$F_j, \ j = 3, \dots, n,$$

in the variables $(z_3, \dots, z_n)$. Finally we are left with one polynomial of the form $a \cdot z_n^M$. If $a \not\equiv 0$, then we have

$$G_1P_1 + \dots + G_nP_n = az_n^M,$$

i.e. we obtain the wanted decomposition. In exactly the same way we can write down monomials z_j^M. Notice that with the degrees of the polynomials grow rapidly at every step, and the final degree M will in general be greater than $||\mathbf{k}|| - n + 1$. But here we do not have to solve any systems of equations.

EXAMPLE 8.4. Let

$$P_1 = a_0x^2 + a_1xy + a_2xz + a_3y^2 + a_4yz + a_5z^2,$$
$$P_2 = b_0x^2 + b_1xy + b_2xz + b_3y^2 + b_4yz + b_5z^2,$$
$$P_3 = c_0x^2 + c_1xy + c_2xz + c_3y^2 + c_4yz + c_5z^2.$$

Making up the resultant R_1 of the polynomials P_1 and P_2 and eliminating x, we have

$$\begin{aligned} R_1 \;=\; & y^4(a_0a_3b_1^2 - 2a_0a_3b_0b_3 + a_0^2b_3^2 - a_0a_1b_1b_3 + a_1^2b_0b_3 - a_1a_3b_0b_3 + \\ & + a_3^2b_0^2) + y^3z(a_0a_4b_1^2 + 2a_0a_3b_1b_3 - 2a_0a_4b_0b_3 - 2a_0a_3b_0b_4 + \\ & + 2a_0^2b_3b_4 - a_0a_1b_1b_4 - a_0a_1b_2b_3 - a_0a_2b_1b_2 + 2a_1a_2b_0b_3 + a_1^2b_0b_4 - \\ & - a_1a_3b_0b_2 - a_2a_3b_0b_1 - a_1a_4b_0b_3 + 2a_3a_4b_0^2) + y^2z^2(a_0a_3b_2^2 + \\ & + 2a_0a_4b_1b_2 + a_0a_5b_1^2 - 2a_0a_3b_0b_5 - 2a_0a_4b_0b_4 - 2a_0a_5b_0b_3 + \\ & + 2a_0^2b_3b_6 + a_0^2b_4^2 - a_0a_1b_1b_5 - a_0a_1b_2b_4 - a_0a_2b_1b_4 - a_0a_2b_2b_3 + \\ & + a_1^2b_0b_5 + 2a_1a_2b_0b_4 + a_2^2b_0b_3 + a_1a_4b_0b_2 + a_1a_5b_0b_2 + a_2a_3b_0b_2 + \\ & + a_2a_4b_0b_1 + 2b_0^2a_3a_5 + b_0^2a_4^2) + yz^3(a_0a_4b_2^2 + 2a_0a_5b_1b_2 - 2a_0a_4b_0b_5 - \\ & - 2a_0a_5b_0b_4 + 2a_0^2b_4b_5 - a_0a_1b_2b_5 - a_0a_2b_1b_5 - a_0a_2b_2b_4 + 2a_1a_2b_0b_5 + \\ & + a_2^2b_0b_4 - a_1a_5b_0b_2 - a_2a_4b_0b_2 - a_2a_5b_0b_1 + 2b_0^2a_4a_5) + z^4(a_0a_5b_2^2 - \\ & - 2a_0a_5b_0b_5 + a_0^2b_5^2 - a_0a_2b_2b_5 + a_2^2b_0b_5 - a_2a_5b_0b_2 + b_0^2a_5^2). \end{aligned}$$

The resultant R_2 of the polynomials P_1 and P_2 differs from R_1 in that the b_j are replaced by the c_j. Furthermore,

$$F_1'P_1 + F_2'P_2 = R_1, \quad F_1''P_1 + F_2''P_3 = R_2.$$

Forming now the resultant of the polynomials R_1 and R_2 and eliminating y, we obtain $G_1P_1 + G_2P_2 = az^8$. From here we get

$$G_1(F_1'P_1 + F_2'P_2) + G_2(F_1''P_1 + F_2''P_3) = az^8$$

and hence

$$P_1(G_1F_1' + G_2F_1'') + G_1F_2'P_2 + G_2F_2''P_3 = az^8.$$

We have thus found the entries of the matrix $\mathbf{A}$. Notice here that the degree in z is equal to 8, but by the Macaulay theorem we can find polynomials a_{jk} of degree 4.

We shall now show how we can simplify the calculations by considering the system (8.2) as a system with parameter.

Let our system have the form

$$\begin{aligned} f_1 = f_1(\mathbf{z}, w) &= 0, \\ \dots\dots\dots\dots& \\ f_n = f_n(\mathbf{z}, w) &= 0, \\ f_{n+1} = f_{n+1}(\mathbf{z}, w) &= 0, \end{aligned} \tag{8.11}$$

where $\mathbf{z} \in \mathbf{C}^n$ and $w \in \mathbf{C}$. We shall assume that for almost every w the system

$$\begin{aligned} f_1(\mathbf{z}, w) &= 0, \\ \dots\dots\dots& \\ f_n(\mathbf{z}, w) &= 0 \end{aligned} \tag{8.12}$$

is non-degenerate.

For this it suffices to verify that (8.12) is non-degenerate for at least one value of w.

Let $f_1(\mathbf{z}, w) = P_j(\mathbf{z}, w) + Q_j(\mathbf{z}, w)$, where P_j is the highest homogeneous part of f_j in the variables $z_1, \ldots, z_n$, $\deg P_j = k_j$. We then find a matrix

$$\mathbf{A} = ||a_{jk}(\mathbf{z}, w)||_{j,k=1}^{n}$$

such that the conditions

$$z_j^{N_j+1} = \sum_{k=1}^{n} a_{jk}(\mathbf{z}, w) P_k(\mathbf{z}, w), \quad j = 1, \ldots, n,$$

are fulfilled. Observe that the coefficients of the polynomials $a_{jk}(\mathbf{z}, w)$ are rational functions in w, and if P_k are independent of w, then a_{jk} are also independent of w.

Further, applying Theorem 8.1 to the polynomials $f_{n+1}^j(\mathbf{z}, w)$, $j = 1, 2, \ldots, k_1 \ldots k_n$, we obtain expressions $S_j(w)$ which are rational functions in w (if P_k are independent of w, then $S_j(w)$ are polynomials in w). We recall that

$$S_j(w) = \sum_{\mathbf{a} \in E_{\mathfrak{f}}} f_{n+1}^j(\mathbf{a}, w),$$

where $E_{\mathfrak{f}}$ is the set of roots of system (8.12) with w is fixed. (These roots also depend on w.) Using the Newton formulas (2.1) we can now find the function

$$R(w) = \prod_{\mathbf{a} \in E_{\mathfrak{f}}} f_{n+1}(\mathbf{a}, w).$$

In the terminology of Tsikh (see [**11**, sec. 21]) this function is the resultant of the polynomial f_{n+1} with respect to the system

$$f_j = 0, \quad j = 1, \ldots, n.$$

In general, $R(w)$ is a rational function in w, but if P_k are independent of w, then $R(w)$ is a polynomial. This function has the property that $R(w) = 0$ if and only if the system (8.11) has a solution for this given w (see [**11**, sec. 21]). So if we remove the denominator from $R(w)$ then we obtain the desired resultant.

We now describe this method for the case of three equations in $\mathbf{C}^3$:

$$\begin{aligned} f_1(x, y, w) &= 0, \\ f_2(x, y, w) &= 0, \\ f_3(x, y, w) &= 0. \end{aligned}$$

First we choose two equations and a parameter. Suppose the system

$$\begin{aligned} f_1(x, y, w) &= 0, \\ f_2(x, y, w) &= 0 \end{aligned}$$

is non-degenerate for some value of w. We write down f_1 and f_2 as polynomials in x, y and select the highest homogeneous parts:

$$f_1 = P_1 + Q_1, \quad f_2 = P_2 + Q_2.$$

Then we construct the matrix Res for P_1 and P_2 (the entries of this matrix will be polynomials in w), and we find

$$\Delta_j, \ \tilde{\Delta}_j, \ \Delta(j, s)$$

from which we get

$$a_{11},\ a_{12},\ a_{21},\ a_{22},\ \det \mathbf{A}.$$

Next we apply Theorem 8.1 to the functions f_3^j, $j = 1, 2, \dots, k_1k_2$, and find the power sums $S_j(w)$. Finally we use the Newton formulas to construct the resultant $R(w)$.

EXAMPLE 8.5. Let there be given a system

$$\begin{aligned} f_1 &= a_0x + a_1y + a_2, \\ f_2 &= b_0x^2 + b_1xy + b_2y^2 + b_3x + b_4y + b_5, \\ f_3 &= c_0x^2 + c_1xy + c_2y^2 + c_3x + c_4y + c_5. \end{aligned}$$

We shall assume that the coefficients of these polynomials are polynomials in w. Also let the system $f_1 = 0$, $f_2 = 0$ be non-degenerate for some w.

Simple computations give $\text{Res} = \begin{pmatrix} a_0 & a_1 & 0 \\ 0 & a_0 & a_1 \\ b_0 & b_1 & b_2 \end{pmatrix}$,

$$\tilde{\Delta}_1 = \begin{vmatrix} a_0 & a_1 \\ b_1 & b_2 \end{vmatrix} = a_0b_2 - a_1b_1,$$

$$\tilde{\Delta}_2 = -\begin{vmatrix} a_1 & 0 \\ b_1 & b_2 \end{vmatrix} = -a_1b_2, \quad \tilde{\Delta}_3 = \begin{vmatrix} a_1 & 0 \\ a_0 & a_1 \end{vmatrix} = a_1^2,$$

$$\Delta_1 = \begin{vmatrix} 0 & a_0 \\ b_0 & b_1 \end{vmatrix} = -a_0b_0, \quad \Delta_2 = -\begin{vmatrix} a_0 & a_1 \\ b_0 & b_1 \end{vmatrix} = a_1b_0 - a_0b_1,$$

$$\Delta_3 = \begin{vmatrix} a_0 & a_1 \\ 0 & a_0 \end{vmatrix} = a_0^2,$$

$$\Delta = a_0^2b_2 - a_0a_1b_1 + a_1^2b_0,$$

$$a_{11} = \frac{1}{\Delta}(\tilde{\Delta}_1x + \tilde{\Delta}_2y) = \frac{1}{\Delta}(x(a_0b_2 - a_1b_1) - ya_1b_2),$$

$$a_{12} = \frac{1}{\Delta}\tilde{\Delta}_3 = \frac{1}{\Delta}a_1^2,$$

$$a_{21} = \frac{1}{\Delta}(\Delta_1x + \Delta_2y) = \frac{1}{\Delta}(-xa_0b_0 + y(a_1b_0 - a_0b_1)),$$

$$a_{22} = \frac{1}{\Delta}\Delta_3 = \frac{1}{\Delta}a_0^2.$$

We then have

$$\begin{aligned} x^2 &= a_{11}P_1 + a_{12}P_2, \\ y^2 &= a_{21}P_1 + a_{22}P_2, \end{aligned}$$

where

$$P_1 = a_0x + a_1y, \quad P_2 = b_0x^2 + b_1xy + b_2y^2,$$

$$\det \mathbf{A} = \frac{1}{\Delta}(a_0x - a_1y),$$

$$\Delta(1,3) = a_0, \quad \Delta(2,3) = -a_1,$$

$$J = \begin{vmatrix} a_0 & a_1 \\ 2b_0x + b_1y + b_2 & b_1x + 2b_2y + b_4 \end{vmatrix} =$$

$$= x(a_0b_1 - 2a_1b_0) + y(2b_2a_0 - a_1b_1) + a_0b_4 - a_1b_3,$$

$$Q_1 = a_2, \quad Q_2 = b_3x + b_4y + b_5.$$

Writing down the formula from Theorem 8.2 in our case, we get

$$S_j = \sum_{\substack{k_{sm}\geq 0 \\ k_{11}+k_{12}+k_{21}+k_{22}\leq 2j \\ j=1,2}} \frac{(-1)^{||\mathbf{k}||}(k_{11}+k_{12})!(k_{21}+k_{22})!}{k_{11}!k_{12}!k_{21}!k_{22}!}\times$$

$$\times\mathfrak{M}\left[\frac{f_3^j Q_1^{k_{11}+k_{21}} Q_2^{k_{12}+k_{22}} J_{\mathfrak{f}} \det \mathbf{A} a_{11}^{k_{11}} \dots a_{22}^{k_{22}}}{x^{2(k_{11}+k_{12})+1}y^{2(k_{21}+k_{22})+1}}\right].$$

In order to find S_j we must substitute the monomials x^ty^m, $t+m \leq 4$, into this formula (the result is denoted $S_{t,m}$). We have

$$S_{0,0} = \mathfrak{M}\left[\frac{J_{\mathfrak{f}} \det \mathbf{A}}{xy}\right] = \frac{1}{\Delta}(a_0(2b_2a_0 - a_1b_1) - a_1(a_0b_1 - 2a_1b_0)) = 2,$$

$$S_{1,0} = \mathfrak{M}\left[\frac{xJ_{\mathfrak{f}} \det \mathbf{A}}{xy} - \frac{xQ_1J_{\mathfrak{f}} \det \mathbf{A}a_{11}}{x^3y} - \frac{xQ_2J_{\mathfrak{f}} \det \mathbf{A}a_{12}}{x^3y}\right.$$

$$\left. - \frac{xQ_1J_{\mathfrak{f}} \det \mathbf{A}a_{21}}{xy^3} - \frac{xQ_2J_{\mathfrak{f}} \det \mathbf{A}a_{22}}{xy^3}\right] = \frac{1}{\Delta}a_1(a_1b_3 - a_0b_4) -$$

$$-\frac{1}{\Delta^2}(a_0a_2(a_0b_2 - a_1b_1)(2b_2a_0 - a_1b_1) - a_1a_2(a_0b_2 - a_1b_1)(a_0b_1 -$$

$$-2a_1b_0) - a_1a_2a_0b_2(a_0b_1 - 2a_1b_0) + a_1^2a_0b_3(2b_2a_0 - a_1b_1) -$$

$$-a_1^2b_3(a_0b_1 - 2a_1b_0) + a_1^2a_0b_4(a_0b_1 - 2a_1b_0) -$$

$$-a_2a_1(2a_0b_2 - a_1b_1)(a_1b_0 - a_0b_1) - a_0^2a_1b_4(2b_2a_0 - a_1b_1)) =$$

$$= \frac{1}{\Delta}(a_1^2b_3 - a_0a_1b_4 - 2a_0a_2b_2 + a_1a_2b_1 - 2a_1^2b_3 + 2a_0a_1b_4) =$$

$$= \frac{1}{\Delta}(a_0a_1b_4 + a_1a_2b_1 - a_1^2b_3 - 2a_0a_2b_2).$$

Similarly

$$S_{0,1} = \frac{1}{\Delta}(a_0a_1b_3 + a_0a_2b_1 - a_0^2b_4 - 2a_1a_2b_0).$$

Another method for finding the matrix $\mathbf{A}$, by means of on Gröbner bases, will be considered in Chapter 4.

Now we consider some results of Yuzhakov (see [**157**]). Let

$$f_j(\mathbf{z}) = P_j(\mathbf{z}) + Q_j(\mathbf{z}) = 0, \quad j = 1, \dots, n, \tag{8.13}$$

be a non-degenerate system of algebraic equations. The degree of the homogeneous polynomial P_j is equal to k_j and $\deg Q_j < k_j$, $j = 1, \dots, n$.

We assume that system (8.13) satisfies the following conditions:

(*) the polynomial P_1 is a product of linear factors

$$P_1(\mathbf{z}) = \prod_{\beta_1=1}^{\mu} L_{\beta_1}^{p_{\beta_1}}(\mathbf{z}), \quad \sum_{\beta_1}^{\mu} p_{\beta_1} = k_1,$$

(**) for any $\beta_1 \in \{1, \dots, \mu\}$ the restriction of the polynomial P_2 to the plane

$$T_{\beta_1} = \{\mathbf{z} : L_{\beta_1}(\mathbf{z}) = 0\}$$

is a product of linear factors

$$P_2(\mathbf{z})|_{T_{\beta_1}} = \prod_{\beta_2=1}^{\mu(\beta_1)} L_{\beta_1\beta_2}^{p_{\beta_1\beta_2}}(\mathbf{z}) \not\equiv 0,$$

and similarly,

(* * *) the restriction of P_j to the $(n-j+1)$–dimensional plane

$$T_{\beta_1 \dots \beta_{j-1}} = \left\{\mathbf{z} : L_{\beta_1}(\mathbf{z}) = \dots = L_{\beta_1\dots\beta_{j-1}}(\mathbf{z}) = 0\right\}$$

is a product of linear factors

$$P_j(\mathbf{z})|_{T_{\beta_1\dots\beta_{j-1}}} = \prod_{\beta_j=1}^{\mu(\beta_1,\dots,\beta_{j-1})} L_{\beta_1\dots\beta_j}^{p_{\beta_1\dots\beta_j}}(\mathbf{z}) \not\equiv 0,$$

$$\sum_{\beta_j} p_{\beta_1\dots\beta_j} = k_j, \quad j = 1, \dots, n.$$

It is clear that $\mu(\beta_1, \dots, \beta_{n-1}) = 1$ and $p_{\beta_1\dots\beta_n} = k_n$.

We shall use the notation

$$B = \{\boldsymbol{\beta} = (\beta_1, \dots, \beta_n) : \quad 1 \le \beta_j \le \mu(\beta_1, \dots, \beta_{j-1}), \quad j = 1, \dots, n\}.$$

If a system of the form (8.13) satisfies the conditions (*)–(* * *) then it is nondegenerate.

If the system (8.13) is nondegenerate then it is sufficient to verify the conditions (*)–(***) only for the first $n-2$ polynomials, since the restriction of a homogeneous polynomial to a two–dimensional or a one–dimensional plane is always a product of linear factors. In particular, for a system of three equations it is sufficient to require factorization of only one polynomial.

The systems from Corollary 8.1 and from Theorem 8.2 evidently satisfy the conditions (*)—(* * *).

For any $\boldsymbol{\beta} \in B$ we make the linear change of variables

$$\zeta_1 = L_{\beta_1}(\mathbf{z}),\ \zeta_2 = L_{\beta_1\beta_2}(\mathbf{z}),\ \dots,\ \zeta_n = L_{\beta_1\dots\beta_n}(\mathbf{z}). \tag{8.14}$$

In these variables the polynomials P_j $(j = 1, \dots, n)$ have the form

$$\zeta_j^{r_{jj}} \prod_{q=j+1}^{n} \left[\prod_{i=1}^{r_{jq}} \left(a_{jqi}\zeta_q + \sum_{s=j}^{q-1} b_{jqis}\zeta_s\right)\right] + \sum_{p=1}^{j-1} \zeta_p P_{jp}(\boldsymbol{\zeta}) = \tag{8.15}$$

$$= c_j \zeta_j^{r_{jj}} \cdots \zeta_n^{r_{jn}} + \Psi_j(\boldsymbol{\zeta}),$$

where $r_{jj} + \dots + r_{jn} = k_j$, and P_{jp}, Ψ_j are homogeneous polynomials of degrees $k_j - 1$ and k_j, respectively. Here $c_j\zeta_j^{r_{jj}} \cdots \zeta_n^{r_{jn}}$ is the leading monomial of the polynomial P_j (in the variables $\boldsymbol{\zeta}$) with respect to the following ordering: $\boldsymbol{\zeta}^{\boldsymbol{\alpha}} \succ \boldsymbol{\zeta}^{\boldsymbol{\beta}}$ if $||\boldsymbol{\alpha}|| > ||\boldsymbol{\beta}||$ or $||\boldsymbol{\alpha}|| = ||\boldsymbol{\beta}||$, and $\alpha_1 = \beta_1, \dots, \alpha_{q-1} = \beta_{q-1}, \alpha_q > \beta_q$, for some $q \in \{1, \dots, n-1\}$.

Under the above assumptions the following theorem holds.

THEOREM 8.3 (Yuzhakov). *If system (8.13) satisfies the conditions* $(*)$–$(***)$ *then the sum of the values of a polynomial* $R(\mathbf{z})$ *of degree* m *at the roots* $\mathbf{a}$ *of system (8.13) is expressed by the formula*

$$\sum_{\mathbf{a}} R(\mathbf{a}) = \sum_{\beta \in B} \mathfrak{M}_\beta, \tag{8.16}$$

with

$$\mathfrak{M}_\beta = \mathfrak{M}\left[\sum_{\alpha \in A} \frac{(-1)^{\|\alpha\|} R g_1^{\alpha_1} \cdots g_n^{\alpha_n} J_{\hat{f}}}{\prod\limits_{j=1}^{n} \zeta_j^{r_{1j}(\alpha_1+1)+\cdots+r_{jj}(\alpha_j+1)-1}}\right], \tag{8.17}$$

where $\mathfrak{M}$ *is the linear functional from Theorem 8.1 in the variables* $\boldsymbol{\zeta}$, $\hat{f}_j = f_j/c_j$, $g_j = (\Psi_j + Q_j)/c_j$ *and*

$$A = \{\alpha = (\alpha_1, \dots, \alpha_n) : \alpha_q = \lambda_q + \sigma_q,\ \lambda_n = 0,\ 0 \le \sigma_1 \le m, \tag{8.18}$$

$$\sum_{j=1}^{q} \lambda_j H(r_{j,q+1}) + \sigma_{q+1} \le m + \sum_{j=1}^{q} (r_{jj} + \cdots + r_{jq} - 1)(\sigma_j + 1) + q - 1,\ j = 1, \dots, n\},$$

with the function satisfying $H(0) = 0$ *and* $H(x) = 1$ *for* $x > 0$. *If* $r_{j,j+1} = \cdots = r_{jn} = 0$, *then we must take* $\alpha_j = \sigma_j$ *(i.e.* $\lambda_j = 0$*) in (8.18).*

The elimination of unknowns from system (8.13) is facilitated if we know part of its roots. In fact, if we know q roots $\mathbf{a}^{(1)}, \dots, \mathbf{a}^{(q)}$, then we can subtract the sum

$$\left(a_j^{(1)}\right)^p + \cdots + \left(a_j^{(q)}\right)^p, \quad p = 1, \dots, k - q,$$

from the power sums of order p of j–th coordinates of all roots ($k = k_1 \cdots k_n$ is the number of all roots of system (8.13)). Then we obtain the power sums (of order p) of j–th coordinates of the remaining roots. By the Newton formulas (2.1) we can then find the polynomial whose zeros are j–th coordinates of the remaining roots.

LEMMA 8.2. *Let the polynomial* Q *and the homogeneous polynomial* P *satisfy the conditions* $\deg Q < \deg P = k$, *and* $P \ne 0$ *on a compact set* $K \subset \mathbf{C}^n$. *Then for each* $\varepsilon > 0$ *there exist* $\lambda > 1$, *such that*

$$|Q(\mathbf{z})| < \varepsilon\, |P(\mathbf{z})|, \tag{8.19}$$

for any

$$\mathbf{z} \in K_\lambda = \{\mathbf{z} = \lambda \cdot \boldsymbol{\zeta} = (\lambda\zeta_1, \dots, \lambda\zeta_n) : \boldsymbol{\zeta} \in K\}.$$

Proof. Write

$$M = \min_K |P|, \quad A = \max_K Q^*$$

where Q^* is the sum of absolute values of all terms of the polynomial Q. Then

$$\min_{K_\lambda} |P| = \lambda^k M, \quad \max_{K_\lambda} Q^* \le \lambda^{k-1} A.$$

Therefore, if we take

$$\lambda > \frac{A}{\varepsilon M}$$

on the compact set K_λ, then inequality (8.19) will hold. □

Proof of theorem 8.3. Consider the cycle

$$\Gamma = \{\mathbf{z} \in \mathbf{C}^n : \ |P_1(\mathbf{z})| = \rho_1, \ldots, |P_n(\mathbf{z})| = \rho_n\},$$

and also, for every $\boldsymbol{\beta} \in B$, the cycle

$$\gamma_{\beta} = \{\mathbf{z} \in \mathbf{C}^n : \ |L_{\beta_1}(\mathbf{z})| = R_1, \ldots, |L_{\beta_1\ldots\beta_n}(\mathbf{z})| = R_n\} =$$

$$= \{\boldsymbol{\zeta} : \ |\zeta_1| = R_1, \ldots, |\zeta_n| = R_n\} = \left\{\boldsymbol{\zeta} : \ \left|c_j \zeta_j^{r_{jj}} \cdots \zeta_n^{r_{jn}}\right| = \rho_j, \ j = 1, \ldots, n\right\},$$

where R_j is connected with ρ_j by the equalities

$$|c_j| R_j^{r_{jj}} \cdots R_n^{r_{jn}} = \rho_j, \quad j = 1, \ldots, n. \tag{8.20}$$

The orientation of these cycles is chosen so that the differential form

$$d\tau_1 \wedge \ldots \wedge d\tau_n$$

is non-negative, where

$$P_j(\mathbf{z}) = R_j e^{i\tau_j}, \quad \zeta_j = R_j e^{i\tau_j}, \quad j = 1, \ldots, n,$$

are the parametrizations of Γ and γ_β, respectively.

Construct a tubular neighborhood

$$U_\beta = \left\{\mathbf{z} : \ R_j'' < |L_j| = |L_{\beta_1\ldots\beta_j}(\mathbf{z})| < R_j', \quad j = 1, \ldots, n\right\}$$

around each cycle γ_β $(R_j'' < R_j < R_j')$.

As will be shown below, the numbers ρ_j, R_j', R_j'' can be chosen in such a way that

1) $\cup U_\beta \subset \mathbf{C}^n \setminus \{\mathbf{z} : \ f_1(\mathbf{z}) \cdots f_n(\mathbf{z}) = 0\}$;
2) $\overline{U}_\beta \cap \overline{U}_\alpha = \emptyset$ for every $\boldsymbol{\alpha}, \boldsymbol{\beta} \in B$ and $\boldsymbol{\alpha} \neq \boldsymbol{\beta}$;
3) $\Gamma = \sum_{\beta \in B} \Gamma_\beta$, where $\Gamma_\beta \subset U_\beta$;
4) the cycles Γ_β and γ_β are homological cycles ($\Gamma_\beta \sim \gamma_\beta$) in U_β;
5) on Γ the inequalities

$$|Q_j(\mathbf{z})| < |P_j(\mathbf{z})| = \rho_j, \quad j = 1, \ldots, n,$$

hold;

6) on γ_β the inequalities

$$|g_j(\boldsymbol{\zeta})| < \left|\zeta_j^{r_{jj}} \cdots \zeta_n^{r_{jn}}\right| = \frac{\rho_j}{|c_j|}$$

hold.

By Theorem 5.3 we get the equality

$$\sum_{\mathbf{a}} R(\mathbf{a}) = \frac{1}{(2\pi i)^n} \int_\Gamma R(\mathbf{z}) \frac{d f_1(\mathbf{z})}{f_1(\mathbf{z})} \wedge \ldots \wedge \frac{d f_n(\mathbf{z})}{f_n(\mathbf{z})}.$$

From this and from conditions 3), 4), 1) we obtain (by Stokes' formula) the identity (8.16), where

$$\mathfrak{M}_\beta = \frac{1}{(2\pi i)^n} \int_{\Gamma_\beta} R \frac{d f_1}{f_1} \wedge \ldots \wedge \frac{d f_n}{f_n} = \frac{1}{(2\pi i)^n} \int_{\gamma_\beta} \frac{R J_{\mathfrak{f}}}{f_1 \cdots f_n} d\boldsymbol{\zeta}. \tag{8.21}$$

Condition 6) implies that the quotient

$$\frac{R\,J_{\mathfrak{f}}}{f_1\cdots f_n} = R\frac{J_{\hat{\mathfrak{f}}}}{\hat{f}_1\cdots\hat{f}_n}$$

may be expanded on the compact set γ_β in a multiple geometric series

$$\frac{R\,J_{\mathfrak{f}}}{f_1\cdots f_n} = \frac{R\,J_{\hat{\mathfrak{f}}}}{\prod\limits_{j=1}^{n}\zeta_j^{r_{jj}}\cdots\zeta_n^{r_{jn}}\left(1+\frac{g_j}{\zeta_j^{r_{jj}}\cdots\zeta_n^{r_{jn}}}\right)} =$$

$$= \sum_{\|\alpha\|\geq 0}\frac{(-1)^{\|\alpha\|}Rg_1^{\alpha_1}\cdots g_n^{\alpha_n}J_{\hat{\mathfrak{f}}}}{\zeta_1^{r_{11}(\alpha_1+1)}\cdots\zeta_n^{r_{1n}(\alpha_1+1)+\cdots+r_{nn}(\alpha_n+1)}}.$$

Substituting this series into (8.21), we get

$$\mathfrak{M}_\beta = \sum_{\|\alpha\|\geq 0}\frac{(-1)^{\|\alpha\|}}{(2\pi i)^n}\int_{\gamma_\beta}\frac{R\,\mathrm{g}^\alpha J_{\hat{\mathfrak{f}}}d\zeta}{\zeta^{\mathrm{r}(\alpha+\mathrm{I})}}. \tag{8.22}$$

Using Examples 1.1 and 4.1 we obtain (8.17), in which the summation is carried out over all multiindeces α, $\|\alpha\|\geq 0$.

We now show that for $\alpha\notin A$ the terms of the series (8.22) vanish. Indeed, the term corresponding to the multiindex α can be written as a sum of terms of the form

$$\mathfrak{M}\left(\frac{cRJ_{\hat{\mathfrak{f}}}\theta_1^{\lambda_1}\cdots\theta_{n-1}^{\lambda_{n-1}}Q_1^{\mu_1}\cdots Q_n^{\mu_n}\prod\limits_{j=2}^{n}\left(\sum\limits_{q=1}^{j-1}\zeta_q P_{jq}\right)^{\eta_j}}{\zeta_1^{r_{11}(\alpha_1+1)-1}\cdots\zeta_n^{r_{1n}(\alpha_1+1)+\cdots+r_{nn}(\alpha_n+1)-1}}\right), \tag{8.23}$$

where $\lambda_j+\mu_j+\eta_j=\alpha_j$, $\lambda_n=\eta_1=0$,

$$\theta_j = \zeta_j^{r_{jj}}\prod_{q=j+1}^{n}\prod_{i=1}^{r_{jq}}\left(a_{jqi}\zeta_q+\sum_{s=j}^{q-1}b_{jqis}\zeta_s\right)-c_j\zeta_j^{r_{jj}}\cdots\zeta_n^{r_{jn}}, \tag{8.24}$$

and c is a constant.

If $r_{j,j+1}=\ldots=r_{jn}=0$ then $\theta_j\equiv 0$. In this case we assume that $\alpha_j=\mu_j+\eta_j$ i.e. $\lambda_j=0$ (see formula (8.18)). The degrees of the polynomials θ_j, $j=1,\ldots,n-1$, with respect to all the variables $\zeta_1,\ldots,\zeta_q$ $(q=j,\ldots,n-1)$ are greater than or equal to

$$r_{jj}+\cdots+r_{jq}+H(r_{j,q+1}).$$

The degrees of the polynomials R, $J_{\hat{\mathfrak{f}}}$, θ_j, P_{jq} in all variables $\zeta_1,\ldots,\zeta_n$ are equal to m, $k_1+\cdots+k_n-n$, k_j, k_j-1, respectively, and $\deg Q_j\leq k_j-1$ (recall that $r_{jj}+\cdots+r_{jn}=k_j$).

Comparing the degrees of numerator and denominator in (8.23), and taking into account that the functional $\mathfrak{M}$ is different from 0 only for those terms in which the degree of the numerator agrees with the degree of the denominator in each variable, we find that (8.23) can be different from 0 only for η_j, μ_j and for λ_j that satisfy

$$\mu_1+\cdots+\mu_n\leq m, \tag{8.25}$$

$$\sum_{j=1}^{q} \lambda_j \left[r_{jj} + \cdots + r_{jq} + H(r_{jq+1})\right] + \sum_{j=2}^{q+1} \eta_j \leq \sum_{j=1}^{q} (r_{jj} + \cdots + r_{jq})(\alpha_j + 1) - 1, \tag{8.26}$$

$$j = 1, \ldots, n-1.$$

Adding to the inequalities (8.26) the inequality $\mu_1 + \cdots + \mu_{q+1} \leq m$, which is implied by (8.25), and writing $\sigma_j = \mu_j + \eta_j$, $j = 2, \ldots, n$, $\sigma_1 = \mu_1$, we obtain the inequalities contained in (8.18).

It remains to show that the numbers ρ_j, R'_j, R''_j can be chosen so that conditions 1) – 6) hold. For every $\boldsymbol{\beta} \in B$ the polynomial P_j can be represented in the form

$$P_j = \prod_{\beta_j=1}^{\mu(\beta_1,\ldots,\beta_{j-1})} L_{\beta_1\ldots\beta_j}^{p_{\beta_1,\ldots,\beta_j}} + \sum_{q=1}^{j-1} L_{\beta_1\ldots\beta_q} P_{jq}, \tag{8.27}$$

where P_{jq} is a homogeneous polynomial of degree $k_j - 1$. Fix a closed polydisk

$$F = \{\mathbf{z} : |z_j| \leq R,\ j = 1, \ldots, n\},$$

and denote

$$M_j = \max_{\mathbf{z}\in F,\boldsymbol{\beta},q} |P_{jq}|, \qquad M = \max_{\mathbf{z}\in F,\boldsymbol{\beta},j} \left|L_{\beta_1\ldots\beta_j}\right|,$$

$$M_0 = \min_{\boldsymbol{\beta},j,q,i} |a_{jqi}|, \qquad M' = \max_{\boldsymbol{\beta},j,q,i,s} |b_{jqis}|,$$

where a_{jqi} and b_{jqis} are coefficients from the representation (8.15).

Choose step by step the numbers R'_n, ρ_n, $R''_n, \ldots, R'_1$, ρ_1, R''_1 so that the following inequalities are fulfilled:

$$R'_j \geq \left[\rho_j(1+\varepsilon)\right]^{1/k_j}, \quad R''_j \leq \frac{\rho_j(1-\varepsilon)}{M^{k_j-1}}, \quad j = 1, \ldots, n, \tag{8.28}$$

$$R'_q < \frac{\varepsilon\,\rho_j}{(j-1)M_j}, \tag{8.29}$$

$$R'_q < \frac{\varepsilon\, R''_j}{(1+(j-1)M_0)}, \tag{8.30}$$

$$R'_q < \frac{\delta\, M' R''_j}{(j-1)M_0}, \tag{8.31}$$

where $1 \leq j < q \leq n$, and ε, δ are sufficiently small positive numbers, with

$$(1+\delta)^{k_j} - 1 \leq \varepsilon, \qquad j = 1, \ldots, n, \tag{8.32}$$

and finally, for each $\boldsymbol{\beta} \in B$

$$R'_q < \frac{\varepsilon|c_j|\,(R''_j)^{r_{jj}} \cdots (R''_n)^{r_{jn}}}{(j-1)M_j}, \quad 1 \leq q < j \leq n, \tag{8.33}$$

$$\overline{U}_{\boldsymbol{\beta}} \subset F = \{\mathbf{z} :\ |z_j| \leq R,\ j = 1, \ldots, n\}, \quad R''_j < R_j < R'_j,$$

where R_j are the numbers defined by formula (8.20).

Take now an arbitrary point

$$\mathbf{z} \in \Gamma = \{\mathbf{z} :\ |P_j| = \rho_j,\ j = 1, \ldots, n\}.$$

The equality

$$|P_1| = \left|\prod_{\beta_1=1}^{\mu} L_{\beta_1}^{p_{\beta_1}}\right| = \rho_1$$

and the inequality (8.28) imply that at least for one $\beta_1 \in \{1,\dots,\mu\}$ we must have

$$R_1'' < |L_{\beta_1}(\mathbf{z})| < R_1'.$$

From the representation (8.27) and the inequalities (8.29), for $j = 2$, we get for the given $\mathbf{z}$

$$\rho_2(1-\varepsilon) < \left|\prod_{\beta_2=1}^{\mu(\beta_1)} L_{\beta_1\beta_2}^{p_{\beta_1\beta_2}}\right| < \rho_2(1+\varepsilon).$$

Hence at least for one β_2 we obtain the inequality

$$R_2'' < |L_{\beta_1\beta_2}(\mathbf{z})| < R_2'.$$

Continuing our reasonings analogously (with the help of (8.27), (8.28), (8.29)) we get that for $j = 3,\dots,n$, and for every $\mathbf{z} \in \Gamma$, there exists $\boldsymbol{\beta} \in B$, such that

$$R_j'' < \left|L_{\beta_1\dots\beta_j}(\mathbf{z})\right| < R_j'.$$

Therefore $\mathbf{z} \in U_{\boldsymbol{\beta}}$, and condition 3) is accomplished.

To prove condition 2), we let

$$\boldsymbol{\alpha} = (\alpha_1,\dots,\alpha_n) \neq \boldsymbol{\beta} = (\beta_1,\dots,\beta_n)$$

and $\alpha_1 = \beta_1,\dots,\alpha_{j-1} = \beta_{j-1}$, but $\alpha_j \neq \beta_j$ for some $j \in \{1,\dots,n-1\}$.

Then in the variables (8.14) the function $L_{\beta_1\dots\beta_{j-1}\alpha_j}$ will have the form

$$a_{jqi}\zeta_q + \sum_{s=j}^{q-1} b_{jqis}\zeta_s.$$

Formula (8.30) then shows that for $\mathbf{z} \in \overline{U}_{\boldsymbol{\beta}}$ the inequality

$$\left|L_{\beta_1\dots\beta_{j-1}\alpha_j}\right| > R_j'$$

holds. Hence we have $\mathbf{z} \notin \overline{U}_{\boldsymbol{\beta}}$.

Further, the representations (8.15), (8.24) and the inequalities (8.31), (8.32), (8.33) imply that in $\overline{U}_{\boldsymbol{\beta}}$ we have the inequality

$$|P_j| = \left|c_j\zeta_j^{r_{jj}}\cdots\zeta_n^{r_{jn}} + \theta_j + \sum_{q=1}^{j-1}\zeta_q P_{jq}\right| >$$

$$> |c_j|\left(R_j''\right)^{r_{jj}}\cdots\left(R_n''\right)^{r_{jn}}\left[1-\varepsilon-(1+\delta)^{k_j-r_{jj}}+1\right] > 0.$$

Then, by Lemma 8.2 applied to the compact set

$$K = \cup_{\boldsymbol{\beta}}\overline{U}_{\boldsymbol{\beta}},$$

there exists $\lambda \geq 1$ such that the inequalities (8.19) hold in K_λ. Multiply the numbers R, R_j', R_j, R_j'', M by λ, ρ_j by λ^{k_j} and M_j by λ^{k_j-1}, respectively. Then for the new values R, R_j', R_j'', ρ_j, M_j, M, by virtue of the homogeneity of the polynomials P_j, P_{jq}, $L_{\beta_1\dots\beta_j}$, the inequalities (8.28)–(8.33) remain true. Hence the previous

previous reasoning is still valid, so for the new values ρ_j, R'_j, R''_j the conditions 2), 3), 5) will be fulfilled.

For each $\boldsymbol{\beta} \in B$ we now set

$$P_j(\mathbf{z},t) = L_j^{r_{jj}} \prod_{q=j+1}^{n} \left[\prod_{i=1}^{r_{jq}} \left(a_{jqi}\zeta_q + t\sum_{s=q}^{q-1} b_{jqis}\zeta_s \right) \right] + t\sum_{p=1}^{j-1} \zeta_p P_{jp}(\boldsymbol{\zeta}),$$

where $\boldsymbol{\zeta}$ and $\mathbf{z}$ are connected with each other by formula (8.14).

The inequalities (8.28)–(8.33) imply that we have the inclusion

$$\Gamma_t = \{\mathbf{z} : \ |P_j(\mathbf{z},t)| = \rho_j, \ j = 1, \dots, n, \ 0 \le t \le 1\} \subset U_{\boldsymbol{\beta}} \cup \left(\mathbf{C}^n \setminus \overline{U}_{\boldsymbol{\beta}}\right).$$

From here we get that the (properly oriented) sets

$$\gamma_{\boldsymbol{\beta}}(t) = \Gamma(t) \cap U_{\boldsymbol{\beta}}, \quad 0 \le t \le 1,$$

constitute a family of cycles which realize a homotopy between the cycles $\Gamma_{\boldsymbol{\beta}}$ and $\gamma_{\boldsymbol{\beta}}$ in $U_{\boldsymbol{\beta}}$. Therefore, condition 4) is satisfied.

It remains to examine the conditions 1) and 6). By the representations (8.13), (8.15), (8.25) and formulas (8.14), (8.30)–(8.33) we have in $U_{\boldsymbol{\beta}}$ the inequalities

$$|Q_j| < \varepsilon |P_j|,$$

$$|P_j| \ge \left|c_j\zeta_j^{r_{jj}} \cdots \zeta_n^{r_{jn}}\right| - |\theta_j| - \left|\sum_{q=1}^{j-1} \zeta_q P_{jq}\right| > |c_j| \left(R''_j\right)^{r_{jj}} \cdots \left(R''_n\right)^{r_{jn}} (1 - 2\varepsilon),$$

and hence,

$$|f_j| \ge |P_j| - |Q_j| > |P_j(1-\varepsilon)| > |c_j| \left(R''_j\right)^{r_{jj}} \cdots \left(R''_n\right)^{r_{jn}} (1-\varepsilon)(1-2\varepsilon) > 0.$$

Therefore, $f_j \ne 0$ in $U_{\boldsymbol{\beta}}$ for $\boldsymbol{\beta} \in B$, and condition 1) is settled.

Since

$$|c_j g_j| = |\Psi_j + Q_j| \le |\Psi_j| + \varepsilon |P_j| \le (1+\varepsilon)|\Psi_j| + \varepsilon \left|c_j\zeta_j^{r_{jj}} \cdots \zeta_n^{r_{jn}}\right|,$$

it follows that

$$\left|c_j\zeta_j^{r_{jj}} \cdots \zeta_n^{r_{jn}}\right| - |c_j g_j| \ge$$

$$\ge (1-\varepsilon)\left|c_j\zeta_j^{r_{jj}} \cdots \zeta_n^{r_{jn}}\right| - (1+\varepsilon)\left(|\theta_j| + \left|\sum_{q=1}^{j-1} \zeta_q P_{jq}\right| \right) >$$

$$> |c_j| \left(R''_j\right)^{r_{jj}} \cdots \left(R''_n\right)^{r_{jn}} ((1-\varepsilon)(1+\varepsilon)2\varepsilon) > 0.$$

This completes the proof of Theorem 8.3. □

For calculations it is convenient to have a formula expressing the sum

$$\sum_{\mathbf{a}} R(\mathbf{a})$$

directly in terms of the coefficients of the polynomials R and f_j.

Theorem 8.4 (Yuzhakov [**157**]). *Let*

$$R = \sum_{\|\mu\| \le m} c_\mu \zeta^\mu, \quad g_j = \sum_{\|\alpha\| \le k_j}^{(j)} a_{j\alpha} \zeta^\alpha, \quad j = 1, \dots, n. \tag{8.34}$$

Then $\mathfrak{M}_\beta$ in formula (8.16) is expressed through the coefficients c_μ and $a_{j\alpha}$ by the formula

$$\mathfrak{M}_\beta = \sum_{\mu, \mathbf{S}} (-1)^{\|\rho(\mathbf{S})\|} c_\mu D(\mathbf{S}) \prod_{j=1}^{n} [\rho(S_j) - 1]! \prod_\alpha^{(j)} \frac{(a_{j\alpha})^{s_{j\alpha}}}{s_{j\alpha}!}, \tag{8.35}$$

where $\mathbf{S} = (S_1, \dots, S_n)$, $S_j = \{s_{j\alpha}\}_\alpha^{(j)}$ being sets of nonnegative integers (indexed by the multiindeces α which occur in the sum $\sum^{(j)}$ from (8.34)); is going;

$$\rho(S_j) = \sum_\alpha^{(j)} s_{j\alpha}, \quad \rho(\mathbf{S}) = (\rho(S_1), \dots, \rho(S_n)),$$

$$D(\mathbf{S}) = \det \|r_{jq} \rho(S_j) - \sigma_q(S_j)\|,$$

where r_{jq} is the exponent from (8.15) for $j \le q \le n$, $r_{jq} = 0$ if $q < j$;

$$\sigma_q(S_j) = \sum_\alpha^{(j)} s_{j\alpha} \alpha_q.$$

The summation in $\sum\limits_{\mu, \mathbf{S}}$ is taken over all multiindeces μ for which $c_\mu \ne 0$, and over the families $\mathbf{S} = (S_1, \dots, S_n)$ satisfying the equations

$$\mu_q + \sum_{j=1}^{n} \sigma_q(S_j) = \sum_{j=1}^{n} \rho(S_j) r_{jq}, \quad q = 1, \dots, n. \tag{8.36}$$

Moreover, if $\mathbf{S} = (S_1, \dots, S_n)$ contains the family S_j for which $\rho(S_j) = 0$ then in the corresponding term of formula (8.35) we must replace $[\rho(S_j) - 1]!$ by 1 and in $D(\mathbf{S})$ we must replace the element $r_{jq}\rho(S_j) - \sigma_q(S_j)$ by r_{jq}.

Proof. The expression (8.21) can be written as

$$\mathfrak{M}_\beta = \frac{1}{(2\pi i)^n} \int_{\gamma_\beta} R \frac{\partial \ln \hat{\mathbf{f}}}{\partial \zeta} d\zeta. \tag{8.37}$$

By condition 6) from the proof of the previous theorem we have on γ_β the following expansion

$$\ln \hat{f}_j = \ln \zeta_j^{r_{jj}} \cdots \zeta_n^{r_{jn}} + \ln \left(1 + \frac{g_j}{\zeta_j^{r_{jj}} \cdots \zeta_n^{r_{jn}}} \right) =$$

$$= \ln \zeta^{\mathbf{r}_{(j)}} + \sum_{p=1}^{\infty} \frac{(-1)^{p-1}}{p} \left(\frac{g_j}{\zeta^{\mathbf{r}_{(j)}}} \right)^p,$$

where $\mathbf{r}_{(j)} = (r_{j1}, \dots, r_{jn})$, $r_{j1} = \dots = r_{j,j-1} = 0$.

Plugging in the expression for g_j from (8.34) we get

$$\ln \hat{f}_j = \sum_{q=1}^{n} r_{jq} \ln \zeta_q +$$

$$+\sum_{p=1}^{\infty}\frac{(-1)^{p-1}}{p}\sum_{\rho(S_j)=p} p!\prod_{\alpha}{}^{(j)}\frac{(a_{j\alpha})^{s_{j\alpha}}}{s_{j\alpha}!}\zeta^{\sigma(S_j)-\rho(S_j)\mathbf{r}_{(j)}} =$$

$$=\sum_{q=1}^{n} r_{jq}\ln\zeta_q + \sum_{\rho(S_j)\geq 1}(-1)^{\rho(S_j)-1}[\rho(S_j)-1]!\prod_{\alpha}{}^{(j)}\frac{(a_{j\alpha})^{s_{j\alpha}}}{s_{j\alpha}!}\zeta^{\sigma(S_j)-\rho(S_j)\mathbf{r}_{(j)}},$$

where $\boldsymbol{\sigma}(S_j) = (\sigma_1(S_j),\dots,\sigma_n(S_j))$.

From here we have

$$\frac{\partial\ln\hat{f}_j}{\partial\zeta_q} = \frac{r_{jq}}{\zeta_q}+$$

$$+\sum_{\rho(S_j)\geq 1}(-1)^{\rho(S_j)}[\rho(S_j)-1]![\rho(S_j)r_{jq}-\sigma_q(S_j)]\left[\prod_{\alpha}{}^{(j)}\frac{(a_{j\alpha})^{s_{j\alpha}}}{s_{j\alpha}!}\right]\zeta^{\sigma(S_j)-\rho(S_j)\mathbf{r}_{(j)}-\mathbf{e}_q},$$

where $\mathbf{e}_q = (0,\dots,0,\underbrace{1}_{q},0,\dots,0)$.

We shall consider the summand r_{jq}/ζ_q as a term of the sum

$$\sum_{\rho(S_j)\geq 0}$$

corresponding to the multiindex S_j for which $\rho(S_j) = 0$. Then we can write

$$R\frac{\partial\ln\hat{\mathbf{f}}}{\partial\zeta} =$$

$$=\sum_{\mu,\mathbf{S}}(-1)^{\rho(\mathbf{S})}c_\mu D(\mathbf{S})\left[\prod_{j=1}^{n}[\rho(S_j)-1]!\prod_{\alpha}{}^{(j)}\frac{(a_{j\alpha})^{s_{j\alpha}}}{s_{j\alpha}!}\right]\zeta^{\mu+\sum\limits_{j=1}^{n}[\sigma(S_j)-\rho(S_j)\mathbf{r}_{(j)}]-\mathbf{I}}. \quad (8.38)$$

Substituting (8.38) into (8.37) and taking into account Example 4.1, we obtain formula (8.35), where the summation is performed over all multiindeces $\boldsymbol{\mu}$ and $\mathbf{S}$ satisfying the condition

$$\boldsymbol{\mu}+\sum_{j=1}^{n}(\boldsymbol{\sigma}(S_j)-\rho(S_j)\mathbf{r}_{(j)}) = 0,$$

which is precisely (8.36). □

EXAMPLE 8.6. Eliminate the unknowns y and z from the system

$$\begin{aligned} y^2z^2 - x^3 &= 0,\\ y^2+z^2-2x &= 0,\\ x^2-xy+xz-y^2+yz-z^2 &= 0. \end{aligned} \quad (8.39)$$

This system has 16 roots. We know the following roots: $(0,0,0)$ of multiplicity 8, $(1,1,1)$ and $(1,-1,-1)$ of multiplicity 1. Thus we know 10 roots and we must determine the remaining 6 roots. The x-coordinates of these roots are the solutions to the equation

$$x^6+a_1x^5+a_2x^4+a_3x^3+a_4x^2+a_5x+a_6 = 0, \quad (8.40)$$

where the coefficients a_j are connected with the power sums S_j by the Newton formulas (2.1). The power sums S_j we can find by formulas (8.16) and (8.35), setting $R = x^j$, $j = 1, \dots, 6$. In our case $P_1 = y^2z^2$, $P_2 = y^2 + z^2$, and the set B consists of two elements $(1,1,1)$ and $(2,1,1)$. To them there correspond two triples of linear factors

$$L_1 = y,\ L_{11} = z,\ L_{111} = x;\ L_2 = z,\ L_{21} = y,\ L_{211} = x.$$

Keeping in mind that the equations (8.39) consist of monomials which are symmetric in the variables y, z, we get (after elementary transformations) the formula

$$S_j = -2 + \sum_{\alpha,\beta,t} A_{\alpha\beta t},$$

where the summation is taken over all sets of nonnegative integers

$$\alpha_1,\ \alpha_2,\ \alpha_3,\ \alpha_4,\ \alpha_5,\ \beta_1,\ \beta_2$$

satisfying the inequalities

$$0 \le \beta_2 \le j,\quad 0 \le \beta_1 \le j - \beta_1,\quad 0 \le \alpha_5 \le p - \beta_1 - \beta_2,$$

$$0 \le \alpha_4 \le 2(j - \beta_1 - \beta_2 - \alpha_5),\quad \frac{\alpha_5}{2} \le t \le \frac{j + \beta_1 + \alpha_5}{2} - \frac{\alpha_4}{4},$$

and the equalities

$$\alpha_1 = 2(j + \beta_1 + \alpha_5 - 2t) - \alpha_4,\quad \alpha_2 = 2(j - \alpha_5 - \beta_1 - \beta_2) - \alpha_4,\quad \alpha_3 = 2t - \alpha_5.$$

The terms in $A_{\alpha\beta t}$ are given by the formulas

$$A_{\alpha\beta t} = (-1)^{\alpha_4+\alpha_5} 2^2 \frac{(\alpha_1 + \alpha_2 + \alpha_3 + \alpha_4 + \alpha_5 - 1)!}{\alpha_1!\alpha_2!\alpha_3!\alpha_4!\alpha_5!}(2\alpha_1 - \alpha_2 + 4\alpha_3 + \alpha_4 - 2\alpha_5),$$

if $\beta_1 = \beta_2 = 0$;

$$A_{\alpha\beta t} = 2^{j+3} j \frac{(2j - 2t - 1)!}{(2t)!(2j - 4t)!},$$

if $\beta_2 = j$;

$$A_{\alpha\beta t} = (-1)^{\alpha_4+\alpha_5+\beta_1} 2^{\beta_2+2} \frac{(\alpha_1 + \cdots + \alpha_5 - 1)!}{\alpha_1! \cdots \alpha_5!} \frac{(\beta_1 + \beta_2 - 1)!}{\beta_1!\beta_2!} \times$$
$$\times (\beta_1(\alpha_1 + \alpha_2 + 2(\alpha_3 + \alpha_4 + \alpha_5)) + \beta_2(\alpha_2 + 2\alpha_3 + \alpha_4)),$$

if $\beta_1 + \beta_2 > 0$, $\beta_2 < j$.

Straight-forward calculations now give the following values for the coefficients in formula (8.40):

$$a_1 = 14,\ a_2 = 129,\ a_3 = -496,\ a_4 = 1376,\ a_5 = -1280,\ a_6 = 256.$$

We end this section by quoting (without proof) some results of Tsikh (see [**135**]), in which he replaced the projective space $\mathbf{CP}^n$ by the n-fold product

$$\bar{\mathbf{C}}^n = \mathbf{CP}^1 \times \cdots \times \mathbf{CP}^1.$$

Let $z_j : w_j$ be homogeneous coordinates in j–th factor of $\bar{\mathbf{C}}^n$, and let

$$f_j(z_1, w_1, \dots, z_n, w_n) = 0,\quad j = 1, \dots, n \tag{8.41}$$

be a system of equations, where f_j are polynomials which are homogeneous in each pair of variables z_k, w_k. We consider only those roots $(z_1, w_1, \dots, z_n, w_n)$ of the system (8.41) for which

$$(z_k, w_k) \in \mathbf{C}^2 \setminus \{(0,0)\}, \quad k = 1, \dots, n,$$

and roots with pair wise proportional coordinates determine the same root in the space $\bar{\mathbf{C}}^n$.

Let

$$\mathbf{a} = \left(z_1^0 : w_1^0, \dots, z_n^0 : w_n^0\right)$$

be a root of (8.41) and assume that $w_k^0 \neq 0$, $k = 1, \dots, n$. Then the multiplicity of $\mathbf{a}$ is (by definition) equal to the multiplicity of the root

$$\left(\frac{z_1^0}{w_1^0}, \dots, \frac{z_n^0}{w_n^0}\right)$$

of the system

$$f_j(z_1, 1, \dots, z_n, 1) = 0, \quad j = 1, \dots, n,$$

in $\mathbf{C}^n$.

Recall that the permanent of a matrix

$$\mathbf{A} = (a_{ij})_{i,j=1,\dots,n}$$

is defined to be the number

$$p(\mathbf{A}) = \sum_{\mathbf{J}} a_{1j_1} \cdots a_{nj_n},$$

where the summation is over all permutations $\mathbf{J} = (j_1, \dots, j_n)$ of the numbers $1, 2, \dots, n$.

THEOREM 8.5 (Tsikh). *Let m_{ij} be the degree of homogeneity of the polynomial f_j in the pair of variables z_i, w_i. Then the system (8.41) has either an infinite number of roots or the number of roots (with multiplicities) is equal to the permanent of the matrix*

$$(m_{ij})_{i,j=1,\dots,n}.$$

A proof of this theorem can be found in [**135**] (see also [**136**]).

Consider the following system of equations

$$f_i(\mathbf{z}) = (z_1 - a_{i1})^{m_{i1}} \cdots (z_n - a_{in})^{m_{in}} + Q_i(\mathbf{z}) = 0, \quad i = 1, \dots, n, \tag{8.42}$$

where m_{ij} are positive integers, Q_i are polynomials for which the degrees in the variable z_j is less than m_{ij}, and a_{ij} are complex numbers such that for every $j = 1, \dots, n$ all the numbers $a_{1j}, \dots, a_{nj}$ are different from each other.

It is clear that the system (8.42) has no infinite roots in the space $\bar{\mathbf{C}}^n$. This means that the system

$$\hat{f}_i = w_1^{m_{i1}} \cdots w_n^{m_{in}} f_i\left(\frac{z_1}{w_1}, \dots, \frac{z_n}{w_n}\right) = 0, \quad i = 1, \dots, n,$$

has no roots in the planes $\{w_j = 0\}$, $j = 1, \dots, n$ ($z_j \neq 0$ for all j).

Hence the system (8.42) has a finite number N of roots, and by Theorem 8.5

$$N = p(m_{ij}).$$

The system (8.42) is the analogue in $\bar{\mathbf{C}}^n$ of the Aizenberg system from Corollary 8.1.

THEOREM 8.6 (Tsikh). *Let $R(\mathbf{z})$ be a polynomial of degree μ_i in the variable z_i, $i = 1, \ldots, n$. Let the vectors*

$$\mathbf{z}^{(1)}, \ldots, \mathbf{z}^{(N)}$$

represent all roots of the system (8.42) (counted according to their multiplicities). Then

$$\sum_{j=1}^{N} R(\mathbf{z}^{(j)}) =$$

$$= \sum_{||\alpha||=0}^{\mu} (-1)^{||\alpha||} \sum_{\mathbf{J}} \frac{1}{\beta(\alpha, \mathbf{J})!} \frac{\partial^{||\beta(\alpha,\mathbf{J})||}}{\partial \mathbf{z}^{\beta(\alpha,\mathbf{J})}} \left[R J_{\mathbf{f}} \frac{Q_1^{\alpha_1} \cdots Q_n^{\alpha_n}}{q_{1j_1}^{\alpha_1+1} \cdots q_{nj_n}^{\alpha_n+1}} \right] \Bigg|_{z_k = a_{j_k k}},$$

where $\alpha = (\alpha_1, \ldots, \alpha_n)$ and

$$\beta(\alpha, \mathbf{J}) = (m_{j_1 1}(\alpha_{j_1} + 1) - 1, \ldots, m_{j_n n}(\alpha_{j_n} + 1) - 1)$$

are multiindeces,

$$\mu = \max_{1 \le j \le n} \{\mu_j\}$$

and

$$q_{kj_k} = \frac{f_k - Q_k}{(z_k - a_{kj_k})^{m_{kj_k}}}, \quad k = 1, \ldots, n.$$

For a proof of this theorem we again refer to [**135**] and [**136**].

9. A formula for the logarithmic derivative of the resultant

We shall consider the following system of algebraic equations

$$\begin{aligned} f_1 &= f_1(\zeta, w) = 0, \\ &\ldots \\ f_{n+1} &= f_{n+1}(\zeta, w) = 0 \end{aligned} \tag{9.1}$$

where $\zeta = (\zeta_0, \zeta_1, \ldots, \zeta_n) \in \mathbf{C}^{n+1}$, $w \in \mathbf{C}^1$, and the functions $f_j(\zeta, w)$ are polynomials in ζ, w which are homogeneous in ζ. The total degree of f_j in ζ is equal to k_j and the degree of f_j in w is equal to p_j, $j = 1, \ldots, n+1$. The polynomials f_j can be written

$$f_j(\zeta, w) = \sum_{||\alpha||=k_j} a_\alpha^j(w) \zeta^\alpha,$$

where $\alpha = (\alpha_0, \alpha_1, \ldots, \alpha_n)$ is a multiindex and

$$||\alpha|| = \alpha_0 + \alpha_1 + \ldots + \alpha_n, \quad \zeta^\alpha = \zeta_0^{\alpha_0} \cdots \zeta_n^{\alpha_n},$$

and the a_α^j are polynomials in w of degree $\le p_j$. We shall assume $\zeta \in \mathbf{CP}^n$ with homogeneous coordinates $(\zeta_0, \zeta_1, \ldots, \zeta_n)$. As before, the hyperplane at infinity is given by

$$\Pi = \{\zeta \in \mathbf{CP}^n : \zeta_0 = 0\},$$

and $\mathbf{C}^n$ is identified with the subspace $\{\boldsymbol{\zeta} \in \mathbf{CP}^n : \zeta_0 \neq 0\}$ via the relation

$$\mathbf{z} = (z_1, \dots, z_n) = \left(\frac{\zeta_1}{\zeta_0}, \dots, \frac{\zeta_n}{\zeta_0}\right).$$

It follows then that $\mathbf{CP}^n = \mathbf{C}^n \cup \Pi$.

The main feature of the system (9.1) is that it has a finite set of roots $(\boldsymbol{\zeta}^k, w_k)$ in $\mathbf{CP}^n \times \mathbf{C}^1$, $k = 1, \dots, M$. We recall that the resultant (in w) of the system (9.1) is the polynomial

$$P(w) = \prod_{k=1}^{M} (w - w_k).$$

(Every root of the system (9.1) is counted according to its multiplicity in the local coordinate system.)

Fix a point $w \neq w_k$, $k = 1, \dots, M$. For this w the system (9.1) has no roots, so the subsystems

$$f_j(\boldsymbol{\zeta}, w) = 0, \quad j \neq l,\ j = 1, \dots, n+1,$$

have finite set of roots in $\mathbf{CP}^n$ (see, for example, [**85**, Ch. 4, Th. 9]). These roots will be denoted

$$\boldsymbol{\zeta}^l_{(j)}(w),\ j = 1, \dots, M_l.$$

By the Bézout theorem we have

$$M_l = \prod_{j \neq l} k_j.$$

To every root $\boldsymbol{\zeta}^l_{(j)}(w)$ we associate a local residue as follows: Suppose the zero-coordinate of the point $\boldsymbol{\zeta}^l_{(j)}(w)$ is non-zero, so that

$$\boldsymbol{\zeta}^l_{(j)}(w) = \mathbf{z}^l_{(j)}(w) \in \mathbf{C}^n.$$

We now write $F_k(\mathbf{z}, w) = f_k(1, z_1, \dots, z_n, w)$ and consider the domain

$$D^l_j(w) = \{\mathbf{z} \in U(\mathbf{z}^l_{(j)}(w)) : |F_k(\mathbf{z}, w)| < \varepsilon_k,\ k \neq l\}$$

and the cycle

$$\Gamma^l_j(w) = \{\mathbf{z} \in U(\mathbf{z}^l_{(j)}(w)) : |F_k(\mathbf{z}, w)| = \varepsilon_k,\ k \neq l\},$$

which is oriented so that

$$d(\arg F_1) \wedge \dots \wedge d(\arg F_{l-1}) \wedge d(\arg F_{l+1}) \wedge \dots \wedge d(\arg F_{n+1}) > 0 \quad \text{on } \Gamma^l_j(w).$$

If ω is a meromorphic form

$$\omega = \frac{h d\mathbf{z}}{F[l]} = \frac{h(z) dz_1 \wedge \dots \wedge dz_n}{F_1 \dots F_{l-1} F_{l+1} \dots F_{n+1}},$$

and h is a holomorphic function in $D^l_j(w)$, then the local residue is

$$\operatorname*{res}_{j,l}(\omega) = \operatorname*{res}_{j,l} \frac{h d\mathbf{z}}{F[l]} = \frac{1}{(2\pi i)^n} \int_{\Gamma^l_j(w)} \frac{h d\mathbf{z}}{F[l]}$$

(see Sec. 4). The global residue $\operatorname{Res}_l \omega$ of the form ω is equal to the sum of local residues

$$\operatorname*{Res}_l \omega = \sum_{j=1}^{M_l} \operatorname*{res}_{j,l} \omega,$$

(see Sec. 7). We have thus defined local residues for points $\zeta^l_{(j)}(w)$ in $\mathbb{C}^n$. For a point $\zeta^l_{(j)}(w)$ belonging to Π the local residue is defined similarly, only instead of the zero-coordinate we must take choose another coordinate which is not equal to 0. Observe also that $\operatorname{Res}_l \omega$ is a function of w.

THEOREM 9.1 (Kytmanov [**100**]). *Let $P(w)$ be the resultant (in w) of the system (9.1) having a finite set of roots (ζ^k, w_k) in $\mathbf{CP}^n \times \mathbf{C}^1$, $k = 1, \dots, M$. Then*

$$\frac{P'(w)}{P(w)} = \sum_{l=1}^{n+1} \operatorname*{Res}_l \frac{f_l' df[l]}{f_1 \dots f_{n+1}} = \sum_{l=1}^{n+1} \sum_{j=1}^{M_l} \frac{f_l'}{f_l}(\zeta^l_{(j)}(w), w), \tag{9.2}$$

where f_l' is the derivative of the polynomial f_l with respect to w, and

$$df[l] = df_1 \wedge \dots \wedge df_{l-1} \wedge df_{l+1} \wedge \dots \wedge df_{n+1}.$$

Proof. We shall assume that for $w = 0$ the system (9.1) has no roots, and we write

$$\zeta^l_{(j)} = \zeta^l_{(j)}(0).$$

Since (9.1) has a finite set of roots in $\mathbf{CP}^n \times \mathbf{C}^1$, and the subsystems

$$f_j(\zeta, 0) = 0, \ j \neq l,$$

also have finite sets of roots in $\mathbf{CP}^n$, then after a linear transformation of $\mathbf{CP}^n$ we may assume that Π contains none of the roots of (9.1) or of any of its subsystems.

We first prove formula (9.2) in the following case: All the roots (ζ^k, w_k) belong to $\mathbf{C}^n \times \mathbf{C}^1$ and all the roots $\zeta^l_{(j)}$ are in $\mathbf{C}^n$. We denote these also by $\mathbf{z}^l_{(j)}$. If

$$F_j(\mathbf{z}, w) = P_j(\mathbf{z}, w) + Q_j(\mathbf{z}, w),$$

where $P_j(\mathbf{z}, w)$ is highest homogeneous part of F_j in $\mathbf{z}$ and $\deg Q_j < k_j$, then the systems

$$P_j(\mathbf{z}, 0) = 0, \quad j = 1, \dots, n+1, \quad j \neq l \tag{9.3}$$

are non-degenerate by Lemma 8.1, and hence for all w except a finite number, the systems

$$P_j(\mathbf{z}, w) = 0, \quad j = 1, \dots, n+1, \quad j \neq l \tag{9.4}$$

are also non-degenerate.

It follows that the systems

$$f_j(0, \zeta_1 \dots \zeta_n, w) = 0, \quad j = 1, \dots, n+1, \quad j \neq l \tag{9.5}$$

have roots for each fixed w, except a finite number of values $w = \tilde{w}_s$. But since the system (9.1) has no roots for $\zeta_0 = 0$, the systems (9.5) can have only finitely many roots for fixed w. We conclude that (9.5) has a finite set of roots in $\mathbf{CP}^{n-1} \times \mathbf{C}^1$. (This will be important for carrying out of the induction below.)

Let $|w|$ be sufficiently small, and let

$$R(w) = \prod_{j=1}^{M_{n+1}} F_{n+1}(\mathbf{z}^{l}_{(j)}(w), w)$$

be the resultant of the function F_{n+1} with respect to the system $F_1 = 0, \ldots, F_n = 0$ (see Sec. 8).

As shown in [**11**, Sec. 21, 22], one may write

$$R(w) = \frac{P(w)}{Q(w)},$$

where $P(w)$ is the classical resultant defined above, and Q is some polynomial with no factor in common with P.

In view of the condition imposed on the system (9.1) we have $R(0) \neq 0$, and hence we can consider $\ln R(w)$ and $\ln F_{n+1}(\mathbf{z}, w)$ in some simply connected neighborhood $U(\mathbf{z}^{l}_{(j)})$ of the origin. Then, by Rouché's Theorem 5.3 and Theorem 5.2, we may take $\Gamma^{l}_{j}(0)$ as the cycle of integration in the formula for $\mathrm{res}_{j,l}$. We thus get

$$\ln R(w) = \operatorname*{Res}_{n+1} \frac{\ln F_{n+1} dF[n+1]}{F_1 \ldots F_n},$$

and hence

$$\frac{d \ln R(w)}{dw} = \frac{R'}{R} = \operatorname*{Res}_{n+1} \frac{F'_{n+1} dF[n+1]}{F_1 \ldots F_{n+1}} +$$

$$+ \sum_{l=1}^{n} \operatorname*{Res}_{n+1} \left(\ln F_{n+1} \frac{dF_1}{F_1} \wedge \ldots \wedge \left(\frac{dF_l}{F_l} \right)' \wedge \ldots \wedge \frac{dF_n}{F_n} \right).$$

Since

$$\left(\frac{dF_l}{F_l} \right)' = d \left(\frac{F'_l}{F_l} \right)$$

and

$$\ln F_{n+1} \frac{dF_1}{F_1} \wedge \ldots \wedge \left(\frac{dF_l}{F_l} \right)' \wedge \ldots \wedge \frac{dF_n}{F_n} =$$

$$= (-1)^{l-1} d \left(\ln F_{n+1} \frac{F'_l}{F_l} \right) \wedge \frac{dF}{F}[l] + (-1)^{n+l-1} \frac{F'_l dF[l]}{F_1 \ldots F_{n+1}},$$

we obtain

$$\frac{R'(w)}{R(w)} = \sum_{l=1}^{n+1} (-1)^{n+l-1} \operatorname*{Res}_{n+1} \frac{F'_l dF[l]}{F_1 \ldots F_{n+1}}. \tag{9.6}$$

This formula was proved in [**11**, Sec. 22] for the case of local residues.

Using the theorem on the complete sum of residues in $\mathbf{CP}^n$ we now pass to the residue Res_j, but then there appears a residue at infinity, which will be equal to Q'/Q.

Let us first find $Q(w)$. The system

$$P_j(\mathbf{z}, w) = 0, \quad j = 1, \ldots, n, \tag{9.7}$$

has a finite set of roots in $\mathbf{CP}^{n-1} \times \mathbf{C}^1$, i.e. it has the same form as (9.1), but the number of variables is smaller than in (9.1). Let $\tilde{Q}(w)$ be the classical resultant of the system (9.7). Then it is easy to see that $Q(w) = \tilde{Q}^{k_{n+1}}(w)$.

The rest of the proof is carried out by induction on n. Let $n = 1$, and

$$F_1 = a_0(w)z^{k_1} + a_1(w)z^{k_1-1} + \ldots + a_{k_1}(w),$$
$$F_2 = b_0(w)z^{k_2} + b_1(w)z^{k_2-1} + \ldots + b_{k_2}(w).$$

For $w = 0$ we have $a_0(0) \neq 0$ and $b_0(0) \neq 0$, which means that

$$P(w) = a_0^{k_2} \prod_{j=1}^{k_1} F_2(z^2_{(j)}(w), w)$$

for small $|w|$. Thus $R(w) = P(w)a_0^{-k_2}(w)$, and

$$\frac{R'}{R} = \frac{P'}{P} - \frac{k_2 a_0'}{a_0}.$$

On the other hand, by the theorem on the complete sum of residues, we have

$$\operatorname*{Res}_2 \frac{F_1' dF_2}{F_1 F_2} + \operatorname*{Res}_1 \frac{F_1' dF_2}{F_1 F_2} + \operatorname*{Res}_\infty \frac{F_1' dF_2}{F_1 F_2} = 0,$$

where

$$\operatorname*{Res}_\infty \frac{F_1' dF_2}{F_1 F_2} = \frac{1}{2\pi i} \int_{|z|=r} \frac{F_1' dF_2}{F_1 F_2} =$$
$$= -\frac{1}{2\pi i} \int_{|\varepsilon|=1/r} \frac{(\tilde{F}_1')(\tilde{F}_{2z}')d\varepsilon}{\varepsilon \tilde{F}_1' \tilde{F}_2'},$$
$$\tilde{F}_1 = a_0 + \varepsilon a_1 + \ldots + \varepsilon^{k_1} a_{k_1},$$
$$\tilde{F}_2 = b_0 + \varepsilon b_1 + \ldots + \varepsilon^{k_2} b_{k_2},$$
$$\tilde{F}_1' = a_0' + \varepsilon a_1' + \ldots + \varepsilon^{k_1} a_{k_1}',$$
$$\tilde{F}_2' = b_0' + \varepsilon b_1' + \ldots + \varepsilon^{k_2} b_{k_2}'.$$

It follows that

$$\operatorname*{Res}_\infty \frac{F_1' dF_2}{F_1 F_2} = -\frac{k_2 a_0'}{a_0},$$

and hence

$$\frac{P'}{P} = \operatorname*{Res}_1 \frac{F_1' dF_2}{F_1 F_2} + \operatorname*{Res}_2 \frac{F_2' dF_1}{F_1 F_2}.$$

Now let $n > 1$. Since

$$R = \frac{P}{Q}$$

and

$$\frac{R'}{R} = \frac{P'}{P} - \frac{Q'}{Q},$$

where $Q = \tilde{Q}^{k_{n+1}}$, we find that

$$\frac{R'}{R} = \frac{P'}{P} - \frac{k_{n+1}\tilde{Q}'}{\tilde{Q}}.$$

By the induction hypothesis we have the equality

$$\frac{\tilde{Q}'}{\tilde{Q}} = \sum_{l=1}^{n} \widetilde{\mathrm{Res}}_l \frac{P_l' dP[l, n+1]}{P_1 \dots P_n}$$

for the system (9.7). Here $\widetilde{\mathrm{Res}}_l$ is the global residue with respect to the system $P_j(1, z_2, \dots, z_n, w) = 0$, $j = 1, \dots, n$, $j \neq l$. Consider now the forms

$$\omega_l = \frac{F_l' dF[l]}{F_1 \dots F_{n+1}}, \quad l = 1, \dots, n+1.$$

Let us pass to homogeneous coordinates $(\zeta_0, \zeta_1, \dots, \zeta_n)$, i.e. we make the change of variables $z_j = \zeta_j / \zeta_0$. The polynomial $f(\mathbf{z})$ of degree k is then transformed to

$$f^*(\zeta_0, \zeta_1, \dots, \zeta_n) = \zeta_0^k f(\zeta_1/\zeta_0, \dots, \zeta_n/\zeta_0),$$

and from ω_l we get the new form

$$\omega_l^* = \frac{1}{\zeta_0} \frac{(F_l')^* J_l^* \nu(\zeta)}{F_1^* \dots F_{n+1}^*},$$

where J_l is the Jacobian ($J_l d\mathbf{z} = dF[l]$) and

$$\nu(\zeta) = \sum_{k=0}^{n} (-1)^k \zeta_k d\zeta[k].$$

So we see that the plane Π is singular for the form ω_l^*. (Notice that if w is not contained in the highest homogeneous part of F_l then F_l' has a degree $k_l - 1$ in $\mathbf{z}$, and then the plane Π is nonsingular for ω_l^*). By a theorem of Yuzhakov [**137**] (which is a corollary to the theorem on the complete sum of residues in $\mathbf{CP}^n$ [**80**, Ch. 5]), we then obtain

$$\mathop{\mathrm{Res}}_{n+1} \omega_l = (-1)^{n+l-1} \mathop{\mathrm{Res}}_{l} \omega_l + \mathop{\mathrm{Res}}_{\infty} \omega_l.$$

As $\mathrm{Res}_\infty \omega_l$ we can take, for example, the global residue of the form w_l^* with respect to the functions $\zeta_0, f_1, \dots, [l], \dots, f_n$ (see [**137**]). After performing the integration with respect to ζ_0, we have

$$\mathop{\mathrm{Res}}_{\infty} \omega_l = (-1)^{n-1} \sum_{j \geq 1} \frac{1}{(2\pi i)^{n-1}} \int_{\Gamma'_{j,l}} \frac{P_l' I_l dz_2 \wedge \dots \wedge dz_n}{P_1 \dots P_{n+1}},$$

where

$$\Gamma'_{j,l} = \{\mathbf{z}' \in U(\mathbf{z'}^l_{(j)}) : |P_s(1, z_2, \dots, z_n, 0)| = \varepsilon_s, \ s \neq l, n+1\}$$

with $\mathbf{z}' = (z_2, \dots, z_n)$, and $\mathbf{z'}^l_{(j)}$ are roots of the system

$$P_s(1, z_2, \dots, z_n, 0) = 0, \ s = 1, \dots, n, \ s \neq l,$$

$P_j = P_j(1, z_2, \dots, z_n, w)$, and finally

$$I_l = \begin{vmatrix} \frac{\partial P_1}{\partial z_1} & \dots & \frac{\partial P_1}{\partial z_n} \\ \dots & \dots & \dots \\ [l] & \dots & \dots \\ \dots & \dots & \dots \\ \frac{\partial P_{n+1}}{\partial z_1} & \dots & \frac{\partial P_{n+1}}{\partial z_n} \end{vmatrix}.$$

The Jacobian I_l can be transformed as follows. We divide and multiply the first column by z_1, and multiplying the other columns by z_j we add them to the first. Then we obtain

$$I_l = \frac{1}{z_1}\begin{vmatrix} k_1P_1\frac{\partial P_1}{\partial z_2} & \dots & \frac{\partial P_1}{\partial z_n} \\ \dots & \dots & \dots \\ [l] & \dots & \dots \\ \dots & \dots & \dots \\ k_{n+1}P_{n+1}\frac{\partial P_{n+1}}{\partial z_2} & \dots & \frac{\partial P_{n+1}}{\partial z_n} \end{vmatrix} = \frac{1}{z_1}\sum_{j\neq l} k_jP_j\Delta_j,$$

and setting $z_1 = 1$ we get

$$\frac{P_l' I_l dz'}{P_1 \dots P_{n+1}} = \sum_{j\neq l} \frac{k_j P_l' \Delta_j dz'}{P_1 \dots [j] \dots P_{n+1}}.$$

But since

$$\int_{\Gamma_{j,l}'} \frac{P_l'\Delta_j \mathbf{dz}'}{P_1\dots[j]\dots P_{n+1}} = 0, \quad \text{if } j \neq n+1,$$

we actually have

$$\operatorname*{Res}_{\infty} \omega_l = \sum_{j\geq 1} \frac{k_{n+1}}{(2\pi i)^{n-1}} \int_{\Gamma_{j,l}'} \frac{P_l' dP[l, n+1]}{P_1 \dots P_n}.$$

The induction hypothesis now gives

$$\sum_{l=1}^{n} \operatorname*{Res}_{\infty} \omega_l = k_{n+1}\frac{\tilde{Q}'}{\tilde{Q}},$$

and so the theorem is proved in the case when all roots of (9.1) are finite. To prove the theorem in the general case we make a linear transformation in $\mathbf{CP}^n$ moving all roots of system (9.1) and all the roots $\zeta_{(j)}^l$ to the finite part $\mathbf{CP}^n \setminus \Pi$. Then we apply the already proved assertion and make an inverse transformation. The expressions $F_l'(\mathbf{z})/F_l(\mathbf{z})$ and $f_l/f_l(\boldsymbol{\zeta})$ are equal if $z_j = \zeta_j/\zeta_0$, and therefore the form on the right hand side of formula (9.2) is invariant with respect to projective transformations. □

Let now all roots of system (9.1) be finite. Is it really necessary in (9.2) to take into account all roots of the subsystems, or would the finite roots suffice?

EXAMPLE 9.1. We consider the system

$$\begin{aligned} F_1 &= z_1 - wz_2 + a = 0, \\ F_2 &= z_1 - wz_2 = 0, \\ F_3 &= z_1^2 - wz_1z_2 + z_2^2 = 0, \end{aligned}$$

where $a \neq 0$. Since it has no roots in $\mathbf{CP}^2 \times \mathbf{C}^1$ we have $P \equiv 1$. Solving the first subsystem we obtain $z_1 = 0,\ z_2 = 0$, and $F_1'/F_1 = 0$. The second subsystem has the roots

$$z_1 = \left(a(w^2 - z) \pm aw\sqrt{w^2-4}\right)/2,$$

and

$$z_2 = (aw \pm a\sqrt{w^2-4})/2,$$

so we get $\sum F_2'/F_2 = w$. The third subsystem has no finite roots, and therefore the sum $\sum F_l'/F_l$ over all finite roots of the subsystems is equal to w, which is different from $P'/P = 0$.

The infinite root of the third subsystem is $\zeta_0 = 0$, $\zeta_1 = w\zeta_2$, $\zeta_2 = \zeta_2$. Then $F_3'/F_3 = -w$, and hence all sums $\sum F_l'/F_l$ vanish.

In spite of this example it is in many cases sufficient to take into account only the finite roots.

COROLLARY 9.1. *If all roots of system (9.1) are finite and if w is contained in at most one of the polynomials $f_j(0, \zeta_1, \ldots \zeta_n, w)$, then*

$$\frac{P'(w)}{P(w)} = \sum_{l=1}^{n} \sum_{\{j\}} \frac{f_j'}{f_j}(\zeta_{(j)}^l(w), w),$$

where the summation is taken only over finite roots $\zeta_{(j)}^l(w)$.

Proof. We must show that the sum in (9.2) taken over infinite roots $\zeta_{(j)}^l(w)$ is equal to 0. If w is not contained in the highest homogeneous part of $F_j(\mathbf{z}, w)$, then after passing to homogeneous coordinates we get

$$\frac{F_l'}{F_l}(\mathbf{z}_{(j)}^l(w), w) = \frac{\zeta_0^m (F_l')^*(\zeta_{(j)}^l(w), w)}{f_l(\zeta_{(j)}^l(w), w)},$$

where $m \geq 1$. Since $\zeta_0 = 0$ on Π such a term is equal to zero.

Let now w be contained in highest homogeneous part of $F_l(\mathbf{z}, w)$. The roots $\zeta_{(j)}^l(w)$ lying on Π must satisfy the system

$$f_j(0, \zeta_1, \ldots, \zeta_n, w) = f_j(0, \zeta_1, \ldots, \zeta_n) = 0, \quad j \neq l,$$

i.e. these roots $\zeta_{(j)}^l(w)$ are independent of w. Plugging in any root in f_l we obtain a polynomial in w which must be constant. Indeed, otherwise we could find a solution, which is impossible, since the system (9.1) was assumed to have only finite roots. Thus $f_l(\zeta_{(j)}^l(w)) \equiv \text{const}$, and hence

$$f_l'(\zeta_{(j)}^l(w), w) = 0.$$

It follows that the sum over the infinite roots of the subsystems is equal to zero.

Now we consider the following system of equations

$$\begin{aligned} f_1 &= f_1(\mathbf{z}, w) = 0, \\ \cdots & \quad \cdots\cdots\cdots\cdots \\ f_{n+1} &= f_{n+1}(\mathbf{z}, w) = 0, \end{aligned} \tag{9.8}$$

where $\mathbf{z} \in \mathbf{C}^n$, $w \in \mathbf{C}$, and f_j are polynomials in $\mathbf{z}$ and w. On this system we impose two conditions:

1. For $w = 0$ the system (9.8) should have no roots in $\mathbf{CP}^n$. This implies that all subsystems

$$f_j(\mathbf{z}, 0) = 0, \quad j \neq l,$$

have finite sets of roots in $\mathbf{CP}^n$ (see [**85**, Ch. 4, Th. 9]), so that after some linear-fractional change of variables they become non-degenerate. We may as well at once assume that the system

$$f_1(\mathbf{z},0)=0, \quad \dots, \quad f_n(\mathbf{z},0)=0 \tag{9.9}$$

is non-degenerate (and also that all previous subsystems are non-degenerate).

2. Consider the system

$$\begin{aligned} f_1(\mathbf{z},w) &= 0, \\ &\cdots \\ f_n(\mathbf{z},w) &= 0. \end{aligned} \tag{9.10}$$

We require that if we set $w=0$ the degree of polynomials f_j are not changed. This means that the highest homogeneous parts of all functions f_j contain some monomial without the parameter w. Thus, if $P_j(\mathbf{z},w)$ is the highest homogeneous part of f_j (in variables $z_1,\dots,z_n$), then

$$P_j(\mathbf{z},w)=G_j(\mathbf{z})+wF_j(\mathbf{z},w)$$

and $G_j\not\equiv 0$. If $\deg f_j=k_j$, then $\deg f_j=\deg P_j=\deg G_j=k_j$, and $\deg F_j=k_j$ or $F\equiv 0$. The conditions 1 and 2 also imply that the system

$$G_1(\mathbf{z})=0, \quad \dots, \quad G_n(\mathbf{z})=0 \tag{9.11}$$

has $z=0$ as its only one root. Let S be the unit sphere in $\mathbf{C}^n$, i.e.

$$S=\{\mathbf{z}\in\mathbf{C}^n : |\mathbf{z}|^2=|z_1|^2+\ldots+|z_n|^2=1\}.$$

Then by condition (9.11), for every point $\mathbf{z}^0\in S$, there exists a number j, such that $G_j(\mathbf{z}^0)\neq 0$. Therefore

$$G_j(\lambda\mathbf{z}^0)=\lambda^{k_j}G_j(\mathbf{z}^0), \quad \lambda\in\mathbf{C},$$

and there exist $\varepsilon>0$ such that for $|w|<\varepsilon$ the inequality

$$|G_j(\lambda\mathbf{z}^0)|>|w|\,|F_j(\lambda\mathbf{z}^0,w)|$$

holds for all $\lambda\in\mathbf{C}$. Now using Theorem 5.3, we obtain that for $|w|<\varepsilon$ the system

$$P_1(\mathbf{z},w)=0, \ \dots, \ P_n(\mathbf{z},w)=0$$

has also the sole root $\mathbf{z}=\mathbf{0}$, and hence the system (9.10) is non-degenerate. In fact, the Hilbert theorem [**142**] shows that system (9.10) can be degenerate only for a finite set of values on w.

We thus see that system (9.8) is special case of system (9.1).

EXAMPLE 9.2. Consider the system

$$\begin{aligned} &b_1 z_1^{k_1}-b_{-1}z_2^{k_1}=\nu_1 w, \\ &\cdots\cdots\cdots\cdots\cdots \\ &b_{n-1}z_{n-1}^{k_{n-1}}-b_{-(n-1)}z_n^{k_{n-1}}=\nu_{n-1}w, \\ &b_n z_n^{k_n}-b_{-n}z_1^{k_n}=\nu_n w, \\ &z_1+\ldots+z_n=1. \end{aligned}$$

For $w = 0$ we obtain

$$\begin{aligned} b_1 z_1^{k_1} &= b_{-1} z_2^{k_1}, \\ \dots&\dots\dots \\ b_n z_n^{k_n} &= b_{-n} z_1^{k_n}, \\ z_1 + \ldots + z_n &= 1. \end{aligned}$$

In order to find its roots in $\mathbf{CP}^n$ we must consider the projectivization of every function. We have

$$\begin{aligned} b_1 \zeta_1^{k_1} &= b_{-1} \zeta_2^{k_1}, \\ \dots&\dots\dots \\ b_n \zeta_n^{k_n} &= b_{-n} \zeta_1^{k_n}, \\ \zeta_1 + \ldots + \zeta_n &= \zeta_0. \end{aligned}$$

Solving this we find

$$\zeta_2 = a_2\zeta_1, \quad \zeta_3 = a_3\zeta_1, \quad \ldots, \quad \zeta_n = a_n\zeta_1,$$

and from the next to the last equation we have that $\zeta_1 = 0$ (the coefficients b_j and b_{-j} are parameters and we find roots for almost every value of the parameters). It follows that $\zeta_1 = \zeta_2 = \ldots = \zeta_n = 0$, and from the last equation we then get $\zeta_0 = 0$. This system has therefore no roots in $\mathbf{CP}^n$, so the second condition for system (9.8) is also fulfilled.

Let $R(w)$ be the resultant of the function f_{n+1} with respect to the system (9.10) and define

$$a_k = \frac{1}{k!}\frac{d^k \ln R}{dw^k}(0).$$

From the proof of Theorem 9.1 one deduces the following assertion.

THEOREM 9.2 (Kytmanov). *For the system (9.8) one has the formulas*

$$\frac{R'(w)}{R(w)} = \frac{1}{(2\pi i)^n}\sum_j \int_{\Gamma_j} \sum_{k=1}^{n+1} (-1)^{n+k-1} \frac{f_k'(\mathbf{z}, w)\, df[k]}{f_1(\mathbf{z}, w)\cdots f_{n+1}(\mathbf{z}, w)}$$

and

$$a_k = \frac{1}{(2\pi i)^n}\sum_j \sum_{l=1}^{n+1} \sum_{\substack{||\alpha|| = k \\ \alpha_1 = \ldots = \alpha_{l-1} = 0,\ \alpha_l > 0}} \frac{(-1)^{n+l-1}}{\alpha!} \int_{\Gamma_j} \left(\frac{f_l'}{f_l}\right)^{(\alpha_l - 1)} \left(\frac{df}{f}[l]\right)^{(\alpha)}_{w=0}.$$

Here

$$\Gamma_j = \{\mathbf{z} \in U(\mathbf{z}_{(j)}) : \ |f_k(\mathbf{z}, 0)| = \varepsilon_k, \ k = 1, \ldots, n\},$$

$\mathbf{z}_{(j)}$ *are roots of system (9.9),* $f^{(\alpha)}$ *is equal to* $d^\alpha f / dw^\alpha$ *and*

$$\left(\frac{df}{f}[l]\right)^{(\alpha)} = \left(\frac{df_1}{f_1}\right)^{(\alpha_1)} \wedge \ldots \wedge \left(\frac{df_{l-1}}{f_{l-1}}\right)^{(\alpha_{l-1})} \wedge \left(\frac{df_{l+1}}{f_{l+1}}\right)^{(\alpha_{l+1})} \wedge \ldots \wedge \left(\frac{df_{n+1}}{f_{n+1}}\right)^{(\alpha_{n+1})}.$$

This formula permits us to compute the coefficients of $R(w)$ by using Lemma 2.1.

Assume now that instead of condition 2 the system (9.8) satisfies the following condition.

3. The parameter w is not contained in the highest homogeneous parts of system (9.10), i.e.

$$f_j(\mathbf{z}, w) = P_j(\mathbf{z}) + Q_j(\mathbf{z}, w), \quad j = 1, \dots, n.$$

In this case we have $R(w) = P(w)$ (see the proof of Theorem 9.1 and Corollary 9.1).

COROLLARY 9.2. *For $|w|$ sufficiently small one has*

$$\frac{R'(w)}{R(w)} = \frac{P'(w)}{P(w)} = \sum_{s=1}^{n+1} \sum_{j=1}^{M_s} \frac{f'_s(\mathbf{z}^s_{(j)}(w), w)}{f_s(\mathbf{z}^s_{(j)}(w), w)}.$$

In particular

$$a_1 = \sum_{s=1}^{n+1} \sum_{j=1}^{M_s} \frac{f'_s}{f_s}(\mathbf{z}^s_{(j)}, 0).$$

Here $\mathbf{z}^s_{(j)}(w)$ are roots of the subsystem

$$f_j(\mathbf{z}, w) = 0, \ j \neq s, \ j = 1, \dots, n+1,$$

and $\mathbf{z}^s_{(j)} = \mathbf{z}^s_{(j)}(0)$.

EXAMPLE 9.3. Consider the system

$$b_1 z_1^2 - b_{-1} z_2^2 = \nu_1 w,$$
$$b_2 z_2 - b_{-2} z_1 = \nu_2 w,$$
$$z_1 + z_2 = 1.$$

In order to find $R(w)$ we solve the system

$$b_1 z_1^2 - b_{-1} z_2^2 = \nu_1 w,$$
$$b_2 z_2 - b_{-2} z_1 = \nu_2 w.$$

Then we obtain

$$\mathbf{z}^3_{(1),(2)}(w) =$$

$$\left(\frac{b_{-1} b_{-2} \nu_2 w \pm b_2 (b_1 b_{-1} \nu_2^2 w^2 + \nu_1 w (b_1 b_2^2 - b_{-1} b_{-2}^2))^{\frac{1}{2}}}{b_1 b_2^2 - b_{-1} b_{-2}^2}, \right.$$

$$\left. \frac{b_1 b_2 \nu_2 w \pm b_{-2} (b_1 b_{-1} \nu_2^2 w^2 + \nu_1 w (b_1 b_2^2 - b_{-1} b_{-2}^2))^{\frac{1}{2}}}{b_1 b_2^2 - b_{-1} b_{-2}^2} \right),$$

and hence

$$R(w) = f_3(\mathbf{z}^3_{(1)}(w), w) f_3(\mathbf{z}^3_{(2)}(w), w) =$$

$$= \frac{\nu_1^2 w^2 (b_1 - b_{-1}) - w(2\nu_2 (b_1 b_2 + b_{-1} b_{-2}) + \nu_1 (b_2 + b_{-2})^2) + b_1 b_2^2 - b_{-1} b_{-2}^2}{b_1 b_2^2 - b_{-1} b_{-2}^2},$$

from which we get

$$\frac{R'(w)}{R(w)} = \frac{2\nu_2 w(b_1 - b_{-1}) - 2\nu_2(b_1b_2 + b_{-1}b_{-2}) - \nu_1(b_2 + b_{-2})^2}{\nu_2^2 w^2(b_1 - b_{-1}) - w(2\nu_2(b_1b_2 + b_{-1}b_{-2}) + \nu_1(b_2 + b_{-2}))^2 + b_1b_2^2 - b_{-1}b_{-2}^2}.$$

Solving the first subsystem

$$b_2z_2 - b_{-2}z_1 = \nu_2 w,$$
$$z_1 + z_2 = 1,$$

we find

$$\mathbf{z}^1_{(1)} = \left(\frac{b_2 - \nu_2 w}{b_2 + b_{-2}}, \frac{\nu_2 w + b_{-2}}{b_2 + b_{-2}}\right),$$

$$\frac{f_1'}{f_1}(\mathbf{z}^1_{(1)}) = \frac{-\nu_1(b_2 + b_{-2})^2}{\nu_2^2 w^2(b_1 - b_{-1}) + 2\nu_2 w(b_1b_2 + b_{-1}b_{-2}) - \nu_1 w(b_2 + b_{-2})^2 + b_1b_2^2 - b_{-1}b_{-2}^2}.$$

Solving the second subsystem

$$b_1z_1^2 - b_{-1}z_2^2 = \nu_1 w,$$
$$z_1 + z_2 = 1,$$

we get

$$\mathbf{z}^2_{(1),(2)} = \left(\frac{-b_{-1} \pm \sqrt{b_1b_{-1} + \nu_1 w(b_1 - b_{-1})}}{b_1 - b_{-1}}, \frac{b_1 \pm \sqrt{b_1b_{-1} + \nu_1 w(b_1 - b_{-1})}}{b_1 - b_{-1}}\right),$$

and hence

$$\frac{f_2'}{f_2}(\mathbf{z}^2_{(1)}) + \frac{f_2'}{f_2}(\mathbf{z}^2_{(2)}) =$$

$$= \frac{2\nu_2 w(b_1 - b_{-1}) - 2\nu_2(b_1b_2 + b_{-1}b_{-2})}{\nu_2^2 w^2(b_1 - b_{-1}) + 2\nu_2 w(b_1b_2 + b_{-1}b_{-2}) - \nu_1 w(b_2 + b_{-2})^2 + b_1b_2^2 - b_{-1}b_{-2}^2}.$$

Since $f_3' = 0$ we have

$$\frac{R'(w)}{R(w)} = \frac{f_1'}{f_1}(\mathbf{z}^1_{(1)}) + \frac{f_2'}{f_2}(\mathbf{z}^2_{(1)}) + \frac{f_2'}{f_2}(\mathbf{z}^2_{(2)}).$$

Now we show that if w occurs in the highest homogeneous part then the identity of Corollary 9.2 may cease to hold.

EXAMPLE 9.4. Consider the system

$$z_1 - w = 0$$
$$z_2 - w = 0$$
$$z_1 + z_2(1 + w) - 1 = 0.$$

For the first subsystem we have $z_2 = w_1,\ z_1 = 1 - w - w^2$, and

$$f_1'/f_1 = -1/(1 - 2w - w^2).$$

The second subsystem is $z_1 = w,\ z_2 = (1 - w)/(1 + w)$, with

$$f_2'/f_2 = -(1 + w)/(1 - 2w - w^2).$$

Finally, the third subsystem is $z_1 = w,\ z_2 = w$, and

$$f_3'/f_3 = -w/(1 - 2w - w^2).$$

Hence

$$f_1'/f_1 + f_2'/f_2 + f_3'/f_3 = \frac{-2-2w}{1-2w-w^2},$$

while the resultant $R(w)$ of the function f_2 is

$$R(w) = \frac{1-w}{1+w} - w = \frac{1-2w-w^2}{1+w}.$$

Since

$$\frac{R'(w)}{R(w)} = \frac{-3-2w-w^2}{(1+w)(1-2w-w^2)},$$

we see that the identity is not valid.

But if we write down the resultant of the function f_3, then we see that it is equal to $P(w) = 2w - w^2 - 1$, and here the identity is valid according to Theorem 9.1.

Now we analyze the system of equations

$$\begin{aligned} f_1 &= P_1(\mathbf{z}) = 0, \\ \dots & \quad \dots\dots\dots\dots\dots \\ f_k &= P_k(\mathbf{z}) = 0, \\ f_{k+1} &= P_{k+1}(\mathbf{z}) - \nu_{k+1}w = 0, \\ \dots & \quad \dots\dots\dots\dots\dots \\ f_n &= P_n(\mathbf{z}) - \nu_n w = 0, \\ f_{n+1} &= P_{n+1}(\mathbf{z}) - 1 = 0, \end{aligned} \tag{9.12}$$

where $P_j(\mathbf{z})$ are homogeneous polynomials of degree $k_j \geq 1$. Systems arising from single-route and multi-route mechanisms of catalytic reactions [**150**] may be reduced to systems of the form (9.12).

We assume that the system (9.12) has no roots in $\mathbf{CP}^n$ for $w = 0$, so that Theorem 9.2 can be applied to it. On the cycles $\Gamma_j = \Gamma$ there holds the inequality $|P_j| > |\nu_j w|$ for small $|w|$. Expanding the function $1/(P_j - \nu_j w)$ in a geometric series we thus obtain

$$\frac{R'(w)}{R(w)} = \frac{1}{(2\pi i)^n}\sum_{l=k+1}^{n}(-1)^{n+l}\nu_l \sum_{||\alpha||}\int_\Gamma \frac{w^{||\alpha||}\nu^\alpha dP[l]}{P_1\dots P_k P_{k+1}^{\alpha_{k+1}+1}\dots P_n^{\alpha_n+1}(P_{n+1}-1)},$$

where $\alpha = (\alpha_{k+1}, \dots, \alpha_n)$, $||\alpha|| = \alpha_{k+1} + \dots \alpha_n$, $\nu^\alpha = \nu_{k+1}^{\alpha_{k+1}} \dots \nu_n^{\alpha_n}$.

From here we get

$$sa_s = \frac{1}{(2\pi i)^n}\sum_{l=k+1}^{n}(-1)^{n+l}\nu_l \sum_{||\alpha||=s-1}\int_\Gamma \frac{\nu^\alpha dP[l]}{(P_{n+1}-1)P_1\dots P_n^{\alpha_n+1}}, \tag{9.13}$$

and, since $|P_{n+1}| < 1$ on the cycle Γ, we also have

$$\frac{1}{P_{n+1}-1} = -\sum_{p=0}^{\infty} P_{n+1}^p.$$

But in (9.13) only finitely many of the integrals are non-zero. Indeed, the degree of the numerator is equal to

$$pk_{n+1} + \sum_{j \neq l} k_j,$$

while the degree of the denominator equals

$$\sum_{j=1}^{n} k_j + \sum_{j=k}^{n} \alpha_j k_j.$$

An integral in (9.13) can be different from zero only in case the degrees of numerator and denominator coincide. This happens when

$$(p+1)k_{n+1} = \sum_{j=k}^{n} \alpha_j k_j + k_l.$$

Hence

$$a_s = \frac{1}{(2\pi i)^n} \sum_{l=k+1}^{n} (-1)^{n+l-1} \frac{\nu_l}{s} \sum_{\|\alpha\|=s-1} \int_\Gamma \frac{\nu^\alpha P_{n+1}^{p_l} dP[l]}{P_1 \dots P_n^{\alpha_n+1}},$$

where

$$p_l k_{n+1} = \sum_{j=k+1}^{n} \alpha_j k_j + k_l - k_{n+1}.$$

If k_{n+1} does not divide the right hand side, then $a_s = 0$.

In order to obtain the formulas for a_s in final form we must, as in Sec. 8, use the generalized formula for the Grothendieck residue (7.5). Let

$$z_j^{N+1} = \sum_{s=1}^{n} a_{js}(\mathbf{z}) P_s(\mathbf{z}), \quad \mathbf{A} = \|a_{js}\|$$

$$\text{and} \quad dP[l] = J_l d\mathbf{z}, \quad k_{n+1} = \deg P_{n+1} = 1.$$

THEOREM 9.3 (Kytmanov). *One has the following formulas*

$$a_s = \sum_{l=k+1}^{n} (-1)^{n+l-1} \frac{\nu_l}{s} \sum_{k_{ij}} \frac{\prod_i \left(\sum_j k_{ij} \right)!}{\prod_{i,j} k_{ij}!} \times$$

$$\times \mathfrak{M} \left[\frac{P_{n+1}^{p_l} J_l \det \mathbf{A} \prod_{i,j} a_{ij}^{k_{ij}}}{z_1^{N\beta_1+\beta_1+N} \dots z_n^{N\beta_n+\beta_n+N}} \right].$$

Here

$$p_l = \sum_{j=k+1} \alpha_j k_j + k_l - 1, \ \sum_j k_{ij} = \beta_i, \ \sum_i k_{ij} = \alpha_j,$$

$\mathfrak{M}$ *(as in Sec. 8) is the functional assigning to a Laurent polynomial*

$$\frac{Q(z)}{z_1^{q_1} \dots z_n^{q_n}}$$

its constant term, $s = 1, 2, \ldots$.

In particular

$$a_1 = \sum_{l=k+1}^{n} (-1)^{n+l-1} \nu_l \mathfrak{M}\left[\frac{P_{n+1}^{k_l-1} J_l \det \mathbf{A}}{z_1^N \ldots z_n^N}\right].$$

Example 9.5. Consider the system

$$\begin{aligned} b_1 z_1^2 - b_{-1} z_2^2 &= u, \\ b_2 z_2^2 - b_{-2} z_1^2 &= u, \\ b_3 z_1 - b_{-3} z_3 &= w, \\ b_4 z_3 - b_{-4} z_1 &= w, \\ z_1 + z_2 + z_3 &= 1. \end{aligned}$$

The corresponding system of the form (9.12) is

$$\begin{aligned} (b_1 + b_{-1}) z_1^2 - (b_{-1} + b_2) z_2^2 &= 0, \\ b_3 z_1 - b_{-3} z_3 &= w, \\ b_4 z_3 - b_{-4} z_1 &= w, \\ z_1 + z_2 + z_3 &= 1. \end{aligned}$$

Since $\deg P_4 = 1$ we have

$$a_1 = \mathfrak{M}\left[\frac{(J_3 - J_2) \det \mathbf{A}}{z_1 z_2 z_3}\right],$$

$$J_2 = -2z_2(b_{-1} + b_2)(b_4 + b_{-4}) - 2z_1 b_4 (b_1 + b_{-2}),$$
$$J_3 = 2z_2(b_{-1} + b_2)(b_3 + b_{-3}) + 2z_1 b_{-3}(b_1 + b_{-2}).$$

Computing the entries of the matrix $\mathbf{A}$, we get

$$a_{11} = 0, \quad a_{13} = \frac{b_{-3}^2 (b_4 z_3 + b_{-4} z_1)}{b_3^2 b_4^2 - b_{-3}^2 b_{-4}^2},$$

$$a_{12} = \frac{b_4^2 (b_3 z_1 + b_{-3} z_3)}{b_3^2 b_4^2 - b_{-3}^2 b_{-4}^2},$$

$$a_{21} = -1/(b_2 + b_{-1}),$$

$$a_{22} = \frac{(b_4^2 + b_{-4}^2)(b_1 + b_{-2})(b_3 z_1 + b_{-3} z_3)}{(b_2 + b_{-1})(b_3^2 b_4^2 - b_{-3}^2 b_{-4}^2)}$$

$$a_{31} = 0, \quad a_{32} = \frac{b_{-4}^3 (b_3 z_1 + b_{-3 z_3})}{b_3^2 b_4^2 - b_{-3}^2 b_{-4}^2},$$

$$a_{33} = \frac{b_3^2 (b_4 z_3 + b_{-4} z_1)}{b_3^2 b_4^2 - b_{-3}^2 b_{-4}^2},$$

$$\det \mathbf{A} = \frac{(b_3 z_1 + b_{-3} z_3)(b_4 z_3 + b_{-4} z_1)}{(b_1 + b_{-1})(b_3^2 b_4^2 - b_{-3}^2 b_{-4}^2)}.$$

Hence

$$a_1 = \frac{2(b_3 b_4 + b_{-3} b_{-4})(b_{-1} + b_2)(b_4 + b_{-4} + b_3 + b_{-3})}{(b_{-1} + b_2)(b_3^2 b_4^2 - b_{-3}^2 b_{-4}^2)} =$$

$$= \frac{2(b_3 + b_4 + b_{-3} + b_{-4})}{b_3 b_4 - b_{-3} b_{-4}}.$$

Theorems 9.1 and 9.3 are contained in [**100**].

10. Multidimensional analogues of the Newton formulas

The relation between the coefficients of a polynomial in $\mathbf{C}$ and the power sums of its zeros are given by the Newton formulas (2.1). The proof of these formulas can be carried out by using the logarithmic residue (see Sec. 2). For systems of algebraic equations in $\mathbf{C}^n$ we can also obtain a relation between coefficients and power sums. At first we consider simpler systems than the ones in Sec. 8, 9 (see [**9**]).

Let the following system be given

$$\begin{aligned} f_1 &= z_1^{k_1} + Q_1 = 0, \\ \dots & \quad \dots\dots\dots\dots \\ f_n &= z_n^{k_n} + Q_n = 0, \end{aligned} \tag{10.1}$$

where $Q_1, \dots, Q_n$ are polynomials in $\mathbf{z} = (z_1, \dots, z_n) \in \mathbf{C}^n$, and $\deg Q_j < k_j$, $j = 1, \dots, n$. In other words, the highest homogeneous part of f_j is equal to $z_j^{k_j}$.

The system (10.1) is non-degenerate, and the polynomials Q_j are of the form

$$Q_j = \sum_{||\alpha|| < k_j} a_\alpha(j) \mathbf{z}^\alpha,$$

where, as usual,

$$\alpha = (\alpha_1, \dots, \alpha_n), \quad \mathbf{z}^\alpha = z_1^{\alpha_1} \dots z_n^{\alpha_n}.$$

Setting $a_\alpha(j) = 1$ if $\alpha = (0, \dots, 0, \underset{j}{k_j}, 0, \dots, 0)$, and $a_\alpha(j) = 0$ for the other α with $||\alpha|| = k_j$, we can write

$$f_j(\mathbf{z}) = \sum_{||\alpha|| \le k_j} a_\alpha(j) \mathbf{z}^\alpha.$$

We choose R so that for all $j = 1, \dots, n$ the inequalities $|z_j|^{k_j} > |Q_j|$ hold on the torus

$$\Gamma_R = \{\mathbf{z} \in \mathbf{C}^n : |z_j| = R, \ j = 1, \dots, n\}.$$

Then, by Theorem 5.3, all the roots of system (10.1) lie in the polydisk

$$D_R = \{\mathbf{z} \in \mathbf{C}^n : \ |z_j| < R, \ j = 1, \dots, n\}.$$

We denote by S_α the expression

$$S_\alpha = \frac{1}{(2\pi i)^n} \int_{\Gamma_R} \frac{\mathbf{z}^\alpha J_{\mathfrak{f}} d\mathbf{z}}{f_1 \dots f_n}, \tag{10.2}$$

where $J_{\mathfrak{f}}$ is the Jacobian of the system (10.1). If $\alpha = (\alpha_1, \dots, \alpha_n)$ has nonnegative integer coordinates then by Theorem 5.3 S_α is the power sum of the roots $\mathbf{z}_{(j)}$ of system (10.1), so that

$$S_\alpha = \sum_{j=1}^{M} \mathbf{z}_{(j)}^\alpha = \sum_{j=1}^{M} z_{1(j)}^{\alpha_1} \dots z_{n(j)}^{\alpha_n}, \quad M = k_1 \dots k_n.$$

Later on we shall allow α also to have negative integers as entries. If there exists such an entry $\alpha_j < 0$, then the expression S_α will be called a pseudopower sum.

We notice if $||\alpha|| < 0$ then $S_\alpha = 0$. Indeed, on the torus Γ_R we have

$$\frac{1}{f_j} = \frac{1}{z_j^{k_j} + Q_j} = \sum_{s=0}^{\infty}(-1)^s \frac{Q_j^s}{z_j^{k_j(s+1)}},$$

and therefore

$$S_\alpha = \frac{1}{(2\pi i)^n}\int_{\Gamma_R}\sum_{||\beta||>0}\frac{J_{\mathbf{f}}\mathbf{z}^\alpha(-1)^{||\beta||}Q_1^{\beta_1}\dots Q_n^{\beta_n}\mathbf{dz}}{z_1^{k_1(\beta_1+1)}\dots z_n^{k_n(\beta_n+1)}}.$$

The degree of the numerator in this integral is equal to

$$||\alpha|| + ||\mathbf{k}|| - n + \beta_1(k_1-1) + \ldots + \beta_n(k_n-1)$$

and the degree of the denominator equals

$$k_1(\beta_1+1) + \ldots + k_n(\beta_n+1).$$

Thus, for $||\alpha|| < 0$, the Laurent polynomial under the sign of integration has no term of the form $(z_1\dots z_n)^{-1}$, and it follows that $S_\alpha = 0$. If $\alpha = (0,\dots,0)$, then

$$S_\alpha = S_{(0,\dots,0)} = k_1\dots k_n,$$

but if $||\alpha|| = 0$ with $\alpha \neq (0,\dots,0)$, then $S_\alpha = 0$, because in this case

$$S_\alpha = \frac{1}{(2\pi i)^n}\int_{\Gamma_R}\frac{\mathbf{z}^\alpha J_{\mathbf{f}}\mathbf{dz}}{z_1^{k_1}\dots z_n^{k_n}},$$

and the Jacobian is

$$J_{\mathbf{f}} = k_1\dots k_n z_1^{k_1-1}\dots z_n^{k_n-1} + \sum_\beta c_\beta \mathbf{z}^\beta$$

with $||\beta|| < ||\mathbf{k}|| - n$.

THEOREM 10.1 (Aizenberg – Kytmanov). *For the system (10.1) one has the following (Newton) recursion formulas*

$$S_{\mathbf{k}-\beta} + \sum_{||\beta||<||\alpha^1+\dots+\alpha^n||<||\mathbf{k}||} a_{\alpha^1}(1)\dots a_{\alpha^n}(n)S_{\alpha^1+\dots+\alpha^n-\beta} +$$

$$+\sum_{\alpha^1+\dots+\alpha^n=\beta}(k_1\dots k_n - \Delta_{\alpha^1\dots\alpha^n})a_{\alpha^1}(1)\dots a_{\alpha^n}(n) = 0. \qquad (10.3)$$

Here $\mathbf{k} = (k_1,\dots,k_n)$, $0 \le ||\beta|| < ||\mathbf{k}||$, $\alpha^j = (\alpha_1^j,\dots\alpha_n^j)$ *are nonnegative multiindeces, and*

$$\Delta_{\alpha^1\dots\alpha^n} = \det ||\alpha_j^k||_{j,k=1}^n.$$

If $||\beta|| < 0$ *then formula (10.3) has the form*

$$S_{\mathbf{k}-\beta} + \sum_{||\alpha^1+\dots+\alpha^n||<||\mathbf{k}||} a_{\alpha^1}(1)\dots a_{\alpha^n}(n)S_{\alpha^1+\dots+\alpha^n-\beta} = 0. \qquad (10.4)$$

Proof. Let us consider the integral

$$J = \frac{1}{(2\pi i)^n}\int_{\Gamma_R}\frac{\mathbf{z}^{-\boldsymbol{\beta}}P_1\dots P_n J_{\mathbf{f}}\mathbf{dz}}{P_1\dots P_n}.$$

On the one hand

$$J = \frac{1}{(2\pi i)^n}\int_{\Gamma_R}\frac{J_{\mathbf{f}}}{\mathbf{z}^{\boldsymbol{\beta}}}\mathbf{dz} =$$

$$= \frac{1}{(2\pi i)^n}\int_{\Gamma_R}\frac{1}{\mathbf{z}^{\boldsymbol{\beta}}z_1\dots z_n}\sum \Delta_{\boldsymbol{\alpha}^1\dots\boldsymbol{\alpha}^n}a_{\boldsymbol{\alpha}^1}(1)\dots a_{\boldsymbol{\alpha}^n}(n)\mathbf{z}^{\alpha_1}\dots\mathbf{z}^{\alpha_n}\mathbf{dz} =$$

$$= \begin{cases} 0, & \text{if } \boldsymbol{\beta} \text{ contains at least one negative coordinate,} \\ \displaystyle\sum_{\boldsymbol{\alpha}^1+\dots+\boldsymbol{\alpha}^n=\boldsymbol{\beta}} \Delta_{\boldsymbol{\alpha}^1\dots\boldsymbol{\alpha}^n}a_{\boldsymbol{\alpha}^1}(1)\dots a_{\boldsymbol{\alpha}^n}(n), & \text{if } \boldsymbol{\beta}\geq 0. \end{cases}$$

On the other hand, by the definition of $S_{\boldsymbol{\alpha}}$,

$$J = \sum_{||\boldsymbol{\alpha}^1+\dots+\boldsymbol{\alpha}^n-\boldsymbol{\beta}||\geq 0} a_{\boldsymbol{\alpha}^1}(1)\dots a_{\boldsymbol{\alpha}^n}(n)S_{\boldsymbol{\alpha}^1+\dots\boldsymbol{\alpha}^n-\boldsymbol{\beta}}.$$

If $\boldsymbol{\alpha}^1+\dots+\boldsymbol{\alpha}^n=\boldsymbol{\beta}$, then

$$S_{\boldsymbol{\alpha}^1+\dots+\boldsymbol{\alpha}^n-\boldsymbol{\beta}} = k_1\dots k_n,$$

but if $||\boldsymbol{\alpha}^1+\dots+\boldsymbol{\alpha}^n|| = ||\boldsymbol{\beta}||$, with $\boldsymbol{\alpha}^1+\dots+\boldsymbol{\alpha}^n\neq\boldsymbol{\beta}$, then

$$S_{\boldsymbol{\alpha}^1+\dots+\boldsymbol{\alpha}^n-\boldsymbol{\beta}} = 0.$$

Let $||\boldsymbol{\alpha}^1+\dots+\boldsymbol{\alpha}^n|| \leq ||k||$. If $||\boldsymbol{\alpha}^1+\dots+\boldsymbol{\alpha}^n|| = ||\mathbf{k}||$ then $||\boldsymbol{\alpha}^j|| = k_j$. Since the coefficients satisfy

$$a_{\boldsymbol{\alpha}^j}(j) = \begin{cases} 1, & \text{if } \boldsymbol{\alpha}^j = k_j\mathbf{e}_j, \\ 0, & \text{if } \boldsymbol{\alpha}^j \neq k_j\mathbf{e}_j, \end{cases}$$

we see that

$$J = S_{\mathbf{k}-\boldsymbol{\beta}} + \sum_{||\boldsymbol{\beta}||<||\boldsymbol{\alpha}^1+\dots+\boldsymbol{\alpha}^n||<||\mathbf{k}||} a_{\boldsymbol{\alpha}^1}(1)\dots a_{\boldsymbol{\alpha}^n}(n)S_{\boldsymbol{\alpha}^1+\dots+\boldsymbol{\alpha}^n-\boldsymbol{\beta}}+$$

$$+k_1\dots k_n \sum_{\boldsymbol{\alpha}^1+\dots+\boldsymbol{\alpha}^n=\boldsymbol{\beta}} a_{\boldsymbol{\alpha}^1}(1)\dots a_{\boldsymbol{\alpha}^n}(n).$$

This implies (10.3), and formula (10.4) is proved analogously. □

We present one more version of the formulas (10.3). Suppose $\boldsymbol{\gamma} = \mathbf{k}-\boldsymbol{\beta}$. Then from (10.3) we have

$$S_{\boldsymbol{\gamma}} + \sum_{j=1}^{n}\sum_{k_j-||\boldsymbol{\gamma}||<||\boldsymbol{\alpha}^j||<k_j} a_{\boldsymbol{\alpha}^j}(j)S_{\boldsymbol{\gamma}+\boldsymbol{\alpha}^j-k_j\mathbf{e}_j} + \dots +$$

$$\sum_{j_1<\dots<j_l}\ \sum_{\substack{||\boldsymbol{\alpha}^{j_s}||<k_{j_s} \\ ||\boldsymbol{\alpha}^{j_1}+\dots+\boldsymbol{\alpha}^{j_l}||>k_{j_1}+\dots+k_{j_l}-||\boldsymbol{\gamma}||}} a_{\boldsymbol{\alpha}^{j_1}}(j_1)\dots a_{\boldsymbol{\alpha}^{j_l}}(j_l)S_{\boldsymbol{\gamma}+\boldsymbol{\alpha}^{j_1}+\dots+\boldsymbol{\alpha}^{j_l}-k_{j_1}\mathbf{e}_{j_1}-\dots-k_{j_l}\mathbf{e}_{j_l}}+$$

$$+\dots+\sum_{\boldsymbol{\alpha}^1+\dots+\boldsymbol{\alpha}^n=\mathbf{k}-\boldsymbol{\gamma}}(k_1\dots k_n - \Delta_{\boldsymbol{\alpha}^1\dots\boldsymbol{\alpha}^n})a_{\boldsymbol{\alpha}^1}(1)\dots a_{\boldsymbol{\alpha}^n}(n) = 0. \qquad (10.5)$$

By the formulas (10.5) one can more easily compute power sums for small norms of γ. Let us find, for example, S_γ, if $||\gamma|| = 1$. In this case only the first and the last terms remain in formula (10.5), and we have

$$S_\gamma = - \sum_{\alpha^1+\ldots+\alpha^n=\mathbf{k}-\gamma} (k_1\ldots k_n - \Delta_{\alpha^1\ldots\alpha^n}) a_{\alpha^1}(1)\ldots a_{\alpha^n}(n). \tag{10.6}$$

Suppose $\gamma = (-\beta_1, -\beta_2, \ldots, \beta_{j+1}, \ldots, -\beta_n)$ and $k_j > \beta_s \geq 0$, $s = 1, \ldots, n$. Then

$$\mathbf{k} - \gamma = (k_1 + \beta_1, \ldots, k_{j-1} + \beta_{j-1}, k_j - \beta_j, \ldots, k_n + \beta_n),$$

and since $||\mathbf{k} - \gamma|| = ||\mathbf{k}|| - 1$, we obtain that the sum in (10.6) consists of one term such that $\alpha^s = k_s \mathbf{e}_s$ for $s \neq j$, and $\alpha^j = (\beta_1, \ldots, k_j - \beta_j, \ldots, \beta_n)$. If γ contains two or more positive coordinates then $S_\gamma = 0$, and hence

$$S_\gamma = -k_1 \ldots k_{j-1}\beta_j k_{j+1} \ldots k_n a_{\alpha^j}(j). \tag{10.7}$$

Thus, there are only finitely many non-zero S_γ with $||\gamma|| = 1$. Considering S_γ for fixed $||\gamma||$ we get from (10.5) by induction that the number of non-zero S_γ is finite. Hence formulas (10.5) are really recursive. They permit one to determine power sums S_γ with norm $||\gamma||$ in terms of power sums with smaller norms $||\gamma||$. Formula (10.7) shows that pseudopower sums can be different from 0.

In the paper [9] more general formulas than (10.3) and (10.4) were considered.

Using the formulas (10.3) and (10.4) one can compute the power and pseudopower sums S_α of the roots of a system (10.1). We saw for $n = 1$ that the Newton formulas (2.1) allows one also to compute the coefficients of a polynomial from its power sums. For $n > 1$ it is necessary to know not only the power sums, but also the pseudopower sums, in order to find the coefficients of system (10.1).

THEOREM 10.2 (Aizenberg - Kytmanov). *Let S_α denote the power and pseudopower sums of system (10.1), $||\alpha|| \leq \max_j k_j$. Then the coefficients of the polynomials Q_j are uniquely determined by (10.3) and (10.4).*

Proof. We write down Q_j in the form

$$Q_j = \sum_{s=0}^{k_j-1} Q_{j,s},$$

where

$$Q_{j,s} = \sum_{||\alpha||=s} a_\alpha(j)\mathbf{z}^\alpha.$$

Formula (10.7) implies that the coefficients $a_\alpha(j)$ of the polynomials Q_{j,k_j-1} are equal to

$$a_\alpha(j) = S_\gamma/(k_1 \ldots k_{j-1}(k_j - \alpha_j)k_{j+1} \ldots k_n),$$

where $\gamma = (-\alpha_1, \ldots, -\alpha_{j-1}, k_j - \alpha_j, \ldots, -\alpha_n)$, $||\gamma|| = 1$.

Consider now the polynomials Q_{j,k_j-2}. If in formula (10.3) we take β with $||\beta|| = ||\mathbf{k}|| - 2$, then the first sum contains also the known coefficients of the polynomials Q_{j,k_j-1}, and the power sums S_γ. The second sum contains the coefficients of Q_{j,k_j-1}

and only one coefficient from one polynomial Q_{j,k_j-2}. Indeed, if we are to find $a_\alpha(j)$ with $||\alpha|| = k_j - 2$, then we take in (10.3)

$$\beta = (k_1 + \alpha_1, \dots, k_{j-1} + \alpha_{j-1}, \alpha_j \dots, k_n + \alpha_n), \quad ||\beta|| = ||\mathbf{k}|| - 2.$$

Notice that the coefficients of $a_\alpha(j)$ are not equal to 0 since

$$k_1 \dots k_n > \Delta_{\alpha^1 \dots \alpha^n}.$$

Finally we can also determine the coefficients of the remaining polynomials $Q_{j,s}$ and to do this it is sufficient to know the sums S_α with $||\alpha|| \le \max_j k_j$. □

We cannot construct Q_j from the power sums S_α, $\alpha \ge 0$ alone. For example, if $k_1 > k_2$ then adding the second equation to the first, we get a system which has the same roots and therefore the same power sums, and it is still of the form (10.1). From this and from Theorem 10.2 we conclude that the pseudopower sums S_α are, in general, not functions of the roots. But sometimes this is the case.

EXAMPLE 10.1. Consider the system

$$z_1^2 + a_1 z_1 + b_1 z_2 + c_1 = 0,$$

$$z_2^2 + a_2 z_1 + b_2 z_2 + c_2 = 0.$$

Let S_α be its power sums. We want to find the coefficients a_j, b_j, c_j, $j = 1, 2$. Considering first β with $||\beta|| = 3$, we have from (10.3)

$$a_1 = -\frac{1}{2} S_{(1,0)}, \ a_2 = -\frac{1}{4} S_{(-1,2)}, \ b_1 = -\frac{1}{4} S_{(2,-1)}, \ b_2 = -\frac{1}{2} S_{(0,1)}.$$

Then we take β with $||\beta|| = 2$, and obtain

$$\begin{aligned} c_1 &= \frac{1}{8} S^2_{(1,0)} - \frac{1}{4} S_{(2,0)} + \frac{1}{16} S_{(0,1)} S_{(2,-1)}, \\ c_2 &= \frac{1}{8} S^2_{(0,1)} - \frac{1}{4} S_{(0,2)} + \frac{1}{16} S_{(-1,2)} S_{(1,0)}, \end{aligned}$$

and also the equation

$$\frac{3}{16} S_{(2,-1)} S_{(-1,2)} = S_{(1,1)} + \frac{3}{4} S_{(1,0)} S_{(0,1)}. \tag{10.8}$$

It remains to determine $S_{(2,-1)}$ and $S_{(-1,2)}$. From (10.3) with $||\beta|| = 1$ we get the equations

$$\begin{aligned} S_{(2,1)} - \frac{1}{4} S_{(2,-1)} S_{(0,2)} - \frac{1}{4} S_{(2,0)} S_{(0,1)} + \frac{1}{16} S^2_{(0,1)} S_{(2,-1)} + & \\ + \frac{1}{8} S^2_{(1,0)} S_{(0,1)} - \frac{1}{2} S_{(1,1)} S_{(1,0)} &= 0 \\ S_{(1,2)} - \frac{1}{4} S_{(2,0)} S_{(-1,2)} - \frac{1}{4} S_{(0,2)} S_{(1,0)} - \frac{1}{2} S_{(1,1)} S_{(0,1)} + & \\ + \frac{1}{8} S^2_{(0,1)} S_{(1,0)} - \frac{1}{16} S^2_{(-1,2)} S_{(1,0)} &= 0. \end{aligned}$$

Hence we can determine $S_{(-1,2)}$ and $S_{(2,-1)}$ which are rational functions of the roots. The equation (10.8) establishes a connection between the S_α. Finally, we have

$$
\begin{aligned}
a_1 &= -\frac{1}{2}S_{(1,0)},\\
a_2 &= \frac{2S_{(0,2)}S_{(1,0)} + 4S_{(1,1)}S_{(0,1)} - S^2_{(0,1)}S_{(1,0)} - 8S_{(1,2)}}{8S_{(2,0)} - 2S^2_{(1,0)}},\\
b_1 &= \frac{2S_{(2,0)}S_{(0,1)} + 4S_{(1,1)}S_{(1,0)} - S^2_{(1,0)}S_{(0,1)} - 8S_{(2,1)}}{2S^2_{(0,1)} - 8S_{(0,2)}},\\
b_2 &= -\frac{1}{2}S_{(0,1)},\\
c_1 &= \frac{2S_{(2,0)}S_{(0,2)} - 2S_{(2,1)}S_{0,1)} + S_{(1,1)}S_{(1,0)}S_{(0,1)} - S^2_{(1,0)}S_{(0,2)}}{2S^2_{(0,1)} - 8S_{(0,2)}},\\
c_2 &= \frac{2S_{(2,0)}S_{(0,2)} - 2S_{(1,0)}S_{1,2)} + S_{(1,1)}S_{(0,1)}S_{(1,0)} - S^2_{(0,1)}S_{(0,2)}}{2S^2_{(0,1)} - 8S_{(0,2)}}.
\end{aligned}
$$

So we see that some of the pseudopower sums from (10.3) are determined in terms of power sums, while others are not.

If we want to find the coefficients of (10.1) from the power sums S_α, then we can determine some of the pseudopower sums from (10.3) and the others we can set equal to zero.

EXAMPLE 10.2. Let us consider the system

$$z_1^2 + az_1 + bz_2 + c = 0, \quad z_2 + d = 0.$$

Then (10.3) implies

$$d = -\frac{1}{2}S_{(0,1)},\ a = -S_{(1,0)},\ b = -\frac{1}{2}S_{(2,-1)},$$

$$c = \frac{1}{4}S_{(0,1)}S_{(2,-1)} - \frac{1}{2}S_{(2,0)} + \frac{1}{2}S^2_{(1,0)}S_{(2,-1)}.$$

Here $S_{(2,-1)}$ is an arbitrary number, say $S_{(2,-1)} = 0$.

Now we consider the more general system of equations

$$
\begin{aligned}
f_1 &= P_1 + Q_1 = 0,\\
&\cdot\quad \cdot\quad \cdot\cdot\cdot \\
f_n &= P_n + Q_n = 0,
\end{aligned}
\tag{10.9}
$$

where P_j and Q_j are polynomials in $\mathbf{C}^n$, P_j are homogeneous polynomials of degree k_j and $\deg Q_j < k_j$, $j = 1, \dots, n$. We assume that the system (10.9) is nondegenerate.

By Theorem 5.2 on logarithmic residues the power sums S_α of the roots of system (10.9) are equal to

$$S_\alpha = \frac{1}{(2\pi i)^n}\int_{\Gamma_\mathbf{f}} \frac{\mathbf{z}^\alpha J_\mathbf{f} d\mathbf{z}}{f_1 \dots f_n}.$$

Since (10.9) is nondegenerate then, as in Sec. 8, we also have the matrix

$$\mathbf{A} = ||a_{jk}(\mathbf{z})||^n_{j,k=1},$$

where $a_{jk}(\mathbf{z})$ are polynomials, such that

$$z_j^{s_j} = \sum_{k=1}^{n} a_{jk}(\mathbf{z})P_k(\mathbf{z}), \quad j = 1, \dots, n,$$

with s_j being certain natural numbers (in Sec. 8 we introduced the convenient notation $N_j + 1 = s_j$), such that $s_j \le ||\mathbf{k}|| - n + 1$. Let $q_j(\mathbf{z})$ denote the polynomials

$$q_j(\mathbf{z}) = \sum_{k=1}^{n} a_{jk}(\mathbf{z})Q_k(\mathbf{z}), \quad j = 1, \dots, n,$$

and write $F_j(\mathbf{z}) = z_j^{s_j} + q_j(\mathbf{z})$, so that $\deg q_j < s_j$. Suppose that

$$F_j(\mathbf{z}) = \sum_{||\alpha|| \le s_j} c_\alpha(j)\mathbf{z}^\alpha,$$

and that $c_\alpha(j) = 1$ for $\alpha = s_j\mathbf{e}_j$, and $c_\alpha(j) = 0$ for all the other α of length s_j.

Using the transformation formula (4.2) (or strictly speaking the transformation formula (7.6) for the global residue) we obtain

$$S_\alpha = \frac{1}{(2\pi i)^n} \int_{\Gamma_F} \frac{\mathbf{z}^\alpha J_{\mathfrak{f}} \det \mathbf{A} d\mathbf{z}}{F_1 \dots F_n}.$$

Choosing the cycle $\Gamma_{\mathbf{z}} = \{\mathbf{z} : \ |z_j| = \varepsilon_j, \ j = 1, \dots, n\}$, with $\varepsilon_j > 0$ sufficiently large we have the inequalities $|z_j^{s_j}| > |q_j|$, $j = 1, \dots, n$, on $\Gamma_{\mathbf{z}}$. Using Theorem 5.3 we then get

$$S_\alpha = \frac{1}{(2\pi i)^n} \int_{\Gamma_{\mathbf{z}}} \frac{\mathbf{z}^\alpha J_{\mathfrak{f}} \det \mathbf{A} d\mathbf{z}}{F_1 \dots F_n}.$$

Let now the multiindex α have at least one negative coordinate. The expression

$$S_\alpha = \frac{1}{(2\pi i)^n} \int_{\Gamma_{\mathbf{z}}} \frac{\mathbf{z}^\alpha J_{\mathfrak{f}} \det \mathbf{A} d\mathbf{z}}{F_1 \dots F_n}$$

will then be called a pseudopower sum.

We can show, just as for the system (10.1), that $S_\alpha = 0$ if $||\alpha|| < 0$. Indeed, expanding the function

$$\frac{1}{F_j} = \frac{1}{(z_j^{s_j} + q_j)}$$

in a geometric series

$$\sum_{s=0}^{\infty} (-1)^s \frac{q_j^s}{z^{s_j(s+1)}},$$

we obtain

$$S_\alpha = \frac{1}{(2\pi i)^n} \sum_{||\beta|| \ge 0} (-1)^{||\beta||} \int_{\Gamma_{\mathbf{z}}} \frac{\mathbf{z}^\alpha J_{\mathfrak{f}} \det \mathbf{A} q_1^{\beta_1} \dots q_n^{\beta_n} d\mathbf{z}}{z_1^{s_1(\beta_1+1)} \dots z_n^{s_n(\beta_n+1)}}.$$

The degree of the denominator is equal to

$$\sum_{j=1}^{n} s_j(\beta_j + 1),$$

while the degree of the numerator equals

$$||\alpha|| + \sum_{j=1}^{n} s_j - n + \sum_{j=1}^{n}(s_j - 1)\beta_j,$$

so that the integral can be different from zero only in case $||\beta|| \leq ||\alpha||$. But since $||\alpha|| < 0$ we actually have $S_\alpha = 0$.

If $||\alpha|| = 0$ then

$$S_\alpha = \frac{1}{(2\pi i)^n} \int_{\Gamma_s} \frac{\mathbf{z}^\alpha J_{\mathfrak{f}} \det \mathbf{A} d\mathbf{z}}{z_1^{s_1} \dots z_n^{s_n}}.$$

In particular, for $\alpha = (0, \dots, 0)$ the Bézout theorem gives $S_{(0,\dots,0)} = k_1 \dots k_n$. If $\alpha \neq (0, \dots, 0)$, then S_α can still be different from zero, since

$$S_\alpha = \frac{1}{(2\pi i)^n} \int_{\Gamma_s} \frac{\mathbf{z}^\alpha J_P \det \mathbf{A} d\mathbf{z}}{z_1^{s_1} \dots z_n^{s_n}} \tag{10.10}$$

and the expression $J_P \det \mathbf{A}$ can have more then one monomial in contrast to the case of system (10.1) where we had

$$J_P = k_1 \dots k_n z_1^{k_1 - 1} \dots z_n^{k_n - 1}.$$

(Here J_P is the Jacobian of the system $P_1, \dots, P_n$). Therefore the number of sums S_α that are different from zero coincides with the number of non-zero monomials of the polynomial $J_P \det \mathbf{A}$.

For instance, in the case of Example 8.5 we get

$$\det \mathbf{A} = \frac{1}{\Delta}(a_0 x - a_1 y), \ J_P = x(a_0 b_1 - 2a_1 b_0) + y(2b_2 a_0 - a_1 b_1),$$

and

$$J_P \det \mathbf{A} = \frac{1}{\Delta}(x^2 a_0(a_0 b_1 - 2a_1 b_0) + y^2 a_1(a_1 b_1 - 2b_2 a_0) + 2xy\Delta).$$

Therefore we must compute all the S_α, with $||\alpha|| = 0$, by formula (10.10) before we can apply the Newton formulas.

Consider the integral

$$I = \frac{1}{(2\pi i)^n} \int_{\Gamma_s} \frac{J_{\mathfrak{f}} \det \mathbf{A} d\mathbf{z}}{\mathbf{z}^\beta},$$

and suppose that

$$J_{\mathfrak{f}} \det \mathbf{A} = \sum_{||\alpha|| \geq 0} b_\alpha \mathbf{z}^\alpha.$$

Then we have

$$I = b_{\beta - \mathbf{I}} = b_{(\beta_1 - 1, \dots, \beta_n - 1)}$$

(if $J_{\mathfrak{f}} \det \mathbf{A}$ has no monomial $\mathbf{z}^\alpha$ with this multiindex α, then $I = 0$).

On the other hand

$$I = \frac{1}{(2\pi i)^n} \int_{\Gamma_s} \frac{\mathbf{z}^{-\beta} J_{\mathfrak{f}} \det \mathbf{A} F_1 \dots F_n d\mathbf{z}}{F_1 \dots F_n} =$$

$$= \sum_{\substack{||\alpha^1+\ldots+\alpha^n-\beta||\geq 0 \\ ||\alpha^j||\leq s_j}} c_{\alpha^1}(1)\ldots c_{\alpha^n}(n)S_{\alpha^1+\ldots+\alpha^n-\beta}.$$

If $||\alpha^j|| = s_j$, then $c_{\alpha^j}(j) \neq 0$ only in case $\alpha^j = s_j\mathbf{e}_j$. Therefore

$$I = S_{\mathbf{s}-\beta} + \sum_{||\beta||\leq||\alpha^1+\ldots+\alpha^n-\beta||<||\mathbf{s}||} c_{\alpha^1}(1)\ldots c_{\alpha^n}(n)S_{\alpha^1+\ldots+\alpha^n-\beta},$$

and we have proved the following result.

THEOREM 10.3 (Kytmanov). *For the system(10.9) one has the formula*

$$S_{\mathbf{s}-\beta} + \sum_{||\beta||\leq||\alpha^1+\ldots+\alpha^n-\beta||<||\mathbf{s}||} c_{\alpha^1}(1)\ldots c_{\alpha^n}(n)S_{\alpha^1+\ldots+\alpha^n-\beta} = b_{\beta-\mathbf{I}}. \tag{10.11}$$

We thus see that every collection of leading homogeneous parts $P_1, \ldots, P_n$ gives rise to a Newton formula of its own. The most elementary of these is formula (10.3) for the system (10.1).

Now consider two systems of algebraic equations

$$P_j(\mathbf{z}) = 0, \quad j = 1, \ldots, n, \tag{10.12}$$

and

$$Q_j(\mathbf{z}) = 0, \quad j = 1, \ldots, n. \tag{10.13}$$

The polynomial P_j has degree k_j,

$$k_1 \geq k_2 \geq \ldots \geq k_n,$$

and the polynomial Q_j is of degree d_j,

$$d_1 \geq d_2 \geq \ldots \geq d_n.$$

Moreover, the sets of roots of systems (10.12) and (10.13) are assumed to coincide, and we denote them by E. We shall further assume that all roots of these systems are simple. The question is: What can be said about the relation between these two systems?

Since the set E is finite we can assume that all roots lie in $\mathbf{C}^n$, and hence that the systems (10.12) and (10.13) are non-degenerate.

THEOREM 10.4 (Kytmanov–Mkrtchyan [**103**]). *Under the above conditions on the systems (10.12) and (10.13), we have*

$$k_1 = d_1, \quad k_2 = d_2, \quad \ldots, \quad k_n = d_n.$$

For the proof we need the following lemma.

LEMMA 10.1. *Let*

$$E_i = \{\mathbf{z} \in \mathbf{C}^n : \ P_j(\mathbf{z}) = 0, \ j = 1, \ldots, n, \ j \neq i\}$$

and let $g(\mathbf{z})$ be a polynomial such that $E \subset g^{-1}(0)$, but $E_i \not\subset g^{-1}(0)$, i.e. $g \not\equiv 0$ on E_i. Then

$$\deg g \geq \deg P_i = k_i.$$

Proof. Assume that $\deg g < k_i$. The analytic set E_i is non-empty and of complex dimension at least 1. Hence $E_i \setminus E \neq \emptyset$.

Consider a point $\mathbf{z}^0 \in E_i \setminus E$ for which $g(\mathbf{z}^0) \neq 0$. At this point we also have $P_i(\mathbf{z}^0) \neq 0$ so there exist numbers α and β such that

$$\alpha P_i(\mathbf{z}^0) + \beta g(\mathbf{z}^0) = 0, \quad \alpha \neq 0, \quad \beta \neq 0.$$

Consider now the system of equations

$$P_j(\mathbf{z}) = 0, \quad j = 1, \ldots, n, \; j \neq i, \tag{10.14}$$
$$\alpha P_i(\mathbf{z}) + \beta g(\mathbf{z}) = 0.$$

Since the highest homogeneous parts of this system coincide with those of system (10.12), it follows that system (10.14) has no roots on the hyperplane at infinity $\mathbf{CP}^n \setminus \mathbf{C}^n$. Hence by the Bézout theorem the number of roots of system (10.14) is equal to $k_1 \cdots k_n$. But every point from E is a root of system (10.14). Apart from them the point $\mathbf{z}^0$ is also a root of system (10.14). From here we get that the number of roots of system (10.14) is at least equal to $k_1 \cdots k_n + 1$, a contradiction. □

Proof of theorem 10.4. First of all we show that $k_1 = d_1$. Consider the set

$$E_1 = \{\mathbf{z} \in \mathbf{C}^n : \; P_2(\mathbf{z}) = 0, \ldots, P_n(\mathbf{z}) = 0\}.$$

There exists at least polynomial Q_j which is not identically equal to 0 on E_1. By Lemma 10.1 we then have

$$\deg Q_j = d_j \geq k_1.$$

Interchanging the systems (10.12) and (10.13) we thus deduce that $d_1 = k_1$.

It is not difficult to show that for almost all α and β the system of equations

$$\alpha P_1(\mathbf{z}) + \beta Q_1(\mathbf{z}) = 0,$$
$$P_2(\mathbf{z}) = 0, \ldots, P_n(\mathbf{z}) = 0 \tag{10.15}$$

has no roots at infinity, and hence the system (10.15) is non-degenerate.

To show that $k_2 = d_2$ we replace the system (10.12) by the system (10.15), and the system (10.13) by the system

$$\alpha P_1(\mathbf{z}) + \beta Q_1(\mathbf{z}) = 0, \tag{10.16}$$
$$Q_2(\mathbf{x}) = 0, \ldots, Q_n(\mathbf{z}) = 0,$$

with the same α and β.

Then we define

$$E_2 = \{\mathbf{z} : \; \alpha P_1 + \beta Q_1 = 0, \; P_3 = \ldots = P_n = 0\},$$

and observe that on E_2 one of the polynomials Q_i, $i = 2, \ldots, n$, does not vanish identically. By Lemma 10.1 there exists a polynomial Q_i, $i \geq 2$, for which $\deg Q_i \geq k_2$, and hence $d_2 \geq k_2$. Interchanging the systems (10.15) and (10.16) we then get that $d_2 = k_2$, and so on. □

In the case of multiple roots Theorem 10.4 is false.

EXAMPLE 10.3. It is sufficient to consider the two systems of equations

$$z_1^4 = 0, \quad z_2 = 0$$

and

$$z_1^2 = 0, \quad z_2^2 = 0.$$

THEOREM 10.5 (Kytmanov–Mkrtchyan [**103**]). *For the systems (10.12) and (10.13) we have*

$$Q_j(\mathbf{z}) = \sum_l R_{lj}(\mathbf{z}) P_l(\mathbf{z}),$$

where the summation is taken over such indices l for which $k_l \leq k_j$. Moreover,

$$\deg R_{lj} + \deg P_l \leq Q_j.$$

In particular, Q_n is a linear combination of those P_j for which $k_j = k_n$. If $k_{n-1} > k_n$ then the polynomials P_n and Q_n are proportional.

Proof. Since Q_j vanishes at the roots of system (10.12), and since these roots are all simple and finite, it follows by the M.Nöther theorem (see, for example, [**119**]) that the polynomials Q_j belong to the ideal generated by $P_1, \ldots, P_n$. □

COROLLARY 10.1 (Bolotov [**22**]). *If in the systems (10.12) and (10.13) one has*

$$k_1 = \ldots = k_n = d_1 = \ldots = d_n,$$

and if the highest homogeneous parts of P_j and Q_j coincide for all j, then

$$P_j \equiv Q_j, \quad j = 1, \ldots, n.$$

11. Elimination of unknowns in different variables. Real roots

Here we consider questions connected with elimination of unknowns. First we focus on the following problem: Suppose that we have found the first coordinates of all roots of a system, by eliminating all the unknowns except the first. How can we find the remaining coordinates?

The classical elimination method reduces the system to triangular form. So, having found the first coordinates of the roots, we insert them into the previous equations (which depend on the first and the second variable), thereby again obtaining polynomials in one variable. Solving theses we now find the second coordinates of the roots etc.

When using the method of Sec. 8 we get at once the resultant $R(z_1)$ in the first variable. In order to find the second coordinates of the roots we must then find the resultant in the second variable etc.

Now we describe a method which simplifies this procedure, for instance, in the case when the zeros of $R(z_1)$ are simple. To determine the other coordinates we do not have to solve any equations at all. This method has been presented in [**3**].

Let a system

$$\begin{aligned} f_1(\mathbf{z}) &= 0, \\ &\cdots \\ f_n(\mathbf{z}) &= 0 \end{aligned} \tag{11.1}$$

of algebraic equations be given. We assume that the system (11.1) has a finite set of roots $\mathbf{z}_{(1)}, \ldots, \mathbf{z}_{(M)}$ in $\mathbf{C}^n$. The elimination method provides us with a polynomial

$R(z_1)$ having the zeros $z_{1(1)}, \dots, z_{1(M)}$. We first assume that all zeros of $R(z_1)$ are simple. Then

$$R(z_1) = \prod_{\mu=1}^{M} (z_1 - z_{1(\mu)}) = \sum_{j=0}^{M} b_j z_1^{M-j}, \quad b_0 = 1.$$

Introducing the polynomial

$$P_j^{(t)}(z_1) = \sum_{\mu=1}^{M} z_{j(\mu)}^t \prod_{\substack{\eta=1 \\ \eta\neq\mu}}^{M} (z_1 - z_{1(\eta)}) = \sum_{\beta=1}^{M} a_{j\beta}^{(t)} z_1^{M-\beta},$$

we have

$$\left. \frac{P_j^{(t)}(z_1)}{R'(z_1)} \right|_{z_1 = z_{1(\mu)}} = z_{j(\mu)}^t,$$

and in particular, for $t = 1$

$$\left. \frac{P_j^{(t)}(z_1)}{R'(z_1)} \right|_{z_1 = z_{1(\mu)}} = z_{j(\mu)}.$$

Hence we must find the polynomial $P_j^{(t)}(z_1)$, or rather its coefficients $a_{j\beta}^{(t)}$.

We consider the auxiliary system of functions

$$\varphi_j^{(t)}(\lambda) = \sum_{\mu=1}^{M} \frac{z_{j(\mu)}^t}{1 - \lambda z_{1(\mu)}} \prod_{\nu=1}^{M} (1 - \lambda z_{1(\nu)}),$$

which may be written

$$\varphi_j^{(t)}(\lambda) = \sum_{\mu=1}^{M} z_{j(\mu)}^t \prod_{\nu\neq\mu} (1 - \lambda z_{1(\nu)}) = \sum_{\beta=1}^{M} a_{j\beta}^{(t)} \lambda^{\beta-1}.$$

On the other hand, using the formula for geometric progression, we get for sufficiently small $|\lambda|$

$$\varphi_j^{(t)}(\lambda) = \sum_{\mu=1}^{M} z_{j(\mu)}^t \sum_{m=0}^{\infty} (\lambda z_{1(\mu)})^m \prod_{\nu=1}^{M} (1 - \lambda z_{1(\nu)}) =$$

$$= \sum_{m=0}^{\infty} \lambda^m \left(\sum_{\mu=1}^{M} z_{j(\mu)}^t z_{1(\mu)}^m \right) \prod_{\nu=1}^{M} (1 - \lambda z_{1(\nu)}) =$$

$$= \left(\sum_{m=0}^{\infty} S_{me_1+te_j} \lambda^m \right) (1 + b_1\lambda + \dots + b_M \lambda^M),$$

where $S_{me_1+te_j}$ is a power sum for multiindex $me_1 + te_j = (m, 0, \dots, 0, t, 0, \dots, 0)$, and b_j are the coefficients of the polynomial $R(z_1)$. Comparing this expression with

$\varphi_j^{(t)}(\lambda)$, we get recursive relations for the calculation of $a_{j\beta}^{(t)}$:

$$a_{j\beta}^{(t)} = \sum_{\alpha=0}^{\beta-1} b_\alpha S_{(\beta-\alpha-1)\mathbf{e}_1+t\mathbf{e}_j}. \tag{11.2}$$

In particular,

$$a_{j\beta}^{(1)} = \sum_{\alpha=0}^{\beta-1} b_\alpha S_{(\beta-\alpha-1)\mathbf{e}_1+\mathbf{e}_j}. \tag{11.3}$$

If the system (11.1) has only finite roots (i.e. if it is non-degenerate), then the power sums can be computed by using Theorem 8.1 or the corresponding Newton formulas from Sec. 10. In this way we find $P_j^{(t)}(z_1)$.

THEOREM 11.1 (Aizenberg – Bolotov – Tsikh). *Let the zeros of the polynomial $R(z_1)$ be simple, and let the coefficients of the polynomials $P_j^{(1)}(z_1)$ be given by the formulas (11.3). The j-th coordinates of the roots of system (11.1) are then equal to*

$$z_{j(\mu)} = \left. \frac{P_j^{(1)}(z_1)}{R'(z_1)} \right|_{z_1=z_{1(\mu)}}, \quad \mu = 1, \ldots, M.$$

COROLLARY 11.1. *If the system (11.1) with real coefficients is such that all zeros of $R(z_1)$ are simple, then the number of real roots coincides with the number of real zeros of the polynomial $R(z_1)$.*

Proof. If the system (11.1) has real coefficients then all the power sums S_α of the roots are real. Therefore the polynomials $R(z_1)$ and $P_j^{(t)}(z_1)$ also have real coefficients. If the first coordinate of the root $\mathbf{z}_{(\mu)}$ is real, then Theorem 11.1 shows that the $z_{j(\mu)}$ are also real, and hence $\mathbf{z}_{(\mu)}$ has all its coordinates real.

This corollary allows us to determine the number of real roots of the system (11.1) by using, for example, the Hermite theorem.

EXAMPLE 11.1. Consider the system of example 8.1.

$$\begin{aligned} z_1^2 + a_1 z_1 + b_1 z_2 + c_1 &= 0, \\ z_2^2 + a_2 z_1 + b_2 z_2 + c_2 &= 0. \end{aligned} \tag{11.4}$$

We must compute $S_{(j,1)}$ for $j = 0, 1, 2, 3$:

$$S_{(j,1)} = \mathfrak{M}\left[\sum_{\alpha_1+\alpha_2 \le j+1} (-1)^{\alpha_1+\alpha_2} \frac{J_{\mathfrak{f}} Q_1^{\alpha_1} Q_2^{\alpha_2}}{z_1^{2\alpha_1+1-j} z_2^{2\alpha_2}} \right],$$

where

$$J_{\mathfrak{f}} = 4z_1z_2 + 2z_2a_1 + 2z_1b_2 + a_1b_2 - b_1a_2,$$

$$Q_1 = a_1z_1 + b_1z_2 + c_1, \quad Q_2 = a_2z_1 + b_2z_2 + c_2.$$

Then we obtain
$$\begin{aligned} S_{(0,1)} &= -2b_2,\\ S_{(1,1)} &= a_1b_2 + 3a_2b_1,\\ S_{(2,1)} &= 2c_1b_2 + 4b_1c_2 - 2b_1b_2^2 - a_1^2b_2 - 5a_1a_2b_1,\\ S_{(3,1)} &= a_1^3b_2 + 7a_1^2a_2b_1 + 5a_2b_1^2b_2 + 3a_1b_1b_2^2-\\ &\quad -6a_1b_1c_1 - 7a_2b_1c_1 - 3a_1b_2c_1. \end{aligned}$$

In Example 8.1 we showed that
$$R(z_1) = z_1^4 + 2a_1z_1^3 + (a_1^2 - b_1b_2 + 2c_1)z_1^2+$$
$$+(a_2b_1^2 - 2a_1c_1 - a_1b_1b_2)z_1 + b_1^2c_2 + c_1^2 - b_1b_2c_1.$$
This together with (11.3) implies that
$$P_2^{(1)}(z_1) = -2b_2z_1^3 + 3(a_2b_1 - a_1b_2)z_1^2 + (a_1a_2b_1 - a_1^2b_2 + 4b_1c_2-$$
$$-2c_1b_2)z_1 + 7a_1b_2c_1 + 8a_1b_1c_2 - a_2b_1c_1 - 6a_1b_1c_1.$$
Let us now consider the general case when the resultant $R(z_1)$ can have multiple zeros. Let $z_{1(\mu)}$ denote the different zeros of the resultant, $\mu = 1,\dots,k$, and let each zero $z_{1(\mu)}$ have multiplicity r_μ, $r_1 + \dots + r_k = M$. Then
$$R(z_1) = \prod_{\mu=1}^{k}(z_1 - z_{1(\mu)})^{r_\mu} = \sum_{\alpha=0}^{M} b_\alpha z_1^{M-\alpha}.$$
We introduce the polynomial
$$P_j^{(t)}(z_1) = \sum_{\mu=1}^{k}\left(z_{j(\mu,1)}^t + \dots + z_{j(\mu,r_\mu)}^t\right) \times$$
$$\times(z_1 - z_{1(\mu)})^{r_\mu-1} \prod_{\substack{\eta=1\\ \eta\neq\mu}} (z_1 - z_{1(\eta)})^{r_\eta} = \sum_{\beta=1}^{M} a_{j\beta}^{(t)} z_1^{M-\beta},$$
where $z_{j(\mu,1)},\dots,z_{j(\mu,r_\mu)}$ are the j-th coordinates of the roots with first coordinate equal to $z_{1(\mu)}$. Then we have
$$\frac{d^{r_\nu-1}P_j^{(t)}(z_1)}{dz_1^{r_\nu-1}} \Bigg/ \frac{d^{r_\nu}R(z_1)}{dz_1^{r_\nu}}\Bigg|_{z_1=z_{1(\mu)}} = \frac{1}{r_\nu}\left(z_{j(\nu,1)}^t + \dots + z_{j(\nu,r_\nu)}^t\right). \tag{11.5}$$

The relations (11.5) allow us to write out the power sums of the j-th coordinates of the roots $z_{j(\nu,1)},\dots,z_{j(\nu,r_\nu)}$ if we know $R(z_1)$ and $P_j^{(t)}(z_1)$. Setting $t = 1,\dots,r_\nu$ we compute r_ν different power sums. Then using the Newton recursion formulas (2.1) we find a polynomial P in z_1 having these zeros. Solving the equation $P = 0$ we find the roots $z_{j(\nu,1)},\dots,z_{j(\nu,r_\nu)}$.

One finds the coefficients $a_{j\beta}^{(t)}$ by the same formulas (11.2).

THEOREM 11.2 (Aizenberg–Bolotov–Tsikh). *Suppose we know the first coordinates of the roots of system (11.1). Then by formula (11.5) we can obtain the power sums of the j-th coordinates of the roots having this given first coordinate, $1 < j \le n$. In this way we reduce the problem of finding the j-th coordinates to finding the zeros of a polynomial of one variable whose degree is equal to the number of roots of system*

(11.1) with the given first coordinate. Moreover, the coefficients of the polynomials $R_j^{(t)}$ are found by formula (11.2), and the power sums are found by Theorem 8.1 and Corollary 8.1.

Corollary 11.1 does not hold in the presence of multiple zeros of $R(z_1)$. If the coefficients of system (11.1) depend on some parameters then generically $R(z_1)$ has only simple zeros. Moreover, to determine the number of real zeros we do not need the polynomial $R(z_1)$.

Indeed, let the system (11.1) be non-degenerate, with real coefficients, and $\deg f_j = k_j$. Then by Theorem 8.1 we find the power sums $S_{\mathbf{e}_1}, \dots, S_{(2M-2)\mathbf{e}_1}$, where $M = k_1 \dots k_n$ and $\mathbf{e}_1 = (1, 0, \dots, 0)$. We consider the matrix (see Corollary 3.1)

$$\mathbf{B} = \begin{pmatrix} M & S_{\mathbf{e}_1} & S_{2\mathbf{e}_1} & \dots & S_{(M-1)\mathbf{e}_1} \\ S_{\mathbf{e}_1} & S_{2\mathbf{e}_1} & S_{3\mathbf{e}_1} & \dots & S_{M\mathbf{e}_1} \\ \dots & \dots & \dots & \dots & \dots \\ S_{(M-1)\mathbf{e}_1} & S_{M\mathbf{e}_1} & S_{(M+1)\mathbf{e}_1} & \dots & S_{(2M-2)\mathbf{e}_1} \end{pmatrix}.$$

If the matrix $\mathbf{B}$ is non-singular (i.e. $\det \mathbf{B} \neq 0$) then the polynomial $R(z_1)$ has only simple zeros (since $\det \mathbf{B}$ is the discriminant of $R(z_1)$ (see Sec. 6)). Then by the Hermite theorem the number of real zeros of $R(z_1)$ (i.e. the number of real roots of system (11.1)) is equal to the signature of the quadratic form with the given matrix.

No simple assertions, like the Descartes sign rule or the Budan–Fourier theorem, are known in general for counting the real roots of a system. Therefore this question is usually reduced to analogous questions for the resultant $R(z_1)$.

Corollary 11.1 may be extended to a more general situation. Let $P(\mathbf{z})$ be a polynomial in $\mathbf{C}^n$ with real coefficients. If $\mathbf{z} \in \mathbf{C}^n$ and $\operatorname{Im} \mathbf{z} = 0$ then $\mathbf{z} = \mathbf{x} = (x_1, \dots, x_n) \in \mathbf{R}^n$. We denote by

$$D_{a,b} = \{\mathbf{x} \in \mathbf{R}^n : \ a < P(\mathbf{x}) < b\}$$

the set consisting of level surfaces of the polynomial P, $a < b$, $a, b \in \mathbf{R}$. How can we determine the number of real roots of the system (11.1) lying in $D_{a,b}$? Consider the complex numbers $P(\mathbf{z}_{(1)}), \dots, P(\mathbf{z}_{(M)})$, i.e. the projections in $\mathbf{C}$ of the roots $\mathbf{z}_{(1)}, \dots, \mathbf{z}_{(M)}$ of the system (11.1). By Theorem 8.1 (or corollary 8.1) we can find the power sums S_j^P of the form ($\deg P = k$)

$$S_j^P = \sum_{\substack{k_{sj} \geq 0 \\ \sum k_{sj} \leq jk, \ \beta_s = \sum_{j=1}^n k_{sj}}} \frac{(-1)^{\sum k_{sj}} \prod_{s=1}^n \left(\sum_{j=1}^n k_{sj} \right)!}{\prod_{s,j=1}^n k_{sj}!} \mathfrak{M} \left[\frac{P^j Q^\alpha J_{\mathrm{f}} \det \mathbf{A} \prod_{s,j=1} a_{sj}^{k_{sj}}}{\prod_{j=1}^n z_j^{\beta_j N_j + \beta_j + N_j}} \right].$$

Then by the Newton formulas (2.1) we find a polynomial $R_P(w)$, $w \in \mathbf{C}$, having the zeros $P(\mathbf{z}_{(j)})$, $j = 1, \dots, M$. If $\mathbf{z}_{(j)}$ is a root of (11.1) with some non-real coordinate then $\overline{\mathbf{z}}_{(j)}$ is also a root of (11.1). If $P(\mathbf{z}_{(j)})$ is a real number then so is $P(\overline{\mathbf{z}}_{(j)})$, i.e. the number $P(z_{(j)})$ is a multiple zero of the polynomial $R_P(w)$. We thus have the following result.

THEOREM 11.3 (Kytmanov). *If the system (11.1) is non-degenerate with real coefficients, and if the polynomial R_P has only simple zeros, then the number of real roots of the system (11.1) in the domain $D_{a,b}$ coincides with the number of real zeros of the polynomial R_P in the interval (a,b).*

The number of real zeros of R_P in (a,b) can be determined either by the Sturm method or by the Descartes sign rule (see Sec. 3). The existence of multiple zeros can be checked by using the discriminant. This theorem permits us to find the number of real roots of system (11.1) in a strip $\{\mathbf{x} \in \mathbf{R}^n,\ a < x_1 < b\}$, in a ball $\{\mathbf{x} \in \mathbf{R}^n : |\mathbf{x}| < r\}$, in an ellipsoid and so on.

Another method for counting real roots was considered in [**124**]

EXAMPLE 11.2. (continuation of Example 11.1).

For the system (11.4), $a_j, b_j, c_j \in \mathbf{R}$, we consider the polynomial $P(\mathbf{x}) = x_1^2 + x_2^2$. We want to determine the number of real roots of (11.4) in a disk or inside a ring. First we compute $S_{(j,k)}$, $j+k \le 4$ by means of the formulas (8.9). We obtain

$$S_{(1,0)} = -2a_1,\ S_{(0,1)} = -2b_2,\ S_{(2,0)} = 2b_1b_2 + 2a_1^2 - 4c_1,$$
$$S_{(1,1)} = a_1b_2 + 3a_2b_1,\ S_{(0,2)} = 2a_1a_2 + 2b_2^2 - 4c_2,$$
$$S_{(3,0)} = -2a_1^3 - 3a_2b_1^2 - 3a_1b_1b_2 + 6a_1c_1,$$
$$S_{(0,3)} = -2b_2^3 - 3b_1a_2^2 - 3b_2a_1a_2 + 6b_2c_2,$$
$$S_{(2,1)} = 2c_1b_2 + 4b_1c_2 - 2b_1b_2^2 - a_1^2b_2 - 5a_1a_2b_1,$$
$$S_{(1,2)} = 2c_2a_1 + 4a_2c_1 - 2a_2a_1^2 - b_2^2a_1 - 5b_1b_2a_2,$$
$$S_{(4,0)} = 2a_1^4 + 4a_1^2b_1b_2 + 2b_1^2b_2^2 - 8a_1^2c_1 - 4b_1^2c_2 + 4c_1^2 + 8a_1a_2b_1^2 - 4b_1b_2c_1,$$
$$S_{(3,1)} = a_1^3b_2 + 7a_1^2a_2b_1 + 5a_2b_1^2b_2 + 3a_1b_1b_2^2 - 6a_1b_1c_1 - 7a_2b_1c_1 - 3a_1b_2c_1,$$
$$S_{(1,3)} = b_2^3a_1 + 7b_2^2b_1a_2 + 5b_1a_1^2a_1 + 3b_2a_2a_1^2 - 6b_2a_2c_2 - 7b_1a_2c_2 - 3b_2a_1c_2,$$
$$S_{(0.4)} = 2b_2^4 + 4b_2^2a_1a_2 + 2a_1^2a_2^2 - 8b_2^2c_2 - 4a_2^2c_1 + 4c_2^2 + 8b_2b_1a_2^2 - 4a_1a_2c_2,$$
$$S_{(2.2)} = 8a_1a_2b_1b_2 + 2a_1^3a_2 + 2b_1b_2^3 + a_1^2b_2^2 + 3a_2^2b_1^2 - 6a_1a_2c_1 - 2a_1^2c_2 - 6b_1b_2c_2 - 2b_2^2c_1 + 4c_1c_2.$$

We must keep in mind that when we have found the sum $S_{(j,k)}$, then for symmetry reasons we obtain the sum $S_{(k,j)}$ by the permutation $a_1 \to b_2,\ a_2 \to b_1,\ b_2 \to a_1,\ c_1 \to c_2,\ c_2 \to c_1$. Then we must compute the sums S_j^P, $j = 1,2,3,4$, corresponding to degrees P^j. But in this case it is easier to use either the multidimensional Newton formulas (10.3) or the following observation: From (11.4) we get $z_1^2 + z_2^2 = -(a_1+a_2)z_1 - (b_1+b_2)z_2 - (c_1+c_2)$ and summing over all roots of system (11.4), we have

$$S_1^P = -(a_1+a_2)S_{(1,0)} - (b_1+b_2)S_{(0,1)} - 4(c_1+c_2),$$

i.e.

$$S_1^P = 2a_1(a_1+a_2) + 2b_2(b_1+b_2) - 4(c_1+c_2).$$

By the same reasoning we can also find S_j^P, $j = 2,3,4$.

These expressions are rather awkward, but after applying the Newton formulas they usually simplify.

To S_j^P we now apply formulas (2.1). (In order to avoid confusion we denote by γ_j the coefficients of the desired polynomial R_P, i.e. $R_P(z) = z_1^4 + \gamma_1 z_1^3 + \gamma_2 z_1^2 + \gamma_3 z_1 + \gamma_4$.)

Since $\gamma_1 + S_1^P = 0$, we have $\gamma_1 = 4(c_1+c_2) - 2a_1(a_1+a_2) - 2b_2(b_1+b_2)$. Similarly, $S_2^P + S_1^P\gamma_1 + 2\gamma_2 = 0$ implies that

$$\gamma_2 = 6(c_1+c_2)^2 + (a_1+a_2)^2\times$$
$$\times(a_1^2 - b_1b_2 + 2c_1) + (b_1+b_2)^2(b_2^2 - a_1a_2 + 2c_2) + 3(a_1+a_2)\times$$
$$\times(b_1+b_2)(a_1b_2 - a_2b_1) - 6a_1(a_1+a_2)(c_1+c_2) - 6b_2(b_1+b_2)(c_1+c_2).$$

Further we have $S_3^P + S_2^P\gamma_1 + S_1^P\gamma_2 + 3\gamma_3 = 0$, and hence

$$\gamma_3 = 4(c_1+c_2)^3+$$
$$+(a_1+a_2)^3(a_1b_1b_2 - 2a_1c_1 - a_2b_1^2) + (b_1+b_2)^3(a_1a_2b_2 - a_2^2b_1-$$
$$-2b_2c_2) + (a_1+a_2)^2(b_1+b_2)(a_1a_2b_1 - a_1^2b_2 - 2b_2c_1 + 4b_1c_1)+$$
$$+(b_1+b_1)^2(a_1+a_2)(a_2b_1b_2 - a_1b_2^2 - 2a_1c_2 + 4a_2c_1) - 6a_1(a_1+a_2)\times$$
$$\times(c_1+c_2)^2 - 6b_2(b_1+b_2)(c_1+c_2)^2 + 6(a_1+a_2)(b_1+b_2)(c_1+c_2)\times$$
$$\times(a_1b_2 - a_2b_1) + 2(a_1+a_2)^2(c_1+c_2)(2c_1 - b_1b_2 + a_1^2)+$$
$$+2(b_1+b_2)^2(c_1+c_2)(2c_2 - a_1a_2 + b_2^2).$$

Finally, $S_4^P + S_3^P\gamma_1 + S_2^P\gamma_2 + S_1^P\gamma_3 + 4\gamma_4 = 0$ gives

$$\gamma_4 = (a_1+a_2)^4\times$$
$$\times(b_1^2c_2 + c_1^2 - b_1b_2c_1) + (b_1+b_2)^4(a_2^2c_1 + c_2^2 - a_1a_2c_2) + (c_1+c_2)^4+$$
$$+(a_1+a_2)^3(b_1+b_2)(4a_1b_1c_1 + a_2b_1c_1 + a_1b_2c_1 - 6a_1b_1c_2)+$$
$$+(b_1+b_2)^3(a_1+a_2)(4b_2a_2c_2 + b_1a_2c_2 + b_2a_1c_2 - 6b_2b_2c_1)+$$
$$-2a_1(a_1+a_2)(c_1+c_2)^3 - 2b_2(b_1+b_2)(c_1+c_2)^3+$$
$$+(a_1+a_2)^2(b_1+b_2)^2(a_1^2c_2 - a_1a_2c_1 - b_1b_2c_2 + b_2^2c_1 - 2c_1c_2)+$$
$$+(a_1+a_2)^2(c_1+c_2)^2(2c_1 - 5a_1^2 - b_1b_2) + (b_1+b_2)^2(c_1+c_2)^2\times$$
$$\times(2c_2 - 5b_2^2 - a_1a_2) + (a_1+a_2)^2(b_1+b_2)(c_1+c_2)(2b_2c_1 + 4a_1^2b_2-$$
$$-3a_1a_2b_1 - b_1b_2^3) + (b_1+b_2)^2(a_1+a_2)(c_1+c_2)(2a_1c_2 + 4b_2^2a_1-$$
$$-3b_1b_2a_2 - a_2a_1^3) + 6(a_1+a_2)(b_1+b_2)(c_1+c_2)^2(a_1b_2 - a_2b_1).$$

We have thus constructed the polynomial $R_P(z)$ provided it has only simple zeros (which may be checked with the discriminant). Then by Theorem 11.3 the number of real zeros of $R_P(z)$ in the interval (a,b) coincides with the number of real roots of system (11.4) inside the ring $\{(x_1,x_2):\ a < x_1^2 + x_2^2 < b\}$.

In conclusion we present a result of Tarkhanov [**132**], which in some cases gives the number of real roots in a domain $D \subset \mathbf{R}^n$.

Let the system (11.1) have real coefficients.

THEOREM 11.4 (Tarkhanov). *If D is a bounded domain in $\mathbf{R}^n$ with piecewise smooth boundary, and if the Jacobian $J_\mathfrak{f} > 0$ in D, then the number of real roots of system (11.1) in domain D is equal to*

$$\frac{1}{2^n}\sum_{\Gamma_0} \operatorname{sign} J_\mathfrak{f}, \tag{11.6}$$

where $\Gamma_0 = \partial D \cap \{\mathbf{x}\in\mathbf{R}^n:\ |f_2| = \varepsilon,\dots,|f_n| = \varepsilon\}$.

The condition $J_{\mathbf{f}} > 0$ is quite restrictive. Usually it will imply that the mapping $\mathbf{f}$ is one-to-one, i.e. the system (11.1) has only one real root. In general, when $J_{\mathbf{f}}$ is allowed to change its sign in D the formula (11.6) gives the index of the mapping $\mathbf{f}$ in D (i.e. the sum of roots in D where every root is multiplied by $\operatorname{sign} J_{\mathbf{f}}$).

We provide also a more complicated formula, in which the Jacobian $J_{\mathbf{f}}$ may change its sign, permitting us to compute power sums of any form (see [**101**]).

Let D be a bounded domain in $\mathbf{R}^n$ with piecewise smooth boundary ∂D. Consider the system

$$\begin{aligned} f_1(\mathbf{x}) &= 0, \\ &\cdots \\ f_n(\mathbf{x}) &= 0, \end{aligned} \tag{11.7}$$

where the f_j are real valued smooth functions. We shall assume that the system (11.7) has a finite set $E_{\mathbf{f}}$ of simple roots in D, with no roots on the boundary ∂D. With the help of the Poincaré index formula we can compute the index of system (11.7) (i.e. the number of roots counted according to the sign of the Jacobian). Then we invoke the Tarkhanov Theorem 11.4 to compute this index in closed form.

We wish to determine the number of roots of system (11.7), and we first consider the situation when the Jacobian does not change sign in $\bar{D}$. For definiteness we shall assume $J_f \geq 0$ on $\bar{D}$. Consider the differential form (for the Poincaré index)

$$\omega(\mathbf{x}) = c_n \sum_{k=1}^{n} (-1)^{k-1} \frac{x_k}{|x|^n} dx[k],$$

where the constant c_n is defined so that

$$\int_{\{|\mathbf{x}|=1\}} \omega(\mathbf{x}) = nc_n \int_{\{|\mathbf{x}|<1\}} dx = 1.$$

It is known that the form $\omega(\mathbf{x})$ is closed in $\mathbf{R}^n \setminus \{0\}$.

LEMMA 11.1. *Suppose φ is a real-valued function from $\mathbf{C}^1(\bar{D})$. Then*

$$\sum_{\mathbf{x} \in E_{\mathbf{f}}} \varphi(\mathbf{x}) = \int_{\partial D} \varphi\omega(\mathbf{f}) - \int_{D} d\varphi \wedge \omega(\mathbf{f}).$$

The proof of Lemma 11.1 is standard. The integrals over the domain D are well defined since they are absolutely convergent (we recall that all roots of system (11.7) are simple, so near these roots the mapping

$$\mathbf{f} = (f_1, \ldots, f_n)$$

is a diffeomorphism). Applying the Stokes formula in the domain

$$D_\varepsilon = D \setminus \{\mathbf{x} : \ |\mathbf{f}(\mathbf{x})| < \varepsilon\}$$

and using the identity

$$\int_{\{|\mathbf{y}|=\varepsilon\}} \omega(\mathbf{y}) = 1$$

we obtain the required formula.

Now we allow the Jacobian $J_{\mathbf{f}}$ to change sign in D.

THEOREM 11.5 (Kytmanov). *If $f_j \in \mathbf{C}^2(\bar{D})$, $j = 1, \ldots, n$, and $\varphi \in \mathbf{C}^1(\bar{D})$, then*

$$\sum_{\mathbf{x}\in E_{\mathbf{f}}} \varphi(\mathbf{x}) = c_{n+1} \int_{\partial D} \varphi I_{\mathbf{f}} \sum_{k=1}^{n} (-1)^{k-1} f_k df[k] -$$

$$-c_{n+1} \int_D I_{\mathbf{f}} d\varphi \wedge \sum_{k=1}^{n} (-1)^{k-1} f_k df[k] - \tag{11.8}$$

$$-2\varepsilon c_{n+1} \int_D \varphi \Delta_{\mathbf{f}} (|f|^2 + \varepsilon^2 J_{\mathbf{f}}^2)^{-\frac{n+1}{2}} dx,$$

where $|\mathbf{f}|^2 = f_1^2 + \ldots + f_n^2$, $\varepsilon > 0$,

$$\Delta_f = \begin{vmatrix} f_1 & \dfrac{\partial f_1}{\partial x_1} & \ldots & \dfrac{\partial f_1}{\partial x_n} \\ \ldots & \ldots & \ldots & \ldots \\ f_n & \dfrac{\partial f_n}{\partial x_1} & \ldots & \dfrac{\partial f_n}{\partial x_n} \\ J_{\mathbf{f}} & \dfrac{\partial J_{\mathbf{f}}}{\partial x_1} & \ldots & \dfrac{\partial J_{\mathbf{f}}}{\partial x_n} \end{vmatrix}$$

and

$$I_{\mathbf{f}} = \begin{cases} \dfrac{2^s(-1)^s}{(2s-1)!} \dfrac{d^s}{d\beta^s} \ln \dfrac{\varepsilon J_{\mathbf{f}} + \sqrt{\varepsilon^2 J_f^2 + \beta}}{-\varepsilon I_f + \sqrt{\varepsilon^2 I_f + \beta}}, & n = 2s, \\ \dfrac{2(-1)^s}{s!} \dfrac{d^s}{d\beta^s} \left(\dfrac{1}{\sqrt{\beta}} \arctan \dfrac{\varepsilon J_{\mathbf{f}}}{\sqrt{\beta}} \right), & n = 2s+1. \end{cases}$$

for $\beta = |\mathbf{f}|^2$. The integrals in this formula converge absolutely.

Proof. Consider (as in [**87**]) the auxiliary system of equations

$$\begin{aligned} f_1(\mathbf{x}) &= 0, \\ &\ldots \\ f_n(\mathbf{x}) &= 0, \\ f_{n+1}(\mathbf{x}, x_{n+1}) &= x_{n+1} J_{\mathbf{f}}(\mathbf{x}) = 0. \end{aligned} \tag{11.9}$$

The roots of the system (11.9) are points in $\mathbf{R}^{n+1}$ of the form $(x_1, \ldots, x_n, 0)$ where $(x_1, \ldots, x_n) \in E_{\mathbf{f}}$. Moreover, the Jacobian of the system (11.9) is equal to $J_{\mathbf{f}}^2 \geq 0$. This means that Lemma 11.1 can be applied to (11.9). The domain of integration we take to be $D_1 = D \times [-\varepsilon, \varepsilon] \subset \mathbf{R}^{n+1}$ ($\varepsilon > 0$). Then we get

$$\sum_{\mathbf{x}\in E_{\mathbf{f}}} \varphi(\mathbf{x}) = \int_{\partial D_1} \varphi \tilde{\omega}(\tilde{\mathbf{f}}) - \int_{D_1} d\varphi \wedge \tilde{\omega}(\tilde{\mathbf{f}}),$$

where $\tilde{\mathbf{f}} = (f_1, \ldots, f_n, f_{n+1})$, $f_{n+1} = x_{n+1} J_{\mathbf{f}}$, and

$$\tilde{\omega}(\tilde{\mathbf{f}}) = c_{n+1} (|\mathbf{f}|^2 + f_{n+1}^2)^{-\frac{n+1}{2}} \left(\sum_{k=1}^{n} (-1)^{k-1} f_k df[k] \wedge df_{n+1} + (-1)^n f_{n+1} \mathbf{df} \right).$$

Since $\partial D_1 = \partial D \times [-\varepsilon, \varepsilon] + D \times [\varepsilon] - D \times [-\varepsilon]$ we have

$$\int_{D\times[\pm\varepsilon]} \varphi\tilde{\omega}(\tilde{\mathbf{f}}) =$$

$$= \pm\varepsilon c_{n+1} \int_D \varphi \left(\sum_{k=1}^{n} (-1)^{k+1} f_k df[k] \wedge dJ_{\mathbf{f}} + (-1)^n J_{\mathbf{f}} d\mathbf{f} \right) (|\mathbf{f}|^2 + \varepsilon^2 J_{\mathbf{f}}^2)^{-\frac{n+1}{2}} =$$

$$= \pm\varepsilon c_{n+1} \int_D \varphi \Delta_{\mathbf{f}} (|\mathbf{f}|^2 + \varepsilon^2 J_{\mathbf{f}}^2)^{-\frac{n+1}{2}} d\mathbf{x}.$$

We also see that

$$\int_{\partial D\times[-\varepsilon,\varepsilon]} \varphi\tilde{\omega}(\tilde{\mathbf{f}}) = \int_{\partial D} \varphi J_{\mathbf{f}} \sum_{k=1}^{n} (-1)^{k-1} f_k df[k] \int_{-\varepsilon}^{\varepsilon} (|\mathbf{f}|^2 + x_{n+1}^2 J_{\mathbf{f}}^2)^{-\frac{n+1}{2}} dx_{n+1},$$

and similarly

$$\int_{D_1} d\varphi \wedge \tilde{\omega}(\tilde{\mathbf{f}}) = \int_D J_{\mathbf{f}} d\varphi \wedge \sum_{k=1}^{n} (-1)^{k-1} f_k df[k] \int_{-\varepsilon}^{\varepsilon} (|\mathbf{f}|^2 + x_{n+1}^2 J_{\mathbf{f}}^2)^{-\frac{n+1}{2}} dx_{n+1}.$$

Computing the integral

$$\int_{-\varepsilon}^{\varepsilon} (|\mathbf{f}|^2 + x_{n+1}^2 J_{\mathbf{f}}^2)^{-\frac{n+1}{2}} dx_{n+1}$$

we now obtain formula (11.8). □

Letting $n = 1$ in (11.8), we have the following result.

COROLLARY 11.2. *If $D = (a, b) \subset \mathbf{R}^1$, then*

$$\sum_{x\in E_f} \varphi(x) = \frac{\varepsilon}{\pi} \int_a^b \frac{(f')^2 + ff''}{f^2 + \varepsilon^2 (f')^2} \varphi dx - \tag{11.10}$$

$$-\frac{1}{\pi} \int_a^b \varphi' \arctan \frac{\varepsilon f'}{f} dx + \frac{1}{\pi} \left(\varphi' \arctan \frac{\varepsilon f'}{f} \right) \Bigg|_a^b .$$

In particular, if $f(x) = \sin \pi x$ and $\varepsilon = 1/\pi$ then formula (11.10) becomes the classical Euler–Maclaurin formula with remainder term R in integral form

$$\sum_{k\in(a,b)} \varphi(k) = \int_a^b \varphi(x) dx + R.$$

If $\varphi = 1$, then formula (11.8) (which occurs in [**87**]) gives the number of roots of the system (11.7) in the domain D.

If we use formula (11.8) to compute the power sums

$$S_j^{\varphi} = \sum_{x\in E_{\mathbf{f}}} \varphi^j(\mathbf{x}), \quad j = 1, 2, \ldots, M,$$

(where M is the number of roots of system (11.7) in the domain D), then by the Newton recursion formulas (2.1) we can construct a polynomial $R_{\varphi}(w)$, $w \in \mathbf{C}^1$, whose zeros are

$$\varphi(\mathbf{x}), \ \mathbf{x} \in E_{\mathbf{f}}.$$

COROLLARY 11.3. *Assume that R_φ has only simple zeros. Then the number of roots of the system (11.7) in the set*

$$\{\mathbf{x} \in D : \; a < \varphi(\mathbf{x}) < b\}$$

coincides with the number of zeros of the polynomial R_φ in the interval (a, b).

CHAPTER 3

APPLICATIONS IN MATHEMATICAL KINETICS

An important field of applications for the methods we have developed is that of mathematical kinetics. The basic objects of study in this field are the equations of chemical kinetics, and these are nonlinear as a rule. In the simplest but commonly encountered case their right hand sides are algebraic functions, which are formed according to the mechanism of a chemical reaction. Recently, the interest in the nonlinear problems of mathematical kinetics has grown considerably in connection with the vast accumulation of experimental facts revealing critical effects of different kinds (see e.g. [**30, 89, 78, 79, 73, 149, 150, 151**]). An essential point in the analysis of the appropriate mathematical models is to determine all steady states, i.e. to find all real roots of certain systems of nonlinear algebraic equations in some bounded domain. In connection with the equations of chemical kinetics this problem was considered for the first time in a number of papers [**4, 5, 6, 7, 8, 34, 35**], which used both traditional approaches as well as methods based on modified elimination procedures for systems of nonlinear algebraic equations [**11**].

Let us consider the general form of the kinetic equations that correspond to a closed homogeneous isothermal system. The transformation scheme or the mechanism of a complex reaction can be represented in the form

$$\sum_{i=1}^{n} \alpha_{ji} Z_i \rightleftarrows \sum_{i=1}^{n} \beta_{ji} Z_i, \qquad j = 1, \ldots, m,$$

where Z_i stands for the i-th intermediate, α_{ji} and β_{ji} are stoichiometric coefficients (nonnegative integers), n is the number of reagents, m is the number of stages (elementary reactions). To every stage there corresponds a numerical characteristic, the rate of the stage:

$$w_j = w_j^+ - w_j^-, \quad w_j^{+(-)} = w_j^{+(-)}(z_1, \ldots, z_n),$$

where $w_j^{+(-)}$ is the rate of the forward (reverse) reaction at the j-th step, and z_i is the concentration of the intermediate Z_i.

The equations for the steady states are given by

$$\sum_{j=1}^{m} \gamma_{ji} w_j(\mathbf{z}) = 0, \quad i = 1, \ldots, n,$$

where $\gamma_{ji} = \beta_{ji} - \alpha_{ji}$, and $\mathbf{z} = (z_1, \ldots, z_n)$. Introducing the matrix $\mathbf{B} = ||\gamma_{ji}||$, and the vector $\mathbf{w} = (w_1, \ldots, w_m)$, the stationary equations of chemical kinetics can thus be written in matrix form

$$\mathbf{B}^T \mathbf{w}(\mathbf{z}) = 0.$$

This system must be supplemented by appropriate linear relations which represent the conservation laws. These determine a linear dependence between the lines and columns in the matrix **B**:

$$\mathbf{A}\mathbf{z} = \mathbf{a},$$

where **A** is a matrix whose entries are proportional to the molecular weights, and **a** is a vector of mass balances. The mass conservation laws for atoms of a given are at each stage expressed by the relation

$$\mathbf{B}\mathbf{A} = 0.$$

The balance relations together with the non-negativity condition for the concentrations $z_i \geq 0$, $i = 1, \ldots, n$, determine the domain of definition, the so-called reaction polyhedron D, for the kinetic equations. In general the construction of D is a separate problem [**78**]. We shall mainly consider the case when D is a simplex, i.e.

$$D = \left\{ \mathbf{z} : \sum_{i=1}^{n} z_i = 1, \; z_i \geq 0 \right\}.$$

The steady state equations together with the balance relations form a system of generally nonlinear equations, whose solutions in D are to be found. The nonlinearity is determined by the character of the dependence on $\mathbf{w}(\mathbf{z})$. In the simplest case corresponding to the mass conservation law, one has

$$w_j^{+(-)} = k_j^{+(-)} \prod_{i=1}^{n} z_i^{\alpha_{ji}(\beta_{ji})},$$

where $k_j^{+(-)}$ are the rate constants at the j-th stage of the forward and reverse reactions, so the stationary equations are algebraic in this case.

It can be shown [**30**] that the polyhedron D is a ω-invariant set for the dynamical system

$$\mathbf{z}' = \mathbf{B}^T \mathbf{w}(\mathbf{z}),$$

i.e. for any initial data $\mathbf{z}(0) \in D$, the solutions $\mathbf{z}(t)$ stay in D for all $t > 0$. This ensures the existence of at least one solution to the stationary system. However, this condition holds also for more general kinds of dependencies of $w_j(\mathbf{z})$.

Scientific findings have shown [**30**] that already simple nonlinear models of chemical kinetics may possess several steady states. General conditions that give rise to critical effects related to structural peculiarities of the chemical transformation scheme can be found, for example, in [**30, 33, 140**]. Moreover, it is important to find the solutions themselves (and all of them) in the given domain D. Standard numerical methods for solving nonlinear equation systems, based on various iteration procedures, are suitable only when the initial data are sufficiently close to a root of the system.

Additional difficulties arise when searching for solutions for which the stability is of "saddle" type character. Methods based on variation of parameters are quite applicable here [**16, 24, 95**]. However, such methods do not guarantee that all steady state solutions in the given domain are found. During the variation of parameters there may appear "isols" in the corresponding system of equations.

A large number of examples of such situations in models of automatic control are given in [**86**]. Zeldovich [**160**] was first to point out the possibility of "isols" in the chemical kinetic equations under nonisothermal conditions.

In order to solve the stated problem we propose to use an approach based on elimination of unknowns from a system of nonlinear algebraic equations [**11**]. As a result of this elimination we obtain a polynomial in one variable, and for these there are well developed methods for finding all roots in a given segment. In this chapter we carry out this scheme for various kinds of equations of chemical kinetics. Some special systems have already been considered above (see Examples 8.2, 9.1, 9.2, 9.4, system (9.4)). The questions of finding the number of solutions in a given domain are analyzed in Sec. 12 for so-called short transformation schemes. The problem of determining all steady states is settled in Sec. 13. Finally, the specific problem of elimination of unknowns from a system of chemical kinetic equations (construction of the kinetic polynomial) is studied in Sec. 14 and 15.

12. Short schemes

In the papers [**31, 32, 46, 48, 49, 51**] mechanisms for chemical reactions with a minimal number (two) of independent reagents were proposed. These so-called short schemes do not contain autocatalytic stages but they do admit non-uniqueness of solutions.

12.1. THE PARALLEL SCHEME

One such short scheme is the three-stage scheme of the general form

$$
\begin{gathered}
mZ \underset{-1}{\overset{1}{\rightleftarrows}} mX, \\
nZ \underset{-2}{\overset{2}{\rightleftarrows}} nY, \qquad (12.1) \\
pX + qY \overset{3}{\rightarrow} (p+q)Z,
\end{gathered}
$$

where Z, X, Y are intermediates, and m, n, p, q are positive integers (stoichiometric coefficients). The digits above and below the arrows indicate the number of reactions. The scheme (12.1) is typical for catalytic reactions. The substance Z (the catalyst) is used in the first and second stages and is then stored. In (12.1) there is a stage of interaction between the different substances X and Y. This is a necessary condition for the existence of several steady states [**30**]. This fact will be proved below as we develop our approach.

Here we analyze in detail how the number of steady states varies with different relations between m, n, p, q and with different reversibility of the stages in (12.1). Simple mechanisms containing autocatalytic stages (reactions of the type $A + X \to 2X$) were considered in [**29, 47, 50, 72, 90**]. The assumption of autocatalysis allows one also to construct simple models of autooscillations [**47**]. In this section we restrict our attention to mechanisms of type (12.1), focusing on the cases of one, three or five steady states. Here the nonlinearity of the corresponding kinetic model comes solely from the nonlinearity of the given transformation scheme.

Mathematical model. To the scheme (12.1) there corresponds the following kinetic model

$$\begin{aligned} \dot{x} &= mk_1 z^m - mk_{-1} x^m - pk_3 x^p y^q, \\ \dot{y} &= nk_2 z^n - nk_{-2} y^n - qk_3 x^p y^q, \end{aligned} \tag{12.2}$$

where x, y and $z = 1 - x - y$ are the concentrations of the intermediates Z, X, Y respectively, and k_i is the rate constant of the corresponding step (k_1, k_2, $k_3 > 0$, $k_{-1}, k_{-2} \geq 0$). The solutions of (12.2) are defined in the reaction simplex

$$S = \{(x,y) : \; x \geq 0, \; y \geq 0, \; x + y \leq 1\}. \tag{12.3}$$

The fact that S is a ω-invariant set for the dynamical system (12.2) guarantees the existence of at least one stationary point.

Steady states. For system (12.2) the steady states (s.s.) are defined as the solutions to the system of algebraic equations

$$\begin{aligned} mk_1 z^m - mk_{-1} x^m - pk_3 x^p y^q &= 0, \\ nk_2 z^n - nk_{-2} y^n - qk_3 x^p y^q &= 0, \\ x + y + z &= 1. \end{aligned} \tag{12.4}$$

The solutions of (12.4) having physical sense lie in the reaction simplex (12.3), and it is in this domain S that we shall estimate the number of solutions of system (12.4). We shall distinguish between interior s.s. and boundary s.s. An interior s.s. is strictly contained in S, i.e. $x > 0$, $y > 0$, $x + y < 1$. For a boundary s.s. we have either $x = 0$, $y = 0$ or $x + y = 1$. The reaction rate, which for (12.1) is given by the expression

$$W = k_3 x^p y^q, \tag{12.5}$$

always satisfies $W > 0$ for interior s.s. but may vanish for boundary s.s.

Let us first consider some particular cases.

1^0. $\mathbf{k_{-i} = 0, i = 1, 2}$. The system (12.4) takes the form

$$\begin{aligned} mk_1 z^m - pk_3 x^p y^q &= 0, \\ nk_2 z^n - qk_3 x^p y^q &= 0, \\ x + y + z &= 1. \end{aligned} \tag{12.6}$$

For (12.6) there always exist two boundary s.s.

$$x_1 = 0, \; y_1 = 1, \; z_1 = 0, \tag{12.7}$$

$$x_2 = 1, \; y_2 = 0, \; z_2 = 0, \tag{12.8}$$

having zero stationary rate (12.5). We shall examine whether or not there are any interior s.s. of system (12.6).

1.1. $\mathbf{m = n}$. If the parameters are related by

$$qk_1 = pk_2, \tag{12.9}$$

then there exists a curve $y = y(x)$ of interior s.s., connecting the two boundary s.s. (12.7) and (12.8) in S. The equation of this curve is implicitly given by

$$F(x,y) = mk_1(1-x-y)^m - pk_3x^py^q = 0. \tag{12.10}$$

Using the formula

$$\frac{\partial y}{\partial x} = -\frac{\partial F}{\partial x} \Big/ \frac{\partial F}{\partial y}$$

it is easy to show that

$$\frac{\partial y}{\partial x} < 0.$$

Moreover, if

$$m = n = p = q, \tag{12.11}$$

then by (12.10) the function $y = y(x)$ can be written down explicitly as

$$y = \frac{1-x}{1+\alpha x}, \quad \alpha = \sqrt[m]{k_3/k_1}. \tag{12.12}$$

The value of the stationary rate is represented by

$$W = W(x) = k_3x^m\left(\frac{1-x}{1+\alpha x}\right)^m.$$

The function $W(x)$ on the segment [0,1] is non-negative and vanishes only at the end points 0 and 1, so $W(x)$ must have a maximum on the interval (0,1). In particular, for $m = n = p = q = 1$ the maximum point x_* for $W(x)$ is given by

$$x_* = -1/\alpha + \sqrt{1 + 1/\alpha^2}$$

and the corresponding maximal rate is

$$W_{\max} = W(x_*) = k_3x_*\frac{1-x_*}{1+\alpha x_*}.$$

The value y_* corresponding to x_* is calculated according to (12.12). If the equality (12.9) does not hold, i.e. if $pk_2 \neq qk_1$ then we have no interior s.s.

1.2. m > n. We will show that in this case, apart from the two boundary s.s., there may also occur two interior s.s. Multiplying the first and the second equations in (12.6) by q and p respectively, and subtracting term by term we have

$$z = \sqrt[m-n]{\frac{npk_2}{mqk_1}} = a. \tag{12.13}$$

Using the balance relation (12.6) we get the following nonlinear equation for the stationary value x

$$f(x) = x^p(b_1 - x)^q = b_2, \tag{12.14}$$

where

$$b_1 = 1 - a, \quad b_2 = \frac{mk_1}{pk_3}a^m.$$

It is clear that a necessary condition for the existence of interior s.s. is imposed by the inequalities $b_1 > 0$ and $a < 1$, i.e.

$$npk_2 < mqk_1. \tag{12.15}$$

A necessary and sufficient condition for the existence of interior s.s. is obtained from the requirement $f(x^*) \geq b_2$ when $f'(x^*) = 0$:

$$p^p q^q \left(\frac{b_1}{p+q}\right)^{p+q} \geq b_2. \tag{12.16}$$

So in order to have two interior s.s. condition (12.15) is necessary while the inequality (12.16) is necessary and sufficient. Notice that to solve (12.14) in general, we must resort to numerical methods. However, in some particular cases this equation can be solved analytically. For example, if $p = q$ then

$$x_{1,2} = \frac{1}{2}b_1 \pm \sqrt{\frac{1}{4}b_1^2 + b_2^{1/p}}.$$

The stationary value y is determined from the stationary value x according to the relation $y = b_1 - x$. It is interesting that the stationary rate of the reaction in this case is the same in interior s.s., and it has the value

$$W = \frac{mk_1}{p}z^m = \frac{mk_1}{p}a^m. \tag{12.17}$$

The interesting case $n = 2$, $m = p = q = 1$, corresponding, for example, to the three-stage mechanism of oxidation:

$$\begin{aligned} &(O_2)+2Z \to 2ZO,\\ &(H_2)+Z \to ZH_2,\\ &ZH_2+ZO \to 2Z+H_2O \end{aligned}$$

has been investigated in detail in [**46**]. Here the reagents in brackets we have concentrations that are kept constant.

1.3. m < n. Except for notation this case is analogous to the previous one. All conclusions obtained in 1.2 will be valid here after interchanging $n \rightleftarrows m$, $p \rightleftarrows q$, $k_1 \rightleftarrows k_2$. For example, the necessary and sufficient conditions for the existence of two interior s.s. together with two boundary s.s. can be written (in accordance with (12.15)) as follows

$$\sqrt[n-m]{\frac{mqk_1}{npk_2}} = a < 1, \qquad mqk_1 < npk_2,$$

$$p^p q^q \left(\frac{b_1}{p+q}\right)^{p+q} \geq b_2,$$

where

$$b_1 = 1 - a, \quad b_2 = \frac{nk_2}{qk_3}a^n.$$

The stationary rate of reaction corresponding to the interior s.s. is given, in view of (12.17), by

$$W = \frac{n}{q}k_2 a^n.$$

2^0. $k_{-1} \neq 0$, $k_{-2} = 0$. The system (12.4) takes the form

$$\begin{aligned} mk_1 z^m - mk_{-1}x^m - pk_3 x^p y^q &= 0, \\ nk_2 z^n - qk_3 x^p y^q &= 0, \\ x + y + z &= 1. \end{aligned} \tag{12.18}$$

Here it is obvious that the only boundary s.s. is given by (12.7). As before we shall analyze the number of interior s.s.

2.1. m = n. From (12.18) we have

$$(pk_2 - qk_1)z^m = qk_{-1}z^m, \tag{12.19}$$

so if the parameters satisfy

$$pk_2 \leq pk_1 \tag{12.20}$$

we have no interior s.s. But if

$$pk_2 > pk_1 \tag{12.21}$$

we have from (12.18)

$$z = \sqrt[m]{\frac{mk_{-1}}{pk_2 - qk_1}}\, x = \beta x. \tag{12.22}$$

2.1.1. m > p. Taking into account (12.21), (12.22) we get (from (12.18)) the following nonlinear equation for positive stationary values x:

$$f(x) = bx^{m-p} = (1 - (1+\beta)x)^q = g(x), \tag{12.23}$$

where

$$b = \frac{mk_2\beta^m}{qk_3}.$$

The solution of equation (12.23) is represented by the point of intersection of the two curves $f(x)$ and $g(x)$ in the interval $(0,1)$. It is clear that such an intersection point always exists, and it is unique because $g(x)$ decreases monotonically to 0 and $f(x)$ increases monotonically from 0. We have thus shown that under the parameter relation (12.20) there exists only boundary s.s. whereas the condition (12.21) produces also an interior s.s. with nonzero rate of reaction.

2.1.2. m = p. Taking into account (12.22) from the second equation of (12.18), one easily obtains

$$y = \sqrt[q]{\frac{mk_2\beta}{qk_3}}.$$

The stationary value x is found from the balance relation

$$x = \frac{(1-y)}{(1+\beta)}.$$

It is clear that for the existence of interior s.s. we must require (in addition to (12.21)) that

$$y < 1 \quad \text{i.e.} \quad mk_2\beta < qk_3.$$

2.1.3. m < p. The equation corresponding to (12.23) takes the form

$$x^{p-m}(1-(1+\beta)x)^q = b. \tag{12.24}$$

Introducing the notation

$$\beta_1 = \frac{1}{(1+\beta)}, \quad \beta_2 = b(1+\beta)^{-q}$$

we can represent equation (12.24) in a form analogous to relation (12.14). Therefore a necessary and sufficient condition for the existence of two interior s.s. (in accordance with (12.16)) is given by

$$q^q(p-m)^{p-m}\left(\frac{\beta_1}{q+p-m}\right)^{q+p-m} \geq \beta_2. \tag{12.25}$$

Notice that the necessary condition (12.15) for the existence of interior s.s. is automatically fulfilled in this case, since $\beta_1 > 0$. Moreover, equality in (12.14) and (12.25) implies the existence of non-rough s.s., whereas a strict inequality guarantees the existence of two rough isolated interior s.s.

2.2. m > n. The equations (12.18) imply that

$$x^m = \frac{1}{qmk_1}(qmk_1z^m - nk_2pz^n),$$

from which

$$x = (z^{m-n} - \overline{a})^{1/m} z^{n/m}(k_1/k_{-1})^{1/m}, \quad \overline{a} = \frac{nk_2p}{mk_1q}, \tag{12.26}$$

$$y = \frac{z^{(1-p/m)n/q}}{(z^{m-n}-\overline{a})^{p/(mq)}}\left(\frac{nk_2}{qk_3}\left(\frac{k_{-1}}{k_1}\right)^{p/m}\right)^{1/q}. \tag{12.27}$$

Substituting (12.26) and (12.27) into the identity $1-x-y=z$ we get an equation with respect to z:

$$1-z = a_1(z^{m-n}-\overline{a})^{1/m}z^{n/m} + a_2\frac{z^{(1-p/m)n/q}}{(z^{m-n}-\overline{a})^{p/(mq)}}, \tag{12.28}$$

where

$$a_1 = \sqrt[m]{k_1/k_{-1}}, \quad a_2 = \left(\frac{nk_2}{qk_3}\left(\frac{k_{-1}}{k_1}\right)^{p/m}\right)^{1/q}.$$

It is obvious that a necessary condition for the existence of a positive solution to equation (12.28) is $\overline{a} < 1$, i.e.

$$nk_2p < mk_1q. \tag{12.29}$$

When the inequality (12.29) is satisfied all solutions of (12.28) with a physical sense are located in the interval $(\overline{\alpha}, 1)$, where

$$\overline{\alpha} = \overline{a}^{1/(m-n)}.$$

2.2.1. m ≤ p. In this case (12.28) may be rewritten in the form

$$1-z = a_1(z^{m-n}-\overline{a})^{1/m}z^{n/m} + \frac{a_2}{(z^{m-n}-\overline{a})^{p/(mq)}z^{(p/m-1)n/q}} = f_1 + f_2. \tag{12.30}$$

It is clear that the functions f_1, f_2 given on the interval $(\overline{\alpha},\ 1)$ are monotone (f_1 increases and f_2 decreases). Hence their sum $f_1 + f_2$ can either decrease or have at most one minimum for $z \in (\overline{\alpha},\ 1)$. The intersection points of the line $1 - z$ with the curve $f_1(z) + f_2(z)$ are solutions of equation (12.30). Therefore we either have no interior s.s. or at least two (rough) s.s., but by virtue of the convexity of f_1 and f_2 we can have no more than two such s.s.

2.2.2. m > p. We write equation (12.28), which is analogous to (12.30), in the form

$$1 - z = f_1(z) + f_3(z),$$

where

$$f_3(z) = a_2 \frac{z^{(1-p/m)n/q}}{(z^{m-n} - \overline{a})^{p/(mq)}}. \tag{12.31}$$

It is obvious that the function $f_3(z)$ can have at most one minimum on the interval $(\overline{\alpha},\ 1)$. Therefore the conclusion on the number of s.s. from case 2.2.1. are true here too, i.e. we either have two s.s. ($f_1 + f_3$ is convex), or none.

2.3. m < n. From (12.18) we express (analogously with (12.26), (12.27)) the values of x and y through z:

$$x = \left(\frac{npk_2}{qmk_{-1}}\right)^{1/m} (\overline{b} - z^{n-m})^{1/m} z,$$

$$y = \left(\frac{nk_2}{qk_3}\right)^{1/q} \left(\frac{qmk_{-1}}{npk_2}\right)^{p/(qm)} \frac{z^{(n-p)/q}}{(\overline{b} - z^{n-m})^{p/(mq)}}.$$

In this case the equation for positive values of z takes the form

$$1 - z = b_1(\overline{b} - z^{n-m})^{1/m} z + b_2 \frac{z^{(n-p)/q}}{(\overline{b} - z^{n-m})^{p/(mq)}}, \tag{12.32}$$

where

$$\overline{b} = \frac{qmk_1}{npk_2}, \quad b_1 = \left(\frac{npk_2}{qmk_{-1}}\right)^{1/m}, \quad b_2 = \left(\frac{nk_2}{qk_3}\right)^{1/q} \left(\frac{qmk_{-1}}{npk_2}\right)^{p/(qm)}.$$

In order that all positive solutions of (12.32) belong to the interval (0,1) it is necessary that $\overline{b} < 1$, i.e.

$$qmk_1 < npk_2 \tag{12.33}$$

which is opposite to (12.29). Under condition (12.33) all positive solutions of (12.32) of interest to us lie in the interval $(0,\ \overline{\beta})$, $\overline{\beta} = \overline{b}^{1/(n-m)}$.

2.3.1. n ≥ p. Equation (12.32) can be written as

$$1 - z = g_1(z) + g_2(z),$$

where the function g_1 has a maximum for $z \in (0,\ \overline{\beta})$, and vanishes at the end points of the interval, while the function $g_2(z)$ increases monotonically towards infinity as $z \to \overline{\beta}$. Therefore we have either one or three interior s.s. (odd integer). The question of more than 3 s.s. is here left open.

2.3.2. n < p. Equation (12.32) can be written as

$$1 - z = g_1(z) + g_3(z),$$

where the function g_3 has a minimum for $z \in (0, \overline{\beta})$, and tends to infinity as $z \to 0$ and $z \to \overline{\beta}$. Therefore we will here have an even number of s.s.: a) no interior s.s., b) two interior s.s., c) four interior s.s.

With the stationary equation written in the form (12.32) it is easy to choose the parameters k_j so that we have four interior s.s. We must take k_{-1} sufficiently small, which ensures that b_1 is large and b_2 is small. In this way a situation with four interior s.s. will be realized.

Observe that in this case we can have up to five s.s. — one on the boundary and four interior ones. As will be shown later, for $k_{-2} > 0$ the boundary s.s. can become an interior s.s., so that we get five interior s.s.

It is probable that the number of s.s. cannot exceed five. Moreover, the above-mentioned case of five s.s. is rather exotic: $m < n < p$, i.e. at least $m = 1$, $n = 2$, $p = 3$ and the third stage in (12.1) is more than three-molecular. If we restrict to

$$m \le 2, \quad n \le 2, \quad p + q \le 3$$

then a situation with more than three s.s. would seem improbable.

3⁰. $\mathbf{k_{-2} \neq 0, k_{-1} = 0}$. All conclusions obtained in item 2^0 are true in this case too if we make the permutation $m \rightleftarrows n$, $p \rightleftarrows q$, $k_1 \rightleftarrows k_2$.

4⁰. $\mathbf{k_{-1}, k_{-2} \neq 0}$. In this case the boundary s.s. are absent.

THEOREM 12.1. *If $m = n$ then for $m \ge p$, $m \ge q$ the system (12.4) has only one root in S.*

Proof. The first equation in system (12.4) defines a decreasing curve in S starting at the point (0,1). Indeed, if we rewrite (12.4) in the form

$$P_1(x,y) = -mk_1(1-x-y)^m + mk_{-1}x^m + pk_3x^py^q = 0, \tag{12.34}$$

$$P_2(x,y) = -nk_2(1-x-y)^n + nk_{-2}y^n + qk_3x^py^q = 0, \tag{12.35}$$

then equation (12.34) gives the curve $y = y(x)$ implicitly. Let us show that it is unique and monotonically decreasing in S. The curve $y(x)$ has only one singular point (0,1) in S, because the gradient

$$\operatorname{grad} P_1 = (m^2k_1(1-x-y)^{m-1} + m^2k_{-1}x^{m-1} + p^2k_3x^{p-1}y^q,$$
$$m^2k_1(1-x-y)^{m-1} + pqk_3x^py^{q-1})$$

vanishes in S only if $m > 1$ and $x = 0$, $y = 1$. To examine how many branches of $y(x)$ there are in S, we take $x = \varepsilon$, so that

$$-mk_1(1-\varepsilon-y)^m + mk_{-1}\varepsilon^m + pk_3\varepsilon^py^q = 0$$

or equivalently

$$(-1)^{m+1}mk_1(y-1+\varepsilon)^m + mk_{-1}\varepsilon^m + pk_3\varepsilon^py^q = 0.$$

If m is odd then

$$pk_3\varepsilon^py^q + mk_{-1}\varepsilon^m = -mk_1(y-1+\varepsilon)^m. \tag{12.36}$$

On the left hand side of this equation stands an increasing function (for $y > 0$), while the right hand side is a decreasing function. Hence (12.36) has at most one solution. If m is even, say $m = 2s$, then

$$pk_3\varepsilon^p y^q + mk_{-1}\varepsilon^{2s} = mk_1(y - 1 + \varepsilon)^{2s}. \tag{12.37}$$

On the segment $[0,\ 1-\varepsilon]$ equation (12.37) can have only one root, for the left side of (12.37) is increasing whereas the right side decreases.

Consider the intersection points of the curve $y(x)$ with the axis Ox. For $y = 0$ we get the equation

$$mk_1(1-x)^m = mk_{-1}x^m,$$

i.e. for even m

$$1 - x = \pm\left(\frac{k_{-1}}{k_1}\right)^{1/m} x,$$

and for odd m

$$1 - x = \left(\frac{k_{-1}}{k_1}\right)^{1/m} x.$$

Consequently

$$x_{1,2} = \left(1 \pm \left(\frac{k_{-1}}{k_1}\right)^{1/m}\right)^{-1},$$

and therefore either $x_1 > 1$ or $x_2 > 1$, i.e. the curve $y(x)$ has only two points of intersection with the sides of the triangle S. Hence in the interior of S there is only one branch of the curve, and this curve is decreasing, since

$$y'_x = \frac{m^2k_1(1-x-y)^{m-1} + m^2k_{-1}x^{m-1} + p^2k_3x^{p-1}y^q}{m^2k_1(1-x-y)^{m-1} + pqk_3x^py^{q-1}} < 0.$$

If we multiply (12.34) by k_2, (12.35) by k_1, and subtract, then we have

$$P = mk_1k_{-2}y^m - mk_2k_{-1}x^m + x^py^q(k_1q - k_2p) = 0.$$

Hence this curve $y = y(x)$ contains the point $(0,0)$ and

$$\operatorname{grad} P = (-mk_2k_{-1}x^{m-1} + pk_3x^{p-1}y^q(k_1q - k_2p)\ ,$$
$$mk_1k_{-2}y^{m-1} + qk_3x^py^{q-1}(k_1q - k_2p)).$$

If $k_1q - k_2p > 0$ then the gradient vanishes only for $y = 0$, and hence $x = 0$ $(m > 1)$. If $k_1q - k_2p < 0$ then again we must have $x = y = 0$ in case $\operatorname{grad} P = (0,0)$. For $k_1q - k_2p = 0$ the same is true. Thus, the only singular point of this curve is (0,0).

Now we analyze how many branches of this curve in our triangle S emanate from the point (0,0). Fixing $x = \varepsilon$ we get

$$P = mk_1k_{-2}y^m + k_3\varepsilon^p(k_1q - k_2p)y^q - mk_2k_{-1}\varepsilon^m = 0.$$

If $m \geq q$ then this equation has only one positive solution. If $m < q$ then it can have two solutions. Thus for $m \geq q$ this curve has one branch lying in S.

Let us consider the derivative

$$y'_x = -\frac{-m^2k_2k_{-1}x^{m-1} + pk_3x^{p-1}y^q(k_1q - k_2p)}{m^2k_1k_{-2}y^{m-1} + qk_3x^py^{q-1}(k_1q - k_2p)}.$$

If $k_1q - k_2p > 0$ then the denominator does not vanish, and for the numerator

$$-m^2k_2k_{-1}x^{m-1} + pk_3x^{p-1}y^q(k_1q - k_2p)$$

to vanish we must have

$$k_3x^py^q(k_1q - k_2p) = m^2k_2k_{-1}x^m/p.$$

Inserting this expression in P we get

$$P = mk_1k_{-2}y^m + mk_2k_{-1}x^m(m-p)/p = 0.$$

So if $m \geq p$ then the numerator is non-vanishing ($y'_x > 0$) in our domain and the equation $P = 0$ defines an increasing curve $y(x)$.

If instead $k_1q - k_2p < 0$, then the numerator does not vanish, but we would have $y'_x = \infty$ if

$$m^2k_1k_{-2}y^m + qk_3x^py^q(k_1q - k_2p) = 0.$$

This in turn would mean that

$$k_3x^py^q(k_1q - k_2p) = -m^2k_1k_{-2}y^m/q,$$

$$P = -mk_2k_{-1}x^m - mk_1k_{-2}y^m(m-q)/q,$$

i.e. $y'_x > 0$ for $m \geq q$. If $k_1q = k_2p$ then it is obvious that $y'_x > 0$. $\square$.

Now we restrict our considerations to cases with small values for the stoichiometric coefficients: $m \leq 2$, $n \leq 2$, $p + q \leq 3$.

4.1. m = 1, n = 2, p = q = 1. The system (12.4) takes the form

$$\begin{aligned} k_1(1 - x - y) - k_{-1}x - k_3xy &= 0, \\ 2k_2(1 - x - y)^2 - 2k_{-2}y^2 - k_3xy &= 0. \end{aligned} \tag{12.38}$$

Solving for y in the first equation we have

$$y = \frac{k_1 - k_1x - k_{-1}x}{k_3x + k_1} > 0.$$

From here it follows that

$$0 < x < \frac{k_1}{k_1 + k_{-1}} = \overline{k}_1, \quad 1 - x - y = \frac{(1-x)k_3x + k_{-1}x}{k_3x + k_1}.$$

Substituting y into the second equation of (12.38) we obtain

$$\begin{aligned} P \;=\; & 2k_2k_3^2x^4 + k_4(k_1k_3 + k_3k_{-1} - 4k_2k_3 - 4k_2k_{-1})x^3 + \\ & +(k_1^2k_3 + k_1k_3k_{-1} + 2k_2k_3^2 + 4k_2k_3k_{-1} + 2k_2k_{-1}^2 - k_1^2k_{-2} - 4k_1k_{-1}k_{-2} - \\ & -2k_{-1}^2k_{-2} - k_1k_3^2)x^2 + k_1(4k_1k_{-2} + 4k_{-1}k_{-2} - k_1k_3)x - 2k_{-2}k_1^2 = 0. \end{aligned}$$

Let us now determine the number of zeros of P in the interval $[0, \overline{k}_1]$. This can be done by means of the Budan-Fourier theorem (Theorem 3.6) in combination with the Descartes sign rule and the Hermite theorem (Theorem 3.2) (the Sturm method is too cumbersome).

Let $P(x)$ be a polynomial of degree n with real coefficients. In order to determine the number of zeros of P in the interval (a,b), we form the two number sequences:

$$1)\ P(a),\ P'(a), \ldots, P^{(n)}(a); \qquad 2)\ P(b),\ P'(b), \ldots, P^{(n)}(b).$$

If none of these numbers is equal to zero, then by the Budan–Fourier theorem the difference between the number $V(a)$ of sign changes in system 1) and $V(b)$ in system 2) either equals the number of roots of P in (a,b), or is less than this number by an even integer.

By the Hermite theorem the number of real zeros of a real polynomial P of degree n is equal to the difference between the number of sign repetitions and the number of sign changes in the sequence

$$\Delta_0 = 1,\ \Delta_1 = s_0,\ \Delta_2 = \begin{vmatrix} s_0 & s_1 \\ s_1 & s_2 \end{vmatrix},\ \ldots,\Delta_n = \begin{vmatrix} s_0 & s_1 & \ldots & s_{n-1} \\ s_1 & s_2 & \ldots & s_n \\ \ldots & \ldots & \ldots & \ldots \\ s_{n-1} & s_n & \ldots & s_{2n-2} \end{vmatrix},$$

where s_j are the power sums of the zeros of the polynomial P, and all determinants are non-vanishing. The power sums s_0, s_1, ..., s_n for the polynomial

$$P = x^n + b_1 x^{n-1} + \ldots + b_n = 0$$

are calculated by the recursion formulas (2.1)

$$s_j + s_{j-1}b_1 + s_{j-2}b_2 + \ldots + s_1 b_{j-1} + jb_j = 0, \qquad \text{for } j \le n,$$
$$s_j + s_{j-1}b_1 + \ldots + s_{j-n}b_n = 0, \qquad \text{for } j > n.$$

Let us now apply these theorems to our case. Since $P(0) = -2k_{-2}k_1 < 0$ and

$$P(\overline{k}_1) = \frac{2k_1^2 k_2((k_3 + k_{-1})(k_1 + k_{-1}) - k_1 k_3)^2}{(k_1 + k_{-1})^4} > 0,$$

we see that P has one or three zeros in the given interval. We make up two columns of numbers

1)	2)
$P(0) < 0,$	$P(\overline{k}_1) = 2k_1^2 k_2(k_1 k_{-1} + k_{-1}^2 + k_3 k_{-1})^2 > 0,$
$P'(0) = 4k_1 k_{-2} + 4k_{-1}k_{-2} - k_1 k_3,$	$P'\overline{k}_1 = \overline{k}_1 k_3(k_1(k_1 + k_{-1}) + k_1 k_3) + 4\overline{k}_1^{3} k_2(k_1 k_{-1} + k_{-1}^2 + k_3 k_{-1})^2/k_1^2 + 4\overline{k}_1^{3} k_2(k_1 k_{-1} + k_{-1}^2 + k_3 k_{-1})/k_1 > 0,$
$P^{(2)}(0) = k_1^2 k_3 + k_1 k_3 k_{-1} + 2k_2 k_3^2 + 4k_2 k_3 k_{-1} + 2k_2 k_{-1}^2 - 2k_1^2 k_{-2} - 4k_1 k_{-1} k_{-2} - 2k_{-1}^2 k_{-2} - k_1 k_3^2,$	$P^{(2)}(\overline{k}_1) = 2k_1 k_3^2 (k_1 + k_{-1})^2 - 12k_1 k_2 k_3^2 k_{-1} - 12k_1 k_2 k_3 k_{-1}(k_1 + k_{-1})k_1 k_3(k_1 + k_{-1})^3 + (k_1 + k_{-1})^2(2k_2 k_3^2 + 4k_2 k_3 k_{-1} + 2k_2 k_{-1}^2) - 2k_{-2}(k_1 + k_{-1})^4,$
$P^{(3)}(0) = k_1 k_3 + k_3 k_{-1} - 4k_2 k_3 - 4k_2 k_{-1},$	$P^{(3)}(\overline{k}_1) = 4k_2 k_3 k_1 - 4k_2 k_3 k_{-1} + k_1^2 k_3 + 2k_1 k_2 k_{-1} + k_3 k_{-1}^2 - 4k_2 k_{-1} \times (k_1 + k_{-1}),$
$P^{(4)}(0) > 0,$	$P^{(4)}(\overline{k}_1) > 0.$

In order for P to have three roots in the interval it is necessary that the right column should have no sign change, i.e. $P^{(2)}(\overline{k}_1) > 0$ and $P^{(3)}(\overline{k}_1) > 0$, while the left column should have three sign changes. Since we can assume that k_{-1} and k_{-2} are small in comparison to the other k_j, we have that $P'(0) < 0$, from which we get that $P^{(4)}(0) > 0$ and $P^{(3)}(0) < 0$. This means that the following inequalities must be satisfied: $P^{(2)}(\overline{k}_1) > 0$, $P^{(3)}(\overline{k}_1) > 0$ and $P^{(2)}(0) > 0$, $P^{(3)}(0) < 0$, or $P'(0) > 0$, $P^{(2)}(0) < 0$, or $P'(0) > 0$, $P^{(3}(0) < 0$. These inequalities are compatible, since for $k_{-1} = k_{-2} = 0$,

$$P^{(2)}(1) > 0, \ P^{(3)}(1) > 0,$$

$$P^{(2)}(0) = k_1^2 k_3 + 2k_2 k_3^2 - k_1 k_3^2, \quad P^{(3)}(0) = k_1 k_2 - 4k_2 k_3.$$

Therefore the domain of three roots can be described by the inequalities $k_1 < 4k_2$ and $k_1^2 + 2k_2 k_3 > k_1 k_3$. In other words, if k_1, k_2, k_3 satisfy these inequalities then we can always choose small k_{-1} and k_{-2} so that the other inequalities hold.

Now let us consider the number of real zeros of the polynomial P. In order for three zeros to lie in the interval $[0, \overline{k}_1]$ it is necessary and sufficient that P has four real zeros, i.e. $\Delta_2 > 0$, $\Delta_3 > 0$ and $\Delta_4 > 0$. Since P has at least one zero on $[0, \overline{k}_1]$ it has at least two zeros on the real axis, so if $\Delta_4 > 0$ we must necessarily have $\Delta_2 > 0$ and $\Delta_3 > 0$. Let

$$P = a_0 x^4 + a_1 x^3 + a_3 x + a_4.$$

We have

$$\Delta_4 = 18a_1^3 a_2 a_3 a_4 + a_1^2 a_2^2 a_3 + 144 a_0 a_1^2 a_2 a_4^2 + 18 a_0 a_1 a_2 a_3^3 + 16 a_0 a_2^4 a_4 +$$
$$+144 a_0^2 a_2 a_3^2 a_4 + 256 a_0^3 a_4^3 - 27 a_1^4 a_4^2 - 4 a_1^3 a_3^3 - 6 a_1^2 a_3^3 a_4 - 4 a_1^2 a_2^2 a_4 -$$
$$-80 a_0 a_1 a_2^2 a_3 a_4 - 192 a_0^2 a_1 a_3 a_4^2 - 4 a_0 a_2^2 a_3^2 - 128 a_0^2 a_4^2 a_2^2 - 27 a_0^2 a_3^4.$$

So the domain of three zeros is given by the inequalities

$$P^{(2)}(\overline{k}_1) > 0, \quad P^{(3)}(\overline{k}_1) > 0,$$

$$P^{(2)}(0) > 0, \quad P^{(3)}(0) > 0 \quad \Delta_4 > 0.$$

Let us show that these inequalities are compatible. Suppose $k_{-1} = 0$, $k_{-2} = 0$. Then we can replace the inequalities $k_1 - 4k_2 < 0$ and $k_1^2 + 2k_2 k_3 - k_1 k_3 > 0$ by the single inequality $k_1 < 2k_2$ representing a somewhat smaller domain. In this way all inequalities (except $\Delta_4 > 0$) will be satisfied. The inequality $\Delta_4 > 0$ is equivalent to the following:

$$Q = k_3(k_1 - 4k_2)^2 (k_1^2 + 2k_2 k_3 - k_1 k_3)^2 + 4k_1^2 k_3^2 (4k_2 - k_1)^3 -$$
$$-36 k_1^2 k_2 k_3 (4k_2 - k_1)(k_1^2 + 2k_2 k_3 - k_1 k_3) - 8k_2 (k_1^2 + 2k_2 k_3 - k_1 k_3)^3 -$$
$$-108 k_1^4 k_2^2 k_3 > 0.$$

Expanding the polynomial Q in powers of k_3 we obtain

$$Q = k_3^3 k_1^2 (2k_2 - k_1)^2 + 2k_3^2 k_1^2 (-3k_1^3 + 4k_1^2 k_2 + 28 k_1 k_2^2 - 32 k_2^3) +$$
$$+k_3 k_1^4 (k_1^2 + 52 k_1 k_2 - 28 k_2^2) - 8 k_1^6 k_2.$$

Since the coefficient of k_3^3 is positive we have $Q > 0$ for sufficiently large k_3. Moreover, Q has only one zero for fixed k_1 and k_2, since the coefficient of k_3 is negative ($k_1 < 2k_2$). Hence $\Delta_4 > 0$ for sufficiently large k_3. Therefore we get that $\Delta_4 > 0$ for sufficiently small k_{-1} and k_{-2}. Thus the domain of three zeros is not empty.

4.2. m = 1, n = 1, p = 2, q = 1. The system (2.4) takes the form

$$\begin{aligned} k_1(1-x-y)-k_{-1}x-2k_3x^2y &= 0,\\ k_2(1-x-y)-k_{-2}y-k_3x^2y &= 0. \end{aligned}$$

Solving for y in the first equation we obtain

$$y=\frac{k_1-k_1x-k_{-1}x}{2k_3x^2+k_1},$$

and hence

$$\begin{aligned} P &= k_3x^3(k_1+k_{-1}-2k_2)+k_3x^2(2k_2-k_1)+\\ &+x(k_2k_{-1}+k_1k_{-2}+k_{-1}k_{-2})-k_1k_{-2}. \end{aligned}$$

We need to determine the number of zeros of P in the segment $[0,\overline{k}_1]$, where

$$\overline{k}_1=\frac{k_1}{k_1+k_{-1}}.$$

Since $P(0)=-k_1k_{-2}<0$ and

$$P(\overline{k}_1)=\overline{k}_1^3k_2(k_{-1}(k_1+k_{-1})^2+2k_2k_1k_{-1})/k_1^2>0,$$

we see that P has one or three zeros in this segment. We have

1) $P(0)<0$,	2) $P(\overline{k}_1)>0$
$P'(0)>0$,	$P'(\overline{k}_1)=k_3k_1^2(k_1-2k_2)+4k_2k_3k_1k_{-1}+$ $+k_3k_1^2k_{-1}+(k_1+k_{-1})^2(k_2k_{-1}+k_2(k_1+k_{-1})^2)$,
$P^{(2)}(0)=2k_2-k_1$,	$P^{(2)}(\overline{k}_1)=k_1(k_1-2k_2)+2k_1k_{-1}+2k_1k_{-1}$,
$P^{(3)}(0)=k_1+k_{-1}-2k_2$,	$P^{(3)}(\overline{k}_1)>0$.

In order that the left column should have three sign changes it is necessary and sufficient that $k_1-2k_2>0$, and then the right column will have no sign change. Therefore a necessary condition for the existence of three zeros is $k_1-2k_2>0$. Now we seek a necessary and sufficient condition for all zeros of P to be real. As in the previous case it is given by the inequality $\Delta_3>0$:

$$\begin{aligned} \Delta_3 &= k_3(2k_2-k_1)^2(k_2k_{-1}+k_1k_{-2}+k_{-1}k_{-2})^2+\\ &+18k_1k_3k_{-2}(k_1+k_{-1}-2k_2)(k_1-2k_2)(k_2k_{-1}+k_1k_{-2}+k_{-1}k_{-2})-\\ &-4k_1k_3^2k_{-2}(k_1-2k_2)^3-4(k_1+k_{-1}-2k_2)(k_2k_{-1}+k_1k_{-2}+\\ &+k_{-1}k_{-2})^3-27k_1^2k_3k_{-2}^2(k_1+k_{-1}-2k_2)^2>0. \end{aligned}$$

This domain is not empty, since for $k_{-2}=0$

$$\Delta_3=k_2^2k_3k_{-1}(k_1-2k_2)^2-4k_3^3k_{-1}^3(k_1+k_{-1}-2k_2),$$

and if k_1-2k_2 is sufficiently large then $\Delta_3>0$, so that for small k_{-2} and large k_1-2k_2 the inequality $\Delta_3>0$ holds. If $k_{-1}=0$, then $\Delta_3<0$ and we have one root.

Notice that k_2 is bounded in this domain when the other k_i are kept fixed, since $k_2<k_1/2$; and k_1 is bounded when the other k_i are fixed, since if we expand Δ_3 in powers of k_1 then the coefficient of k_1^4 is

$$-4k_3^2k_{-2}-4k_{-2}^3-8k_2k_{-2}^2<0.$$

Also k_3 is bounded for the other k_i fixed, since

$$\Delta_3 = -4k_1k_3^2k_{-2}(k_1-2k_2)^3 + k_3((2k_2-k_1)^2(k_2k_{-1}+k_1k_{-2}+k_{-1}k_{-2})^2+$$

$$+18k_1k_{-2}(k_1+k_{-1}-2k_2)(k_1-2k_2)(k_2k_{-1}+k_1k_{-2}+k_{-1}k_{-2})-$$

$$-27k_1^2k_{-2}(k_1+k_{-1}-2k_2)^2) - 4(k_1+k_{-1}-2k_2)(k_2k_{-1}+k_1k_{-2}+k_{-1}k_{-2})^2.$$

Here the leading coefficient is less than zero. Moreover, this domain can have the form

or

but as a whole it is not bounded, as shown by the above reasoning.

4.3. $\mathbf{m=1, n=2, p=2, q=1}$. The system (12.4) takes the form

$$\begin{aligned} k_1(1-x-y) - k_{-1}x - 2k_3x^2y &= 0, \\ 2k_2(1-x-y)^2 - 2k_{-2}y^2 - k_3x^2y &= 0. \end{aligned}$$

From the first equation we get

$$y = \frac{k_1 - k_1x - k_{-1}x}{2k_3x^2 + k_1},$$

and inserting this into the second equation we have

$$P(x) = 8k_3^2k_2x^6 + 2k_3^2x^5(k_1+k_{-1}-8k_2) + 2k_2x^4\times$$

$$\times(4k_3k_2 - 4k_2k_{-1} - k_3k_1) + k_3x^3(8k_2k_{-1}+k_1k_{-1}+k_1^2)+$$

$$+x^2(2k_2k_{-1}^2 - k_3k_1^2 - 2k_{-2}^2k_{-2} - 2k_{-2}k_1^2 4k_{-2}k_{-1}k_1)+$$

$$+4k_1k_{-2}x(k_1+k_{-1}) - 2k_1^2k_{-2} = 0.$$

We have to find the number of zeros of P in the segment $[0, \overline{k}_1]$, where

$$\overline{k}_1 = \frac{k_1}{k_1 + k_{-1}}.$$

Since $P(0) < 0$ and $P(\overline{k}_1) > 0$, we know that P can have one, three or five zeros in this segment. We have

1) $P(0) < 0,$
$P'(0) > 0,$
$P^{(2)}(0) = 2k_2k_{-1}^2 - k_1k_3^2 - 2k_{-2}(k_1 + k_{-1})^2,$
$P^{(3)}(0) > 0,$
$P^{(4)}(0) = 4k_2k_3 - 4k_2k_{-1} - k_1k_3,$
$P^{(5)}(0) = k_1 + k_{-1} - 8k_2,$
$P^{(6)}(0) > 0,$

2) $P(\overline{k}_1) > 0$
$P'(\overline{k}_1) > 0,$
$P^{(2)}(\overline{k}_1) = 8k_2k_3^2k_1^2(k_1^2 - 8k_1k_{-1} + 6k_{-1}^2) + 8k_1^3k_3^2(k_1 + k_{-1})^2 - 2k_{-2}(k_1 + k_{-1})^6 + 24k_1k_2k_3k_{-1}(k_1 + k_{-1})^2(k_{-1} - k_1) + 2k_1^2k_3(k_1 + k_{-1})^4 + 2k_2k_{-1}^2(k_1 + k_{-1})^4,$
$P^{(3)}(\overline{k}_1) > 0,$
$P^{(4)}(\overline{k}_1) = k_2k_3(6k_1^2 - 8k_1k_{-1} + k_{-?}^2) + 4(k_1 + k_{-1})(k_1k_3 - k_2k_1),$
$P^{(5)}(\overline{k}_1) > 0,$
$P^{(6)}(\overline{k}_1) > 0.$

$P^{(5)}(0)$ and $P^{(4)}(0)$ cannot both be positive, since if $k_1 + k_{-1} - 8k_2 > 0$, then

$$8k_2k_3 - 8k_2k_{-1} - 2k_1k_3 < k_1k_3 + k_{-1}k_3 - 8k_2k_{-1} - 2k_1k_2 =$$

$$= k_3(k_{-1} - k_1) - 8k_2k_{-1} < 0$$

So the left column can have either three or five sign changes, for

$$2k_2k_{-1}^2 - k_1k_3^2 - 2k_{-2}(k_1 + k_{-1})^2 < 0.$$

If we suppose that $k_{-1}, k_{-2} = O(k_1, k_2)$ then we can obtain five sign changes. Hence the number of zeros can be determined by forming the power sums $s_0, s_1, \ldots, s_8$ and the determinants $\Delta_0, \Delta_1, \ldots, \Delta_6$.

Let us show that the polynomial P may have at least three zeros. Suppose $k_{-1} = k_{-2} = 0$. Then P has a zero $x = 0$ of multiplicity 2 and a zero $x = 1$ of multiplicity 1. We investigate the polynomial Q on the segment [0,1]:

$$Q = 8k_3k_2x^3 + 2k_3x^2(k_1 - 4k_2) + k_1^2.$$

$Q(0) = 0,$	$Q(1) > 0,$
$Q'(0) = 0,$	$Q'(1) > 0,$
$Q^{(2)}(0) = k_1 - 4k_2,$	$Q^{(2)}(1) > 0,$
$Q^{(3)}(0) > 0,$	$Q^{(3)}(1) > 0.$

We see that for $k_1 - 4k_2 < 0$ there are two sign changes, and for $k_1 > 4k_2$ we have no sign change and no zeros. Therefore the inequality $k_1 - 4k_2 < 0$ is a necessary condition for having more than one zero. Being a polynomial of odd degree Q has at least one real zero. Hence the number of real zeros is determined by the sign of Δ_3: If $\Delta_3 > 0$ then Q has three zeros, and if $\Delta_3 < 0$ then Q has one zero. We have

$$\Delta_3 = 32k_3^3k_1^2(4k_2 - k_1)^3 - 27 \cdot 64k_3^2k_2^2k_1^4 = 32k_3^2k_1^2(k_3(4k_2 - k_1)^3 - 54k_1^2k_2^2) > 0,$$

so if k_3 is sufficiently large this condition is fulfilled and Q has two zeros on the segment $[0, 1]$. Then for small nonzero k_{-1} and k_{-2} the polynomial P will have three zeros.

For finding the zeros we must (as a rule) solve the nonlinear system (12.4) numerically. It is only in particular cases that we can write the solutions of (12.4) in closed form. For example, letting $m = n = p + q$, we get from (12.4)

$$(qk_1 - pk_2)z^m = qk_{-1}x^m - pk_{-2}y^m. \tag{12.39}$$

Substituting (12.39) into the second equation of (12.4), we have

$$\alpha_1 x^{m-p}y^{-q} - \alpha_2 y^{m-q}x^{-p} = \alpha_3, \tag{12.40}$$

where $\alpha_1 = qmk_{-1}k_2$, $\alpha_2 = qmk_1k_{-2}$, $\alpha_3 = qk_3(qk_1 - pk_2)$. Equation (12.40) can also be written

$$u^{p+q} - \gamma_1 u^p - \gamma_2 = 0,$$

where $u = x/y$, $\gamma_1 = \alpha_2/\alpha_1$, $\gamma_2 = \alpha_3/\alpha_1$. Suppose in addition that $p = q$. Then

$$x = cy, \quad c = \sqrt[p]{\frac{1}{2}\gamma_1 + \sqrt{\frac{1}{4}\gamma_1 + \gamma_2}},$$

and since the second equation of (12.4) gives

$$mk_2(1 - (1+c)y)^m = (mk_{-2} + qk_3c^p)y^m,$$

we finally get

$$y = \left(1 + c + \left(\frac{k_{-2}}{k_2} + \frac{qk_3c^p}{mk_2}\right)^{1/m}\right)^{-1}.$$

In general, when there is a possibility of having several s.s., we must either solve the initial system of equations numerically, or use elimination of unknowns to reduce it to a single equation in one unknown, and then look for the zeros of the obtained polynomial. An account of elimination methods for more general situations will be given in Sec. 13.

Conclusions. The analysis of the number of s.s. for a three-stage mechanism of the general form (12.1) shows (Theorem 12.1) that a difference between the kinetic exponents of the first and second stages (i.e. $m \neq n$) is a necessary condition for having several s.s. We considered various cases corresponding to irreversibility of the adsorption stages (the first and second stages), with different relations between the stoichiometric coefficients m, n, p, q. We indicated the cases with one, three, five s.s. For parallel schemes (12.1) the case of five s.s. appears to be rather exotic (for example $m < n < p$, i.e. $m = 1$, $n = 2$, $p = 3$). Moreover, it seems likely that for the given mechanism the number of s.s. does not exceed 5 for any m, n, p, q. So the most typical situations, in which several s.s. occur, are the ones with three s.s.

12.2. THE SEQUENTIAL SCHEME

There exists one more important type of three-stage mechanisms which admits several s.s. [**51**]

$$mZ \underset{-1}{\overset{1}{\rightleftarrows}} mX,$$

$$nX \underset{-2}{\overset{2}{\rightleftarrows}} nY, \tag{12.41}$$

$$pX + qY \xrightarrow{3} (p+q)Z,$$

where m, n, p, q are stoichiometric coefficients positive integers), Z is a catalyst, and X, Y are intermediates. The scheme (12.41) can be called sequential in accordance with the character of the first two stages.

The scheme (12.41) contains a stage of interaction between the different substances X and Y (stage 3). Its presence in the detailed mechanism is a necessary condition for the plurality of s.s. But it is not by itself sufficient for the appearance of several s.s. As the investigations of the parallel scheme have shown, some special relations must hold between the kinetic exponents of the various stages. In this section we shall describe, in terms of the stoichiometric coefficients m, n, p, q, those mechanisms of type (12.41) for which the kinetic models have more than one s.s., and we shall also give estimates for the number of these s.s.

Mathematical model. To the scheme (12.41) there corresponds the following nonstationary kinetic model

$$\begin{aligned} \dot{y} &= nk_2x^n - nk_{-2}y^n - qk_3x^pq^q, \\ \dot{z} &= -mk_1z^m + mk_1x^m + (p+q)k_3x^py^q, \end{aligned} \tag{12.42}$$

where $x = 1 - y - z$, y and z are the concentrations of X, Y and Z respectively, and k_i are the rate constants of the corresponding reactions.

The solutions to the two ordinary nonlinear differential equations (12.42) are defined in the triangle

$$S_r = \{(x,y,z): \ x,y,z \geq 0, \ x+y+z = 1\}, \tag{12.43}$$

which, as before, we shall call the reaction simplex. The system (12.42) has the property that S_r is a ω-invariant set for it. This ensures the existence of at least one s.s. belonging to S_r. As the reaction rate we take the rate of stage 3, i.e.

$$W = k_3x^py^q. \tag{12.44}$$

This corresponds to the assumption that a product emerges in stage 3. In an interior s.s. the stationary rate W_0 is strictly positive, whereas W may vanish in a boundary s.s.

Steady states. The s.s. of system (12.42) are defined to be the solutions of the system of equations

$$\begin{aligned} nk_2x^n - nk_{-2}y^n - qk_3x^py^q &= 0, \\ -mk_1z^m + mk_{-1}x^m + (p+q)k_3x^py^q &= 0, \\ x + y + z &= 1. \end{aligned} \tag{12.45}$$

We rewrite (12.45) in the form

$$\begin{aligned} nk_2x^n - nk_{-2}y^n - qk_3x^py^q &= 0, \\ -mk_1(1-x-y)^m + mk_{-1}x^m + (p+q)k_3x^py^q &= 0. \end{aligned} \tag{12.46}$$

The problem is to find the number of roots of the nonlinear algebraic system (12.46) in the triangle

$$S = \{(x,y): \ x \geq 0, \ y \geq 0, \ x+y \leq 1\}, \tag{12.47}$$

which, just as the triangle S_r (12.43), we shall call the reaction simplex. Observe that for k_{-1}, $k_{-2} > 0$ the system (12.46) has no boundary s.s.

THEOREM 12.2. *For $p \leq n$ the system (12.46) has only one interior s.s.*

Proof. Consider the first equation from (12.46):

$$P_1(x,y) = nk_2x^n - nk_{-2}y^n - qk_3x^py^q = 0. \tag{12.48}$$

The curve $y = y(x)$ which is defined by equation (12.48) has (0,0)as its only singular point, since

$$\operatorname{grad} P_1 = (n^2k_2x^{n-1} - pqk_3x^{p-1}y^q \ , \ -n^2k_{-2}y^{n-1} - q^2k_3x^py^{q-1}),$$

which vanishes if $x = 0$, $y = 0$ and $n > 1$.

We now analyze how many branches originate from the point (0,0) and lie in the quadrant ($x > 0$, $y > 0$). The point of intersection of our curve with the line $x = \varepsilon$ will be determined by the equation

$$nk_2\varepsilon^n - nk_{-2}y^n - qk_3\varepsilon^py^q = 0,$$

which always has one positive root, since there is only one sign change in the coefficient sequence.

So from the point (0,0) there emanates one branch of the curve $y(x)$ into the positive quadrant of the (x, y) plane. The derivative of this branch is

$$y'_x = \frac{nk_2x^{n-1} - qpk_3x^{p-1}y^q}{n^2k_{-2}y^{n-1} + q^2k_3x^py^{q-1}},$$

where the denominator is always greater than 0. For y'_x to vanish we must have

$$n^2k_2x^{n-1} = qpk_3x^{p-1}y^q.$$

If $x \neq 0$ this means that

$$n^2k_2x^n = qpk_3x^py^q,$$

or

$$qk_3x^py^q = \frac{n^2k_2x^n}{p}.$$

From the equation of the curve we thus find that

$$nk_2x^n - nk_{-2}y^n - \frac{n^2k_2x^n}{p} = 0,$$

and hence

$$n\left(1 - \frac{n}{p}\right)k_2x^n - k_{-2}y^n = 0.$$

If $p \leq n$ then $y = 0$, and $x = 0$, so in this case the curve $y(x)$ has no extremal points and is therefore increasing.

Consider the curve given by the second equation from (12.46):

$$P_2(x,y) = -mk_1(1 - x - y)^m + mk_{-1}x^m + (p + q)k_3x^py^q = 0.$$

This curve $y(x)$ has the only singular point (0,1) in S, since the gradient

$$\operatorname{grad} P_2 = (m^2k_1(1 - x - y)^{m-1} + (p + q)pk_3x^{p-1}y^q + m^2k_{-1}x^{m-1},$$
$$m^2k_1(1 - x - y)^{m-1} + (p + q)qk_3x^py^{q-1}),$$

vanishes in the triangle S only if $m > 1$ and $x = 0$, $y = 1$. Taking $x = \varepsilon$ we get

$$mk_1(1-\varepsilon-y)^m + mk_{-1}\varepsilon^m + (p+q)k_3\varepsilon^p y^q = 0,$$

or

$$(-1)^{m+1}mk_{-1}(y-1+\varepsilon)^m + mk_{-1}\varepsilon^m + (p+q)k_3\varepsilon^p y^q = 0.$$

If m is odd then

$$(p+q)k_3\varepsilon^p y^q + mk_{-1}\varepsilon^m = -mk_1(y-1+\varepsilon)^m.$$

The left hand side of this equation is an increasing function for $y > 0$, whereas the right hand side is decreasing. Hence this equation has no more than one solution.

If $m = 2s$ is even, then we get an equation of the form

$$(p+q)k_3\varepsilon^p y^q + mk_{-1}\varepsilon^m = mk_1(y-1+\varepsilon)^m.$$

On the segment $[0, 1-\varepsilon]$ the left hand side is increasing, while the right hand side is a decreasing function. So in this segment there is at most one root. Consider now the intersection points of the curve $y(x)$ with the axis Ox. For $y = 0$ we get the equation

$$mk_1(1-x)^m = mk_{-1}x^m,$$

i.e. for even m

$$1 - x = \pm\sqrt[m]{k_{-1}/k_1}x,$$

and for odd m

$$1 - x = \sqrt[m]{k_{-1}/k_1}x.$$

From here $x_{1,2} = \left(1 \pm \sqrt[m]{\frac{k_{-1}}{k_1}}\right)^{-1}$, and hence either $x_1 < 1$, $x_2 < 1$ or $x_1 x_2 < 1$, i.e. the curve $y(x)$ has only two points of intersection with the sides of the triangle S. There is thus only one branch of the curve $y(x)$ inside S, and this curve is decreasing, since

$$y'_x = -\frac{m^2k_1(1-x-y)^{m-1} + m^2k_{-1}x^{m-1} + (p+q)pk_3x^{p-1}y^q}{m^2k_1(1-x-y)^{m-1} + (p+q)qk_3x^p y^{q-1}} < 0.$$

In conclusion, for $p \le n$ the first curve (given by $P_1 = 0$) increases, while the second ($P_2 = 0$) decreases. The system (12.46) therefore has only one interior s.s. □

So for $p \le n$, $k_{-1}, k_{-2} > 0$ we have only one interior s.s. and no boundary s.s. So cases with many s.s. must be looked for among mechanisms with $p > n$. Below we will estimate the number of s.s. in different particular cases for which the relation $p > n$ holds. Recall that in the parallel scheme the inequality $m \ne n$ was a basic necessary condition for plurality of s.s. Here we have instead the condition $p > n$. The asymmetry of this latter condition is explained by the asymmetry of the scheme (12.41), in contrast to the scheme (12.1). In (12.41) the substances X and Y are involved in the given scheme in different ways, whereas in (12.1) X and Y play identical roles.

Now we consider some particular cases corresponding to different reversibility assumptions at the stages 1 and 2 in the mechanism (12.41).

1⁰. $\mathbf{k_{-1} = k_{-2} = 0}$. The system (12.45) takes the form

$$\begin{aligned} nk_2x^n - qk_3x^py^q &= 0,\\ mk_1z^m - (p+q)k_3x^py^q &= 0,\\ x+y+z &= 1. \end{aligned} \tag{12.49}$$

System (12.49) has one boundary root $x = 0$, $y = 1$, $z = 0$. The other roots satisfy the system

$$\begin{aligned} nk_2x^{n-p} - qk_3y^q &= 0,\\ mk_1z^m - (p+q)k_3x^py^q &= 0,\\ x+y+z &= 1, \end{aligned} \tag{12.50}$$

from which we get

$$y^q = \frac{nk_2}{qk_3}x^{n-p}, \quad qmk_1z^m = (p+q)nk_2x^n,$$

and hence

$$y = \left(\frac{nk_2}{qk_3}\right)^{1/q} x^{(n-p)/q} = c_1x^{(n-p)/q},$$

$$z = \left(\frac{nk_2(p+q)}{qmk_1}\right)^{1/m} x^{n/m} = c_2x^{n/m}.$$

Then we have

$$x + c_1x^{(n-p)/q} + c_2x^{n/m} - 1 = 0,$$

and making the substitution $x \to x^{mq}$, we obtain

$$x^{mq} + c_1x^{(n-p)m} + c_2x^{nq} - 1 = 0. \tag{12.51}$$

the condition $n - p < 0$ then gives

$$x^{m(p+q-n)} + c_2x^{nq+pm-mn} - x^{(p-n)m} + c_1 = 0,$$

and, in view of the Descartes theorem, this equation has at most two roots.

Let us verify that the case of 2 roots is indeed possible. Letting $k_1 \to \infty$ we get

$$P(x) = x^{m(p+q-n)} - x^{(p-n)m} + c_1 = 0. \tag{12.52}$$

We find the zero

$$x_0 = \left(\frac{p-n}{p+q-n}\right)^{1/(mq)}$$

for the derivative of the function $P(x)$, and in order for equation (12.52) to have exactly two positive roots, it is necessary and sufficient that

$$P(x_0) = \left(\frac{p-n}{p+q-n}\right)^{1+(p-n)/q} - \left(\frac{p-n}{p+q-n}\right)^{(p-n)/q} + \left(\frac{nk_2}{qk_3}\right)^q < 0.$$

This can always be achieved by decreasing k_2. So for any m, n, p, q with $p > n$, the system (12.49) may have either two interior roots, or no such roots.

2^0. $\mathbf{k_{-2} = 0}$. Now the system (12.45) takes the form

$$\begin{aligned} nk_2x^n - qk_3x^py^q &= 0, \\ mk_1z^m - mk_{-1}x^m - (p+q)k_3x^py^q &= 0, \\ x+y+z &= 1. \end{aligned} \tag{12.53}$$

System (12.53) has the boundary s.s. $x = 0$, $z = 0$, $y = 1$. The other roots satisfy the system

$$\begin{aligned} nk_2x^{n-p} - qk_3y^q &= 0, \\ mk_1z^m - mk_{-1}x^m - (p+q)k_3x^py^q &= 0, \\ x+y+z &= 1. \end{aligned} \tag{12.54}$$

From here we get

$$y = \left(\frac{nk_2}{qk_3}\right)^{1/q} x^{(n-p)/q} = c_1x^{(n-p)/q},$$

$$\begin{aligned} z = (mk_1p)^{-1/m}(qmk_{-1}x^m + n(p+q)k_2x^n)^{1/m} = \\ = c_2(qmk_{-1}x^m + n(p+q)k_2c^n)^{1/m}. \end{aligned}$$

Then the equation for x becomes

$$F(x) = x + c_1x^{(n-p)/q} + c_2(qmk_{-1}x^m + n(p+q)k_2x^n)^{1/m} - 1 = 0. \tag{12.55}$$

Let us estimate how many positive roots equation (12.55) can have. We compute the second derivative of F and observe that

$$(x^{(n-p)/q})'' > 0$$

for $x > 0$, while

$$((qmk_{-1}z^m + n(p+q)k_2x^n)^{1/m})'' =$$

$$= \frac{n^3(p+q)^2k_2^2x^{2n-1}(n-m) + q(p+q)k_2k_{-1}m^2n(n-m)(n-m+1)x^{m+n-2}}{m^2(qmk_{-1}x^m + n(p+q)k_2x^n)^{(2m-1)/m}}.$$

It follows that $F'' > 0$ for $n \geq m$. So in this case equation (12.55) can have at most two positive roots. The case of a greater number of roots is possible only for $p > n$, $m > n$.

LEMMA 12.1. *Let*

$$\varphi = t^\alpha(1+t)^\beta,$$

with $\alpha, \beta \geq 0$. Then $\varphi'' \geq 0$ or ≤ 0 for $t > 0$ if and only if either $\alpha + \beta \leq 1$, or $\alpha \geq 1$.

This lemma is verified by straight-forward differentiation.

In order that F should have at least 4 zeros it is necessary for the function

$$\Phi(x) = c_1x^{(n-p)/q} + c_2(qmk_{-1}z^m + n(p+q)k_2x^n)^{1/m}$$

to have at least three extremal points, i.e. Φ' should have at least three zeros. Now, the equation

$$\Phi' = -c_1\frac{p-n}{q}x^{(n-p-q)/q} + \frac{c_2}{m}\cdot\frac{qm^2k_{-1}x^{m-1} + n^2(p+q)k_2x^{n-1}}{(qmk_{-1}x^m - n(p+q)k_2x^n)^{1-1/m}} = 0$$

can be written

$$A + Bx^{n-m} - Cx^{(n-p-q)/q}(D + Ex^{n-m})^{1-1/m} = 0.$$

Introducing the new variable $x^{n-m} = t$ we obtain

$$g(t) = A + Bt - Ct^{(p+q-n)/(q(m-n))}(D + Et)^{1-1/m} = 0.$$

We then apply Lemma 12.1 to $g(t)$ with

$$\alpha = \frac{p+q-n}{q(m-n)}, \quad \beta = \frac{m-1}{m},$$

which gives us $g'' \geq 0$ or ≤ 0 if

$$\frac{p+q-n}{q(m-n)} \geq 1$$

or

$$\frac{p+q-n}{q(m-n)} + \frac{m-1}{m} \leq 1,$$

i.e. $p + q - n - qm + qn \geq 0$ or $pm + qn - mn \leq 0$. But since $p > n$ the second inequality cannot be satisfied. So if $p + q + qn \geq n + qm$ then Φ' cannot have three extremal points, and hence equation (12.55) has at most two roots.

Finally we get: If $p \leq n$ then system (12.53) has no interior s.s.; if $p > n$ and $n \geq m$ or $p + q + qn \geq n + qm$, then (12.53) has at most two interior s.s.

3⁰. $\mathbf{k_{-1} = 0}$. The system (12.45) takes the form

$$\begin{aligned} nk_2x^n - nk_{-2}y^n - qk_3x^py^q &= 0, \\ mk_1z^m - (p+q)k_3x^py^q &= 0, \\ x + y + z &= 1. \end{aligned} \tag{12.56}$$

This system (12.56) has no boundary s.s., i.e. all its s.s. are interior. Estimates of their number will be made below, under the general assumption $k_{-1}, k_{-2} \neq 0$, for different concrete values of m, n, p, q.

We restrict our attention to mechanisms of type (12.41), for which $m, n \leq 2$, $p + q \leq 3$. In view of Theorem 12.2 it is enough to analyze only the case $p > n$.

4⁰. $\mathbf{m = n = q = 1, p = 2}$. Consider the system

$$\begin{aligned} k_2x - k_{-2}y - k_3x^2y &= 0, \\ k_1(1 - x - y) - k_{-1}x - 3k_3x^2y &= 0. \end{aligned} \tag{12.57}$$

Solving for y in the second equation, we get

$$y = \frac{k_1 - k_{-1}x - k_1x}{3k_3x^2 + k_1}.$$

Since $y \geq 0$ we must have

$$k_1 - k_{-1}x - k_1x \geq 0,$$

so that

$$0 \leq x \leq \frac{k_1}{k_1 + k_{-1}} = \overline{k}_1 < 1$$

and

$$0 \leq 1 - x - y =$$

$$= \frac{(1-x)(3k_3x^2+k_1)-k_1(1-x)+k_{-1}x}{3k_3x^2+k_1} =$$

$$= \frac{3k_3x^2(1-x)+k_{-1}x}{3k_3x^2+k_1},$$

for $0 \le x \le 1$. So for $0 \le x \le \bar{k}_1$ the root of system (12.57) will be located inside the triangle S. Substituting y into the first equation, we get

$$\begin{aligned} P = k_3x^3(3k_3+k_1+k_{-1}) - k_1k_3x^2 + x(k_1k_2 &+ k_1k_{-2}+k_{-1}k_{-2}) - k_1k_{-2} = 0,\\ P(0) = -k_1k_{-2} &< 0, \\ P(\bar{k}_1) = k_1k_2/(k_1+k_{-1}) &> 0. \end{aligned} \qquad (12.58)$$

Hence the polynomial P has at least one, but no more than three zeros in the segment $[0,\bar{k}_1]$.

The determination of the number of roots of a polynomial in an interval is based on the Budan–Fourier theorem and the Hermite theorem. In order to apply the first of these, we must calculate the two sequences:

1) $P(0) = -k_1k_{-2} < 0,$

$P'(0) = k_1k_2 + k_1k_{-2} +$
$+k_{-1}k_{-2} > 0,$

$P^{(2)}(0) = -k_1k_3 < 0,$

$P^{(3)}(0) = k_3(3k_2+k_1+k_{-1}) > 0,$

2) $P(\bar{k}_1) = \bar{k}_1k_2 > 0,$

$P'(\bar{k}_1) = 9k_1^2k_2k_3 + k_1^2k_3(k_1+k_{-1}) +$
$+(k_1+k_{-1})^2(k_1k_2+k_{-2}k_1+k_{-1}k_{-2}) > 0,$

$P^{(2)}(\bar{k}_1) = 18k_1^2k_2k_3 +$
$+4k_1k_3(k_1+k_{-1})^2 > 0,$

$P^{(3)}(\bar{k}_1) = 6k_3(3k_2+k_1+k_{-1}) > 0.$

We see that the difference between the numbers of sign changes in the left and right columns is equal to three. By the Budan–Fourier theorem it then follows that this polynomial has either one or three zeros. Now we apply the Hermite theorem. To this end we need to calculate the power sums s_0, s_1, s_2, s_3, s_4 of the zeros of P by means of the Newton formulas (2.1). We have

$$s_0 = 3, \quad s_1 = -b_1, \quad s_2 = b_1^2 - 2b_2, \quad s_3 = -b_1^3 + 3b_1b_2 - 3b_3,$$
$$s_4 = b_1^4 - 4b_1^2b_2 + 4b_1b_3 + 2b_2^2.$$

Then we get

$$\Delta_0 = 1, \quad \Delta_1 = 3, \quad \Delta_2 = 2b_1^2 - 6b_2,$$
$$\Delta_3 = b_1^2b_2^2 + 18b_1b_2b_3 - 4b_1^3b_3 - 4b_2^3 - 27b_3^2.$$

Now, substituting the coefficients of the polynomial P (12.58) into these formulas, we get (up to positive factors):

$$\Delta_0 = 1, \quad \Delta_1 = 3,$$
$$\Delta_2 = k_1^2k_3 - 3(3k_2+k_1+k_{-1})(k_1k_2+k_1k_{-2}+k_{-1}k_{-2}),$$

$$\begin{aligned} \Delta_3 \;=\; & k_1^2k_3(k_1k_2+k_1k_{-2}+k_{-1}k_{-2})^2 + \\ & +18k_1k_3^2k_{-2}(3k_2+k_1+k_{-1})(k_1k_2+k_1k_{-2}+k_{-1}k_{-2}) - \\ & -4k_1^4k_3^2k_{-2} - 4(3k_2+k_1+k_{-1})(k_1k_2+k_1k_{-2}+k_{-1}k_{-2})^3 - \\ & -27k_1^2k_3k_{-2}^2(3k_2+k_1+k_{-1})^2. \end{aligned}$$

The polynomial (12.58) has three real zeros if $\Delta_2 > 0$ and $\Delta_3 > 0$, and one real zero if $\Delta_2 < 0$, $\Delta_3 < 0$, or $\Delta_2 > 0$, $\Delta_3 < 0$, i.e. if $\Delta_3 < 0$. (The case $\Delta_2 < 0$, $\Delta_3 > 0$ is not possible since $\Delta_3 > 0$ implies $\Delta_2 > 0$).

So together with the Budan–Fourier theorem this result shows that for $\Delta_3 > 0$ the polynomial P has three zeros in the segment $[0, \overline{k}_1]$, and for $\Delta_3 < 0$ it has one zero.

The domain of three s.s. is defined by the inequality

$$\begin{aligned} Q &= k_1^2k_3(k_1k_2 + k_1k_{-2} + k_{-1}k_{-2})^2 + \\ &+18k_1k_3^2k_{-2}(3k_2 + k_1 + k_{-1})(k_1k_2 + k_1k_{-2} + k_{-1}k_{-2}) - \\ &-4k_1^4k_3^2k_{-2} - 4(3k_2 + k_1 + k_{-1})(k_1k_2 + k_1k_{-2} + k_{-1}k_{-2})^3 - \\ &-27k_1^2k_3k_{-2}^2(3k_2 + k_1 + k_{-1})^2 > 0. \end{aligned}$$

Let us analyze this domain in more detail. Fixing all the variables in Q, except k_1, we get

$$\begin{aligned} Q &= k_1^4(k_3(k_2 + k_{-2})^2 + 18k_3k_{-2}(k_2 + k_{-2}) - 4k_3^2k_{-2} - 4(k_2 + k_{-2})^3 - \\ &-27k_3k_{-2}^2) + k_1^3(2k_3k_{-1}k_{-2}(k_2 + k_{-2}) + 18k_3k_{-1}k_{-2}^2 + \\ &+18k_3k_{-2}(3k_2 + k_{-1})(k_2 + k_{-2}) - 12(k_2 + k_{-2})^2k_{-1}k_{-2} - \\ &-4(3k_2 + k_{-1})(k_2 + k_{-2})^3 - 54k_3k_{-2}^2(3k_2 + k_{-1})) + k_1^2(k_3k_{-1}^2k_{-2}^2 + \\ &+18k_3k_{-1}k_{-2}^2(3k_2 + k_{-1}) - 12(k_2 + k_{-2})k_{-1}^2k_{-2}^2 - 12(3k_2 + \\ &+k_{-1})(k_2 + k_{-2})^2k_{-1}k_{-2} - 27k_3k_{-2}^2(3k_2 + k_{-1})^2) - k_1(4k_{-1}^3k_{-2}^2 + \\ &+12(3k_2 + k_{-1})(k_2 + k_{-2})k_{-1}^2k_{-2}^2) - 4(3k_2 + k_{-1})k_{-1}^3k_{-2}^3. \end{aligned}$$

The constant term of this polynomial and the coefficient of k_1 are both negative. It is easily checked that the coefficient of k_1^2 is also negative. By the Descartes theorem it follows that Q can have at most two positive zeros. If

$$k_3(k_2^2 + 20k_2k_{-2} - 8k_{-2}^2) - 4k_3^2k_{-2} - 4(k_2 + k_{-2})^3 > 0,$$

then Q has one zero, so the intersection with the domain of three s.s. looks like

If on the other hand

$$k_3(k_2^2 + 20k_2k_{-2} - 8k_{-2}^2) - 4k_3^2k_{-2} - 4(k_2 + k_{-2}) < 0$$

then either Q has no zeros or two zeros, and its qualitative structure has the form

One can also investigate this domain by fixing the other k_i. At any rate it is clear that the domain of three s.s. is unbounded in k_1, and that it has a conic form.

5^0. m = p = 2, n = q = 1. The system (12.46) take the form

$$\begin{aligned} k_2x - k_{-2}y - k_3x^2y &= 0, \\ 2k_1(1 - x - y)^2 - 2k_{-1}x^2 - 2k_{-1}x^2 - 3k_3x^2y &= 0. \end{aligned} \qquad (12.59)$$

Computing y from the first equation we obtain

$$y = \frac{k_2 x}{k_3 x^2 + k_{-2}} > 0, \quad \text{if } x > 0.$$

Inserting this expression for y into the second equation of system (12.59), we have

$$\begin{aligned} P = 2k_3^2 x^6(k_1 - k_{-1}) - k_3^2 x^5(4k_1 + 3k_2) + 2k_3 x^4(k_1 k_3 + 2k_1 k_2 + \\ +2k_1 k_{-2} - 2k_{-1} k_{-2}) - k_3 x^3(4k_1 k_2 + 8k_1 k_{-2} + 3k_2 k_{-2}) + \\ +2x^2(k_1(k_2 + k_{-2})^2 + 2k_3 k_1 k_{-2} - k_{-1} k_{-2}^2) - \\ -4k_1 k_{-2}(k_2 + k_{-2}) + 2k_1 k_{-2}^2 = 0. \end{aligned}$$

Now we examine in what interval we should look for the zeros of P. If $x > 0$, $y > 0$, and $1 - x - y > 0$, then

$$1 - x - y = \frac{-k_3 x^3 + k_3 x^2 - x(k_2 + k_{-2}) + k_{-2}}{k_3 x^2 + k_{-2}} > 0,$$

$$P_1 = -k_3 x^3 + k_3 x^2 - x(k_2 + k_{-2}) + k_{-2},$$

$$\begin{array}{ll} P_1(0) = k_{-2} > 0, & P_1(1) = -k_2 < 0, \\ P_1'(0) = -k_2 - k_{-2} < 0 & P_1'(1) = -4k_3 < 0, \\ P_1^{(2)}(0) = 2k_3 > 0, & P_1^{(2)}(1) = -4k_3 < 0, \\ P_1^{(3)}(0) = -6k_3 > 0, & P_1^{(3)}(1) = -6k_3 < 0. \end{array}$$

By the Budan-Fourier theorem the polynomial P can have either one or three zeros in the unit interval [0,1]. As in the previous example the condition for P_1 to have three zeros is given by the inequality $\Delta_3 > 0$. Here

$$\begin{aligned} \Delta_3 = k_3(k_2 + k_{-2})^2 + 18k_3 k_{-2}(k_2 + k_{-2}) - 4k_3^2 k_{-2} - 4(k_2 + k_{-2})^3 - \\ -27k_3 k_{-2}^2 = k_3(k_2^2 + 20k_2 k_{-2} - 8k_{-2}^2) - 4(k_2 + k_{-2})^2 - 4k_3^2 k_{-2} > 0. \end{aligned}$$

So if $\Delta_3 > 0$ and the zeros of P_1 are $x_1 < x_2 < x_3$, then we must look for zeros of P in the segments $[0, x_1]$ and $[x_2, x_3]$. If $\Delta_3 < 0$ then P_1 has one zero x_1, and we must seek for zeros of P in the interval $[0, x_1]$. In any case $P(x_i) < 0$, because

$$P(x_i) = -2k_{-1} x_i^2 (k_3 x_i^2 + k_{-2})^2 - 3k_2 k_3 x_i^2 (k_3 x_i^2 + k_{-2}) < 0.$$

It is equally simple to show that $P'(x_i) < 0$.

We conclude that the polynomial P has at least one zero, but no more than 5 zeros in the above intervals, i.e. the possible numbers of s.s. are 1,3 and 5.

Since the construction of plurality domains is cumbersome in general we consider only the particular case $k_{-2} = 0$. In this case (12.59) has at most three roots in S. Then we have

$$P = 2k_3^2 x^6(k_1 - k_{-1}) - k_3^2 x^5(4k_1 + 3k_2) + 2k_3 x^4(k_1 k_3 + 2k_1 k_2) - k_3 x^3 4k_1 k_2 + 2x^2 k_1 k_2,$$

so $x = 0$ is a zero of multiplicity 2. If we assign a small non-zero value to k_{-2} then this zero splits in two, whereby one of the zeros moves into the domain S, and the second moves outside S. Let us investigate the polynomial

$$\begin{aligned} Q \quad = \quad & 2k_3^2 x^4(k_1 - k_{-1}) - k_3^2 x^3(4k_1 + 3k_2) + \\ & +2k_1 k_3 x^2(k_3 + 2k_2) - 4k_1 k_2 k_3 x + 2k_1 k_2^2 \end{aligned}$$

in the interval $(0,1)$. We have in this case

$$1-x-y=(-k_3x^2+k_3x-k_2)/(k_3x)>0,$$

so that

$$-k_3x^2+k_3x-k_2>0, \quad k_3x^2-k_3x+k_2<0$$

and

$$x_{1,2}=\left(k_3\pm\sqrt{k_3^2-4k_2k_3}\right)\Big/(2k_3).$$

Hence, if $k_3>4k_2$ then the roots $x_{1,2}$ will be real, and moreover,

$$0<x_1=\frac{1}{2k_3}\left(k_3-\sqrt{k_3^2-4k_2k_3}\right)<x_2=\frac{1}{2k_3}\left(k_3+\sqrt{k_3^2-4k_2k_3}\right)<1.$$

For the polynomial Q in the segment $[x_1,x_2]$ we have

1) $Q(x_1)<0$,
$Q'(x_1)<0$,
$Q''(x_1)=4k_1k_3+24k_{-1}k_2k_3+$
$+(12k_3k_{-1}+9k_2k_3)\times$
$\times\sqrt{k_3^2-4k_2k_3}-16k_1k_2k_3-$
$-9k_2k_3^2-12k_{-1}k_3^2$,
$Q^{(3)}(x_1)=-4k_{-1}k_3-3k_2k_3-$
$-4(k_1-k_{-1})\sqrt{k_3^2-4k_2k_3}<0$,
$Q^{(4)}(x_1)>0$,

2) $Q(x_2)<0$,
$Q'(x_2)<0$,
$Q''(x_2)=4k_1k_3+24k_{-1}k_2k_3-$
$-(12k_3k_{-1}+9k_2k_3)\times$
$\times\sqrt{k_3^2-4k_2k_3}-16k_1k_2k_3-$
$-9k_2k_3^2-12k_{-1}k_3^2$,
$Q^{(3)}(x_2)=-4k_{-1}k_3-3k_2k_3-$
$-4(k_1-k_{-1})\sqrt{k_3^2-4k_2k_3}<0$
$Q^{(4)}(x_2)>0$.

In order for Q to have zeros in the segment $[x_1,x_2]$ it is necessary that $Q''(x_1)>0$. If $Q^{(3)}(x_2)<0$ then

$$4(k_1-k_{-1})(k_3^2-4k_2k_3)<(4k_{-1}k_3+3k_2k_3)\sqrt{k_3^2-4k_2k_3},$$

$$Q''(x_2)<-(8k_{-1}k_3+6k_2k_3)\sqrt{k_3^2-4k_2k_3}+4k_{-1}(k_3^2-4k_2k_3)+$$

$$+24k_{-1}k_2k_3-9k_1k_3^2-12k_{-1}k_3^2=-(8k_{-1}k_3+6k_2k_3)\sqrt{k_3^2-4k_2k_3}+$$

$$+8k_{-1}k_2k_3-9k_2k_3^2-8k_{-1}k_3^2.$$

Since $k_2<k_3/4$ it follows that $Q''(x_1)<0$, and hence the right column 2) always has one sign change, while the left column 1) has either one (if $Q''(x_1)<0$) or three sign changes (if $Q''(x_1)>0$). Therefore Q can have two roots in $[x_1,x_2]$. there will then also be two roots outside the segment $[x_1,x_2]$, for the leading coefficient of Q is positive and Q is a polynomial of fourth degree.

Now we form the power sums

$$s_0=4, \quad s_1=-b_1, \quad s_2=b_1^2-2b_2, \quad s_3=-b_1^3+3b_1b_2-3b_3,$$

$$s_4=b_1^4-4b_1^2b_2+4b_1b_3-4b_4+2b_2^2,$$

$$s_5=-b_1^5+5b_1^3b_2-5b_1^2b_3+5b_1b_4-5b_1b_2^2+5b_2b_3,$$

$$s_6=b_1^6-6b_1^4b_2-6b_1^3b_3-6b_1^2b_4+9b_1^2b_2^2-12b_1b_2b_3+b_2b_4-2b_2^3+3b_3^2.$$

Then we get

$$\Delta_0=1, \qquad \Delta_1=4,$$

$$\Delta_2 = 3b_1 - 8b_2 = c(16k_1^2k_3 + 72k_1k_2k_3 + 27k_2^2k_3 - -64k_1^2k_2 + 32k_1k_{-1}k_3 + 64k_1k_{-2}k_2).$$

Taking the condition $k_3 > 4k_2$ into account, we see that $\Delta_2 > 0$.

Thus the polynomial Q has two zeros in the segment $[x_1, x_2]$ if

$$k_3 > 4k_2, \quad Q''(x_1) > 0 \quad \text{and} \quad \Delta_3 > 0.$$

It follows that for small k_{-2} the polynomial P may have three zeros.

Conclusions. For the system (12.45) a necessary condition for the existence of several s.s. is the inequality $p > n$ (Theorem 12.2). We have analyzed particular cases corresponding to irreversibility of the adsorption stages (the first second stages) and to different combinations of the stoichiometric coefficients m, n, p, q. We have also distinguished the situations with one, three or five interior s.s.

Just as in the case of the parallel scheme above, it seems plausible that also the sequential scheme can have no more than 5 s.s. for any values of

$$m, \ n, \ p, \ q.$$

And again the case of five s.s. appears to be quite exotic. At least it can occur only in a very limited domain of parameters k_i. Hence, if in an experiment more than three s.s. (5 or more) with a nonzero reaction rate were discovered, then it would be necessary to describe the process by more complicated schemes than (12.41), containing for example a larger number of stages and reagents.

For a qualitative interpretation of three s.s. it could be sufficient to use of the simple three-stage schemes considered here.

13. The search for all stationary solutions

In the previous section, where we estimated the number of s.s. for kinetic models of short schemes, it was quite sufficient to use elementary mathematical methods. For real transformation schemes whose corresponding systems consist of three or more nonlinear equations it becomes necessary to use the complete arsenal of mathematical tools for analyzing nonlinear system.

In Chapter 2 we proposed an elimination algorithm based on multidimensional logarithmic residues as an effective approach to this problem [**11**].

The classical Kronecker method [**142**] permitting to eliminate all variables except one from a system of nonlinear algebraic equations is well known. This method is useful for proving a number of theorems in theoretical mathematics, but it is inconvenient for practical calculations if the number of equations exceeds 2. A general procedure for solving two algebraic equations, based on the classical method of resultants, is described in [**159**]. However, this approach is limited to the case of 2 equations, since it becomes too cumbersome for systems of three equations. The elimination method based on multidimensional residue theory (see [**11**]) is more efficient. It leads to fewer calculations, and in the case of 3 equations allows one to eliminate the unknown quantities without the use of a computer. This modified elimination algorithm provides the resultant in symbolic form, which is particularly useful when the given system is to be repeatedly solved for different values of the parameters.

Consider the following (rather general) scheme of oxidation on a catalyst Z [**96**]:

$$\begin{array}{ll} 1) & O_2+2Z \rightleftarrows 2ZO, \\ 2) & H_2+2Z \rightleftarrows 2ZH, \\ 3) & H_2+ZO \rightleftarrows Z+H_2O, \\ 4) & ZO+ZH \rightleftarrows Z+ZOH, \\ 5) & ZOH+ZH \rightarrow 2Z+H_2O, \\ 6) & H_2+2ZOH \rightarrow 2Z+2H_2O. \end{array} \tag{13.1}$$

Our mechanism (13.1) is modified in comparison to [**96**]: The possibility of adsorption of a product (H_2O) on the catalyst (for example Pt) is taken into account, i.e. stage 3 is reversible. This stage is monomolecular, in contrast to the bimolecular stages 5 and 6.

Under the assumption of conservation of partial pressures in the gas phase (H_2, O_2, H_2O) the following steady state equation may be written for the mechanism (13.1) (with intermediates Z, ZO, ZH, ZOH):

$$\begin{array}{lcl} 2k_1z^2 - 2k_{-1}x^2 - k_4xy + k_{-4}zu - k_3x + k_{-3}z & = & 0, \\ 2k_2z^2 - 2k_{-2}y^2 - k_4xy + k_{-4}zu - k_5yu & = & 0, \\ k_4xy - k_{-4}zu - k_5yu - 2k_6u^2 & = & 0, \\ x + y + z + u & = & 1, \end{array} \tag{13.2}$$

where x, y, z, u are the concentrations of the intermediates ZO, ZH, Z, ZOH respectively, and the partial pressures O_2, H_2, H_2O enter as factors in the rate constants k_i of the respective stages. The system (13.2) consists of 4 algebraic equations of the second order, the first 3 of which are nonlinear. Therefore, in general (13.2) may have up to 8 solutions (according to the Bézout theorem). We are only interested in non-negative solutions lying in the reaction simplex

$$S = \{(x,y,z,u): \ x,y,z,u \geq 0, \ x+y+z+u = 1\}. \tag{13.3}$$

Those roots $(x,y,z,u) \in S$ which have some zero coordinates will be called boundary s.s. (they are situated on the boundary of the simplex (13.3)), whereas interior s.s. correspond to roots $(x,y,z,u) \in S$ with positive coordinates $x,y,z,u > 0$. Note that if all $k_i > 0$, then (13.2) has no boundary s.s.

By eliminating $u = 1 - x - y - z$ and making the substitution $z = ty$, we can write system (13.2) as

$$\begin{array}{lcl} f_1 & = & a_1x^2 + a_2xy + a_3y^2 + a_4x + a_5y + a_6 = 0, \\ f_2 & = & b_1x^2 + b_2xy + b_3y^2 + b_4x + b_5y + b_6 = 0, \\ f_3 & = & c_1x + c_2y + c_3 = 0, \end{array} \tag{13.4}$$

where

$$a_1 = 2k_{-1}, \quad a_2 = k_4 + k_{-4}t, \quad a_3 = (k_{-4} - 2k_1)t^2 + k_{-4}t,$$

$$a_4 = k_3, \quad a_5 = -(k_3 + k_{-4})t, \quad a_6 = 0, \quad b_1 = k_6,$$

$$b_2 = 2k_6t + 2k_6 - k_5, \quad b_3 = (k_6 - k_2)t^2 + (2k_6 - k_5)t + k_6 + k_{-2} - k_5,$$

$$b_4 = 2k_6, \quad b_5 = -2k_6t + k_5 - 2k_6, \quad b_6 = k_6, \quad c_1 = k_5 - k_4 - k_{-4}t,$$

$$c_2 = (2k_2 - k_4)t^2 + (k_5 - k_{-4})t + k_5 - 2k_{-2}, \quad c_3 = k_{-4}t - k_5.$$

The coefficients a_i, b_i, c_i depend on the new variable t. Now we denote

$$X = c_1 x + x_2 y + c_3, \quad Y = y,$$

so that

$$x = \frac{X - c_2 Y - c_3}{c_1} \quad y = Y.$$

In the new variables X, Y system (13.4) can be written as

$$\begin{aligned} f_1 &= A_1X^2 + A_2XY + A_3Y^2 + A_4X + A_5Y + A_6 = 0, \\ f_2 &= B_1X^2 + B_2XY + B_3Y^2 + B_4X + B_5Y + B_6 = 0, \\ f_3 &= X = 0, \end{aligned} \tag{13.5}$$

where

$$A_1 = a_1, \quad A_2 = a_2c_1 - 2a_1c_2,$$

$$A_3 = a_1c_2^2 - a_2c_1c_2 + a_3c_1^2, \quad A_4 = a_4c_1 - 2a_1c_3,$$

$$A_5 = 2a_1c_2c_3 - a_2c_1c_3 - a_4c_1c_2 + a_5c_1^2, \quad A_6 = a_1c_3^2 - a_4c_1c_3 + a_6c_1^2,$$

$$B_1 = b_1, \quad B_2 = b_2c_1 - 2b_1c_2, \quad B_3 = b_1c_2 - b_2c_1c_2 + b_3c_1^2,$$

$$B_4 = b_4c_1 - 2b_1c_3, \quad B_5 = 2b_1c_2c_3 - b_2c_1c_3 - b_4c_1c_2 + b_5c_1^2,$$

$$B_6 = b_1c_3^2 - b_4c_1c_3 + b_6c_1^2.$$

We are not going to use the concrete form of the coefficients A_i, B_i and a_i, b_i, c_i. The resultant obtained for (13.4) and (13.5) (in symbolic form) will be suitable also for more general systems than (13.2). Since there are only two nonlinear equations in (13.5), one could in principle apply the classical elimination method in this case. However, we shall here illustrate our modified approach which gives an alternative way of obtaining the resultant explicitly.

Consider the first two equations from (13.5)

$$\begin{aligned} f_1(X, Y) = 0, \\ f_2(X, Y) = 0, \end{aligned} \tag{13.6}$$

where $\deg f_1 = \deg f_2 = 2$. Then, following our general elimination scheme from Chapter 2, we write

$$\begin{aligned} P_1 = A_1X^2 + A_2XY + A_3Y^2, \quad Q_1 = A_4X + A_5Y + A_6, \\ P_2 = B_1X^2 + B_2XY + B_3Y^2, \quad Q_2 = B_4X + B_5Y + B_6. \end{aligned}$$

Using the results obtained when solving Example 8.3, we get from the system

$$\begin{aligned} a_{11}P_1 + a_{21}P_2 &= X^3, \\ a_{12}P_1 + a_{22}P_2 &= Y^3, \end{aligned}$$

the expressions

$$a_{11} = \frac{1}{R}\left(\begin{vmatrix} A_1 & A_2 & A_3 \\ B_2 & B_3 & 0 \\ B_1 & B_2 & B_3 \end{vmatrix} X - B_3 \begin{vmatrix} A_2 & A_3 \\ B_2 & B_3 \end{vmatrix} Y\right),$$

$$a_{21} = \frac{1}{R}\left(B_1 \begin{vmatrix} A_1 & A_2 \\ B_1 & B_2 \end{vmatrix} X + \begin{vmatrix} A_1 & A_2 & A_3 \\ B_1 & B_2 & B_3 \\ 0 & B_1 & B_2 \end{vmatrix} Y\right),$$

$$a_{12} = \frac{1}{R}\left(\begin{vmatrix} A_2 & A_3 & 0 \\ A_1 & A_2 & A_3 \\ B_1 & B_2 & B_3 \end{vmatrix} X + A_3 \begin{vmatrix} A_2 & A_3 \\ B_2 & B_3 \end{vmatrix} Y\right),$$

$$a_{22} = \frac{1}{R}\left(-A_1 \begin{vmatrix} A_1 & A_2 \\ B_1 & B_2 \end{vmatrix} X + \begin{vmatrix} A_1 & A_2 & A_3 \\ 0 & A_1 & A_2 \\ B_1 & B_2 & B_3 \end{vmatrix} Y\right),$$

where

$$R = \begin{vmatrix} A_1 & A_2 & A_3 & 0 \\ 0 & A_1 & A_2 & A_3 \\ B_1 & B_2 & B_3 & 0 \\ 0 & B_1 & B_2 & B_3 \end{vmatrix} \not\equiv 0.$$

Calculating a_{ij} and R, we get

$$\begin{aligned} a_{11} &= \frac{1}{R}\left((A_3B_2^2 + A_1B_3^2 - A_2B_2B_3 - A_3B_1B_3)X - B_3(A_2B_3 - A_3B_2)Y\right), \\ a_{12} &= \frac{1}{R}\left(B_1(A_1B_2 - A_2B_1)X + (A_1B_2^2 + A_3B_1^2 - A_1B_1B_3 - A_2B_1B_2)Y\right), \\ a_{21} &= \frac{1}{R}\left((A_2^2B_3 + A_3^2B_1 - A_2A_3B_2 - A_1A_3B_3)X - A_3(A_2B_3 - A_3B_2)Y\right), \\ a_{22} &= \frac{1}{R}\left(A_1(A_2B_1 - A_1B_2)X + (A_1^2B_1 + A_1A_3B_1 - A_1A_2B_2 - A_2^2B_1)Y\right), \end{aligned}$$

$$R = (A_1B_3 - A_3B_1)^2 - (A_1B_2 - A_2B_1)(A_2B_3 - A_3B_2).$$

Furthermore,

$$\Delta_1 = \frac{D(f_1, f_2)}{D(X, Y)}$$

and

$$\Delta_2 = \begin{vmatrix} a_{11} & a_{12} \\ a_{21} & a_{22} \end{vmatrix},$$

$$R_1 = a_{11}Q_1 + a_{21}Q_2, \quad R_2 = a_{12}Q_1 + a_{22}Q_2.$$

If we write

$$d_{ij} = \begin{vmatrix} A_i & A_j \\ B_i & B_j \end{vmatrix} = A_iB_j - A_jB_i,$$

then

$$\begin{aligned}
\Delta_1 &= 2X^2d_{12}+4XYd_{13}+2Y^2d_{23}+X(2d_{15}-d_{24})+Y(d_{25}-2d_{34})+d_{45},\\
\Delta_2 &= \frac{1}{R}(-X^2d_{12}+XYd_{13}-Y^2d_{23}),\\
R_1 &= \frac{1}{R}(X^2(d_{24}d_{23}-d_{34}d_{13})+XY(d_{34}d_{23}+d_{25}d_{23}-d_{35}d_{13})+\\
&\quad +Y^2d_{35}d_{23}+X(d_{26}d_{23}-d_{36}d_{13})+Yd_{36}d_{23}),\\
R_2 &= \frac{1}{R}(-X^2d_{14}d_{12}+XY(d_{14}d_{13}-d_{24}d_{12}-d_{15}d_{12})+\\
&\quad +Y^2(d_{15}d_{13}-d_{25}d_{12})-Xd_{16}d_{12}+Y(d_{16}d_{13}-d_{23}d_{12})).
\end{aligned}$$

Now, for the power sums S_j, $j=1,2,3,4$, we have

$$S_j=\mathfrak{N}\left(X^j\Delta_1\Delta_2\sum_{\alpha_1+\alpha_2\le j}(-1)^{\alpha_1+\alpha_2}R_1^{\alpha_1}R_2^{\alpha_2}\overline{X}^{3\alpha_1}\overline{Y}^{3\alpha_2}\right), \tag{13.7}$$

where the linear functional $\mathfrak{N}$ is equal to 1 on the monomials $X^{3\alpha_1+2}Y^{3\alpha_1+2}\overline{X}^{3\alpha_1}\overline{Y}^{3\alpha_2}$ and 0 on the others (see [**11**, Sec. 26]). A computation shows that

$$\begin{aligned}
S_1 &= \frac{1}{R}d_{23}(2d_{15}-d_{24}),\\
S_2 &= \frac{2}{R}(d_{15}d_{35}-d_{23}d_{26}-d_{34}^2-d_{25}d_{34})+\frac{1}{R^2}d_{23}^2d_{24}^2,\\
S_3 &= -\frac{3d_{36}}{R}(2d_{34}+d_{25})+\frac{d_{23}}{R^2}(3d_{23}d_{24}d_{26}+3d_{24}d_{34}^2+2d_{15}^2d_{35}-\\
&\quad -6d_{14}d_{34}d_{35}-3d_{14}d_{25}d_{35}+3d_{24}d_{25}d_{34})-\frac{d_{23}^3d_{24}^3}{R^3},\\
S_4 &= -\frac{4d_{39}^2}{R}+\frac{1}{R^2}(2d_{23}^2d_{26}^2-16d_{23}d_{15}d_{34}d_{36}+d_{23}d_{24}d_{35}d_{45}-\\
&\quad -4d_{23}d_{16}d_{25}d_{35}-4d_{23}d_{16}d_{25}d_{35}-d_{23}d_{15}d_{35}d_{45}+8d_{23}d_{24}d_{34}d_{36}+\\
&\quad +4d_{23}d_{26}d_{34}^2+4d_{23}d_{25}d_{26}d_{34}+4d_{23}d_{24}d_{25}d_{36}+d_{15}d_{24}d_{35}^2+2d_{15}^2d_{35}^2-\\
&\quad -d_{24}^2d_{35}^2-6d_{15}d_{25}d_{34}d_{35}+5d_{24}d_{25}d_{34}d_{35}-6d_{14}d_{34}d_{35}^2+d_{14}d_{25}d_{35}^2+\\
&\quad +8d_{24}d_{34}^2d_{35}-6d_{15}d_{34}^2d_{35}+2d_{25}^2d_{34}^2+2d_{34}^4+4d_{25}d_{34}^3)-\\
&\quad -\frac{4d_{23}^2d_{24}^2}{R^3}(d_{23}d_{26}+d_{34}^2+d_{25}d_{34})+\frac{d_{23}^4d_{24}^4}{R^4}.
\end{aligned}$$

Here the expressions for S_j are given under the assumption $d_{13}=0$, which causes no loss of generality. For calculating S_j we have used the relations

$$d_{ij}d_{ki}\pm d_{ik}k_{jl}=\pm d_{ij}k_{jk},\quad i\ne j\ne k\ne l.$$

For example, $d_{12}d_{34}-d_{13}d_{24}=d_{14}d_{23}$.

The resultant of system (13.5) can be computed by means of the Newton recursion formulas (2.1):

$$r=6S_4+S_1S_2+6S_1^2S_2-3S_2^2-S_1^4.$$

Inserting the expressions for S_j, we have

$$r = \frac{-24(d_{36}^2 - d_{35}d_{56})}{R},$$

and after one more substitution this becomes

$$r = -\frac{24}{R}((A_3B_6 - A_6B_3)^2 - (A_3B_5 - A_5B_3)(A_5B_6 - A_6B_5)). \qquad (13.8)$$

If we now assign an arbitrary value to d_{13} this expression will not change. The resultant (13.8) is a rational function (a quotient of two polynomials) in one variable t. After cancelling out common factors in numerator and denominator what is left in the numerator will be the desired resultant in t for (13.5).

For (13.4) we have

$$r = -\frac{24R_1^*}{R_2^*},$$

where, in the notation

$$e_{ij} = \begin{vmatrix} a_i & a_j \\ b_i & b_j \end{vmatrix} = a_ib_j - a_jb_i,$$

the numerator and denominator have the form

$$R_1^* = c_1^4(e_{36}^2 - e_{35}e_{56}) + c_1^2c_2(e_{25}e_{56} + e_{34}e_{56} + e_{34}e_{56} + e_{46}e_{35} -$$
$$-2e_{26}e_{36}) + c_1^3c_3(e_{26}e_{35} - e_{35}e_{45} - e_{23}e_{56} - 2e_{34}e_{36}) + c_1^2c_2^2(e_{26}^2 +$$
$$+2e_{16}e_{36} - e_{15}e_{56} - e_{24}e_{56} - e_{25}e_{46}) + c_1^2c_2c_3(3e_{24}e_{36} - e_{25}e_{26} + e_{25}e_{45} +$$
$$+e_{13}e_{56} + e_{34}e_{45} - 2e_{16}e_{35}) + c_1^2c_3^2(e_{23}e_{36} - 2e_{13}e_{36} - e_{45}e_{23} -$$
$$-e_{24}e_{35} + e_{15}e_{35} + e_{34}^2) + c_1c_2^3(e_{14}e_{56} - 2e_{16}e_{56} + +e_{15}e_{46} + e_{24}e_{46}) +$$
$$+c_1c_2^2c_3(3e_{16}e_{25} - 2e_{14}e_{36} - e_{34}^2 - e_{12}e_{56} - e_{24}e_{45} - 2e_{13}e_{46}) +$$
$$+c_1c_2c_3^2(2e_{12}e_{36} - e_{24}e_{34} + e_{25}e_{24} - e_{15}e_{25} + 3e_{13}e_{45} - 2e_{36}e_{16}) +$$
$$+c_1c_3^3(2e_{13}e_{34} - e_{24}e_{23} + e_{23}e_{15} - e_{12}e_{35}) + c_2^2c_3^2(2e_{13}e_{16} - e_{12}e_{45} -$$
$$-e_{14}e_{25}) + c_2c_3^3(e_{12}e_{25} - e_{34}e_{12} - 2e_{15}e_{13} + e_{23}e_{14}) + c_3^4(e_{13}^2 - e_{12}e_{24}),$$
$$R_2^* = e_{13}^2 - e_{12}e_{23}.$$

If R_1^* and R_2^* have no common factors then the equation

$$R_1^*(t) = 0 \qquad (13.9)$$

will not contain X or Y. Thus the method of elimination has been carried out completely for the variables x, y, t of system (13.4), and the resultant is written in the symbolic form (13.9).

Let us indicate some simple generalizations of the just obtained result.

1. The elimination scheme described above may be applied to (13.4), where $f_3 = (c_1x + c_2y + c_3)^2$. In this case the power sums (13.7) for the new system are $\tilde{S}_1 = S_2$, $\tilde{S}_2 = S_4$, where S_2, S_4 are the previously calculated sums (in x^2, x^4). The power sums corresponding to x^6 and x^8 remain to be calculated. In order to simplify the procedure a square should be cancelled in $f_3 = 0$. The roots of (13.5) will have double multiplicity compared to the roots of the new system.

2. All the above conclusions are valid for $f_3 = (c_1x+c_2y+c_3)^n$, $n \geq 3$. Hence, the resultant (13.9) obtained for system (13.4) is accurate also for a system of 3 algebraic

equations, one of which is a power of a linear function in two of the variables and the other two are quadratic forms in these variables.

3. If there is a linear transformation of the quadratic forms in the initial system of 3 algebraic equations which turns one of the functions, f_3 say, into a perfect square, then all the above results are valid for the transformed system.

4. The coefficients a_i, b_i, c_i in (13.4) may contain any powers of the distinguished variable t, because in calculating (13.9) the particular dependence of $a_i(t)$, $b_i(t)$, $c_i(t)$ was not used. Therefore the resultant $R_1^*(t)$ is valid also for systems containing greater powers. In this case the variable t should be such that f_1, f_2 and $f_3 = c_1x + c_2y + c_3$ are quadratic forms with respect to x and y.

It is not quite clear whether a general system of 3 algebraic equations, each of which is of the second order, could be transformed into (13.4) (even by a nonlinear change of variables). It is however quite likely that a general system of 3 quadratic equations always can be transformed into (13.4), with $f_3 = c_1x + c_2y + c_3$ or $f_3 = (c_1x + c_2y + c_3)^2$.

The polynomial $R_1^*(t)$ in (13.9) has been written for system (13.2) with arbitrary coefficients a_i, b_i, c_i. We can write the coefficients of $R_1^*(t)$ in terms of the rate constants k_i, substituting a_i, b_i, c_i into R_1^* and making the necessary algebraic transformations. The final result will not be given here. Quite simple expressions are obtained if $k_6 = 0$, i.e. if step 6 is absent in the scheme (13.1). In this case we have the factorization $R_1^*(t) = t^2P(t)$, and

$$P(t) = P_0t^6 + P_1t^5 + \ldots + P_5t + P_6, \tag{13.10}$$

where

$$\begin{aligned}
P_0 &= k_2^2k_{-4}^2(2k_{-1} + k_3),\\
P_1 &= k_2k_{-4}(4k_2k_5k_{-1} + 2k_2k_3k_5 + k_2k_3k_4),\\
P_2 &= 2k_2^2k_5^2k_{-1} + k_2^2k_3k_5^2 + k_2^2k_3k_4k_5 + k_2k_3k_4k_5k_{-4} - 4k_2k_{-1}k_{-2}k_{-4}^2 -\\
&\quad -2k_2k_3k_{-2}k_{-4}^2 - k_2k_4k_5k_{-3}k_{-4},\\
P_3 &= k_2k_3k_4k_5^2 + k_2k_3k_4k_5k_{-4} - 8k_2k_5k_{-1}k_{-2}k_{-4} - 4k_2k_3k_5k_{-2}k_{-4} -\\
&\quad -2k_2k_3k_4k_{-2}k_{-4} - k_2k_4k_5^2k_{-3} - k_2k_4^2k_5k_{-3},\\
P_4 &= 2k_{-1}k_{-2}^2k_{-4} + k_2k_4^2k_5^2 + k_3k_{-2}^2k_{-4}^2 + k_2k_3k_4k_5^2 + k_4k_5k_{-2}k_{-3}k_{-4} -\\
&\quad -2k_1k_4^2k_5^2 - 4k_2k_5^2k_{-1}k_{-2} - k_3k_4k_5k_{-2}k_{-4} - 2k_2k_3k_4k_5k_{-2} -\\
&\quad -2k_2k_3k_5^2k_{-2} - k_4^2k_5^2k_{-3},\\
P_5 &= 4k_5k_{-1}k_{-2}^2k_{-4} + 2k_3k_5k_{-2}^2k_{-4} + k_3k_4k_{-2}^2k_{-4} + k_4^2k_5k_{-2}k_{-3} +\\
&\quad +k_4k_5^2k_{-2}k_{-3} - k_3k_4k_5^2k_{-2} - k_3k_4k_5k_{-2}k_{-4} - k_4^2k_5^2k_{-3},\\
P_6 &= k_{-2}k_5(2k_{-1}k_{-2}k_5 + k_3k_4k_{-2} + k_3k_5k_{-2} - k_4^2k_5 - k_3k_4k_5).
\end{aligned}$$

If $t = 0$, so that $x = 0$, we get from (13.2) (still with $k_6 = 0$) the two simple roots $r_1 = (0, 0, 0, 1)$ and

$$r_2 = \left(\frac{-k_3}{2k_1}, 0, 0, \frac{2k_{-1} + k_3}{2k_{-1}}\right).$$

The root r_1 is on the boundary of S, and the root r_2 is outside the simplex S. For finding the other 6 roots of system (13.2) it is necessary first to solve the equation

$P(t) = 0$. The remaining coordinates of the roots may then be found using (if necessary) the method described in [4] (see also Ch. 2, Sec. 11).

Let us first consider the number of positive roots of system (13.2). We will have to apply the Descartes theorem and the Budan–Fourier theorem. It is evident, that a positive solution of system (13.2) corresponds to a root t of the equation $P(t) = 0$ precisely if $t \geq t_0$, where $t_0 = \sqrt{k_{-2}/k_2}$ is the only positive root of the equation

$$Q(t) = k_{-4}k_2t^3 + k_2k_5t^2 - k_{-2}k_{-4}t - k_{-2}k_5 = 0. \tag{13.11}$$

Equation (13.11) is obtained from the system (13.2) with $k_6 = 0$. It follows that the number of positive solutions of system (13.2), counted with multiplicity, is equal to the number of zeros of the polynomial $P(t)$ to the right of t_0. It is easy to see that $P(t_0) < 0$. Consequently, in the interval (t_0, ∞) the polynomial $P(t)$ has an odd number (at least 1) of zeros. On the other hand, the total number of zeros does not exceed 3, since in the sequence $P(t_0)$, $P'(t_0)$, ..., $P^{(6)}(t_0)$ the sign cannot change more than three times.

Let us show that for $k_{-3} = 0$ the system (13.2) has at most one positive root. We have

$$\begin{aligned}
P(t_0) &< 0, \\
P'(t_0) &= k_3k_{-2}k_{-4} + k_3k_5k_{-2} + t_0(k_3k_{-2}k_{-4} + k_2k_3k_5 - 2k_1k_4k_5 + \\
&\quad + k_2k_4k_5), \\
P''(t_0) &= 8k_{-1}k_{-2}^2k_{-4}^2 + 4k_3k_{-2}^2k_{-4}^2 + 8k_2k_5^2k_{-1}k_{-2} + 4k_2k_3k_5^2k_{-2} + \\
&\quad + 4k_2k_3k_4k_{-2} + 5k_3k_4k_5k_{-2}k_{-4} + k_2k_3k_4k_5^2 + k_2k_4^2k_5^2 - \\
&\quad - 2k_1k_4^2k_5^2 + t_0(16k_2k_5k_{-1}k_{-2}k_{-4} + 8k_2k_3k_5k_{-2}k_{-4} + \\
&\quad + 4k_2k_3k_4k_{-2}k_{-4} + 3k_2k_3k_4k_5^2 + 3k_2k_3k_4k_5k_{-4}),
\end{aligned}$$

$$P^{(3)}(t_0) > 0, \quad P^{(4)}(t_0) > 0, \quad P^{(5)}(t_0) > 0, \quad P^{(6)}(t_0) > 0,$$

where the equalities are to be understood up to some positive factors. If $P'(t_0) > 0$, then $P''(t_0) > 0$ since $P''(t_0) - k_4k_5P'(t_0) > 0$. So there is only one sign change in this sequence.

Next we show that in the interval (t_0, ∞) the equation $P(t) = 0$ may have three distinct positive roots. Letting $k_{-1} = k_{-2} = k_{-4} = 0$, we have $t_0 = 0$ and $P(t) = tR(t)$, where

$$\begin{aligned}
R(t) &= k_2^2k_3(k_4 + k_5)t^3 + k_2k_4(k_3k_5 - k_5k_{-3} - k_4k_{-3})t^2 + \\
&\quad + k_4k_5(k_2k_3 + k_2k_4 - 2k_1k_4 - k_4k_{-3})t - k_4^2k_5k_{-3}.
\end{aligned}$$

The boundary roots $(0, 0, 0, 1)$ and $(0, 1, 0, 0)$ correspond to the solution $t = 0$. The polynomial $R(t)$ has three positive zeros, if all its roots are real and if there are three sign changes in its coefficient sequence. In order that all the roots of a polynomial $f(t)$ of degree n be simple and real, it is necessary and sufficient that the quadratic form with matrix

$$\begin{pmatrix} n & S_1 & \dots & S_n \\ S_1 & S_2 & \dots & S_{n+1} \\ \dots & \dots & \dots & \dots \\ S_n & S_{n+1} & \dots & S_{2n-2} \end{pmatrix},$$

(where S_j are the power sums of the zeros of the polynomial $f(x)$) be positive. In our case ($n = 3$) the sums S_j are easily calculated by the Newton formulas (2.1) and the necessary and sufficient requirements for the polynomial $R(t)$ to have three distinct positive roots become

$$\begin{aligned} &K_3K_5 - K_5K_{-3} - K_4K_{-3} < 0\\ &K_2K_4 - 2K_1K_4 + K_2K_3 - K_4K_{-3} > 0,\\ &K_1^2 - 3K_0K_2 > 0,\\ &K_1^2K_2^2 + 18K_0K_1K_3 - 4K_0K_2^3 - 27K_0^2K_3^2 - 4K_1^3K_3 > 0, \end{aligned} \tag{13.12}$$

where K_i is the coefficient of t^i in the polynomial $R(t)$. Some of the solutions to the inequalities (13.12) may be found by considering the following equations

$$K_1 = -K_2, \quad K_0 = -K_3, \quad K_1 = -\alpha K_0, \quad \alpha \geq 4. \tag{13.13}$$

Each solution of (13.13) satisfies (13.12). In particular, one may take $k_1 = 0.444$, $k_2 = 3.91$, $k_3 = 1$, $k_4 = 25$, $k_5 = 1$ and $k_{-3} = 0.636$. In this case $R(t)$ has three positive zeros and the system (13.2) has the three positive solutions

$$\begin{aligned} &x_1 = 0.0112, \quad y_1 = 0.5177, \quad z_1 = 0.1922, \quad u_1 = 0.2788,\\ &x_2 = 0.0287, \quad y_2 = 0.1118, \quad z_2 = 0.1431, \quad u_2 = 0.7164,\\ &x_3 = 0.0332, \quad y_3 = 0.0426, \quad z_3 = 0.0951, \quad u_3 = 0.8301, \end{aligned} \tag{13.14}$$

Moreover, (13.2) also has the boundary solution $x = y = z = 0$, $u = 1$ of multiplicity 2. If k_{-i}, $k_6 > 0$ these last roots may produce two more solutions inside S.

As this example shows, the reversibility of the product building step can give rise to multiple steady states.

Thus, in the case studied above there is only one interior s.s. for $k_{-3} = 0$, while for $k_{-3} \neq 0$ three such steady states (13.14) are possible.

Let us consider the case $k_3 = k_{-3} = 0$ separately. (Step 3 is then absent in the scheme (13.1).) We will show that system (13.2) then has a unique positive solution. A special feature of (13.2) in this aces is that the first three equations are homogeneous with respect to the variables x, y, z, u. So, in order to analyze the number of positive solutions we may set $u = 1$ in these equations. Instead of (13.2) we then have the system

$$\begin{aligned} (2k_1 - k_2)z^2 - 2k_{-1}x^2 + k_{-2}y^2 - k_6 &= 0,\\ k_2z^2 - k_{-2}y^2 - k_5y - k_6 &= 0,\\ k_4xy - k_{-4}z - k_5y - 2k_6 &= 0. \end{aligned} \tag{13.15}$$

It may be easily shown that the equations obtained by solving for z in the third equation and substituting it into the first two, give rise to two monotone curves, one increasing and one decreasing, in the first quadrant of the (x, y)-plane. There is hence a unique intersection between the curves, and therefore at most one positive solution of system (13.15).

Let us examine two particular mechanisms which have been discussed in the literature [**96**]:

1) $O_2+2Z \to 2ZO$,	1) $O_2+2Z \to 2ZO$,
2) $H_2+2Z \rightleftarrows 2ZH$,	2) $H_2+2Z \rightleftarrows 2ZH$,
3) $H_2+ZO \to Z+H_2O$,	3) $H_2+ZO \to Z+H_2O$,
4) $ZO+ZH \to ZOH+Z$,	4) $ZO+ZH \to ZOH+Z$,
5) $ZH+ZOH \to 2Z+H_2O$;	5) $H_2+2ZOH \to 2Z+2H_2O$.

Here the enumeration of the stages agrees with (13.1). The uniqueness for the interior s.s. of the kinetic model corresponding to the first mechanism has already been established above in a somewhat more general case. The same conclusion is also valid for the second mechanism, with the corresponding stationary kinetic model (13.2) satisfying $k_{-1} = k_{-3} = k_{-4} = k_5 = 0$. The elimination of variables can in this particular case be done with elementary methods. Indeed, from the first three equations of system (13.2) it is easy to obtain

$$x = \frac{2k_1k_{-2}y^2}{k_1k_3 + k_4(k_2 - k_1)y},$$

$$z^2 = \frac{x(k_3 + k_4y)}{2k_1}, \quad u^2 = \frac{k_3xy}{2k_6}.$$

Since there are no boundary s.s. in this case, so that only positive solutions are available, we must have

$$k_2k_3 + k_4(k_2 - k_1)y > 0.$$

Under this condition it is easily shown that $z'_y, x'_y, u'_y > 0$, and hence the interior s.s. is unique.

Now we add one more stage in (13.1), namely

$$7)\ ZO+2ZH \to 3Z+H_2O.$$

Taken together, the stages 1)–7) provide a fairly complete description of today's understanding of the mechanism for catalytic hydrogen oxidation. Steps 1)–3) and 7) represent one of the possible particular mechanisms. It may be shown here that there is only one positive s.s. for $k_{-3} = 0$, even though the scheme is substantially nonlinear. The occurrence of several s.s. is ensured by the reversibility of step 3). The same phenomenon takes place in the more general case 1)–7). For simplicity we consider irreversible steps 1)–7) having as corresponding system

$$\begin{aligned} 2k_1z^2 - k_4xy - k_3x - k_7xy^2 &= 0, \\ 2k_2z^2 - k_4xy - k_5yu - 2k_7xy^2 &= 0, \\ k_4xy - k_5yu - 2k_6u^2 &= 0, \\ x + y + z + u &= 1. \end{aligned} \qquad (13.16)$$

We add the third equation to the first and second ones, then we divide the second equation by 2, and subtract it from the first one. We thus obtain

$$(2k_1 - k_2)z^2 - k_3x - k_6u^2 = 0,$$

from which we see that system (13.6) has no positive roots for $2k_1 \le k_2$. For $2k_1 > k_2$ there may exist a positive solution, apart from the boundary solution $z = u = x = 0$,

$y = 1$. After simple transformations one obtains

$$\begin{aligned} x &= u(k_5 y + 2k_6 u)/(k_4 y), \\ 2k_1 k_4 z^2 &= u(k_5 y + 2k_6 u)(k_7 y^2 + k_4 y + k_3)/y, \\ u &= \frac{k_5}{2k_6} y \cdot \frac{k_7(2k_1 - k_2)y^2 + k_4(2k_1 - k_2)y - k_2 k_3}{k_7(k_2 - 2k_1)y^2 + k_4(k_2 - k_1)y + k_2 k_3}. \end{aligned} \tag{13.17}$$

These functions x, z, u of the variable y are monotonically increasing in the domain of variation of y (recall that x, z, u are assumed to be positive). On the other hand, the function $z = 1 - z - y - u$ is decreasing, and hence the system (13.16) can only have one positive solution.

Thus, in all considered cases the reversibility of step 3), the water producing step, proves to be responsible for the plurality of s.s., which in turn may result in a complex dynamical behavior of the reaction. Hence, in the multi-stage mechanism at hand the reversibility constitutes a natural "feedback". As seen in the above examples, the full procedure for elimination of variables is sometimes not needed for analyzing the number of possible solutions. For this it may be sufficient to make some indirect considerations, for example, to establishment the monotony of dependencies such as (13.17). However, in cases when several s.s. can occur, the construction of the one-variable resultant may be effective for the determination of their number, and also for a parametric analysis of the solutions. As we have shown in this section, this may be done "manually" without the use of a computer. The main appeal of our approach is that one sometimes manages to perform a qualitative study of the obtained resultant with letter coefficients. As an example of this, we have in this section been able to reveal the crucial role played by the reversibility of the water producing step for the appearance of multiple steady states.

14. The kinetic polynomial. Single-route mechanisms

The concept of kinetic polynomials was introduced in the articles [**111, 112, 155**], and in [**38, 39, 109, 113, 154, 156**] it was further developed and applied to the kinetics of catalytic reactions. The mathematical basis for the new approach was presented in [**37, 40, 99**], and here we provide the complete proofs of the fundamental assertions. On the basis of results from [**98**] we give a simplified algorithm for the determination of the coefficients in the kinetic polynomial, and analyze in detail its use for a typical transformation scheme of catalytic reactions.

In the stationary case the basic kinetic equation for catalytic reactions can be written as

$$\mathbf{\Gamma}^T \mathbf{w}(\mathbf{z}) = \mathbf{0}, \quad \mathbf{A}\mathbf{z} = \mathbf{a}, \tag{14.1}$$

where $\mathbf{\Gamma}$ is a matrix of stoichiometric vectors for the stages of the initial mechanism of the reaction, $\mathbf{w}(\mathbf{z})$ is the rate vector of the stages, $\mathbf{z}$ is the vector of concentrations of the reagents, $\mathbf{A}$ is a matrix with entries proportional to the molecular weights of the reagents, and $\mathbf{a}$ is a vector of balance quantities. The relations $\mathbf{Az} = \mathbf{a}$ correspond to the conservation laws $\mathbf{\Gamma A} = 0$. Moreover, the catalytic transformation scheme is cyclic precisely if there is a linear dependence between the rows of $\mathbf{\Gamma}$ arising from a linear dependence of the columns of $\mathbf{A}$, i.e. there exist non-zero vectors $\boldsymbol{\nu}$ such that

$\nu\Gamma = 0$. The vectors ν are made up from stoichiometric numbers and correspond to different routes in the original transformation scheme.

For a n-stage single-route catalytic reaction the system (14.1) takes the form

$$w_s(\mathbf{z}) = \nu_s w, \quad s = 1, \ldots, n, \quad L(\mathbf{z}) = 1, \tag{14.2}$$

where ν_s is the stoichiometric number for the s-th stage, w is the observable reaction rate, n is the number of intermediates, and $L(\mathbf{z})$ is a linear function. The equations (14.2) are in general nonlinear with respect to $\mathbf{z}$. For example, if $w_s^{\pm}$ are found according to the mass-action law, then (14.1) and (14.2) are systems of algebraic equations. By eliminating $\mathbf{z}$ from (14.2) one obtains a new representation for the stationary equations in terms of the quantity $\mathbf{w}$, which can be measured in experiments. In [**111, 155**] this new representation was called the kinetic polynomial.

In the simplest linear case the kinetic polynomial, then having the form $B_1 w + B_0 = 0$, can be easily found by using methods from graph theory [**70, 145, 146, 147, 148, 153**]. For example, for a linear single-route mechanism [**145**]

$$B_0 = \prod_{i=1}^{n} b_i^- - \prod_{i=1}^{n} b_i^+, \ B_1 = \sum_x D_x, \tag{14.3}$$

where $b_i^{\pm}$ are the weights of the forward and reverse reaction at the i-th stage, and D_x is the weight of the vertex x in the reaction graph (see e.g. [**150**]). The coefficient B_0 in (14.3) corresponds to the kinetic equation of a brutto-reaction (in the case of a stage, one for which the mass-action law is valid). In equilibrium (when $w = 0$) one has $B_0 = 0$.

The equations (14.1) and (14.2) are in general nonlinear, and we shall apply the special elimination method which we developed in Chapter 2. The kinetic polynomial can be represented in the following form ($w_s^{\pm}$ are found in accordance with the mass-action law):

$$R(w) = B_N w^N + B_{N-1} w^{N-1} + \ldots + B_2 w^2 + B_1 w + B_0 = 0. \tag{14.4}$$

The degree of the polynomial $R(w)$ is determined by the nonlinearity of system (14.2) and its coefficients are certain functions of the rate constants. $R(w)$ is the resultant of system (14.2) with respect to w (see, for example, [**137**]).

We thus consider the system of equations

$$b_s^+ \prod_{i=1}^{n} z_i^{\alpha_i^s} - b_s^- \prod_{i=1}^{n} z_i^{\beta_i^s} = \nu_s w, \ s = 1, \ldots, n, \ z_1 + \ldots + z_n = 1, \tag{14.5}$$

where the constants $b_s^{\pm}$ correspond to the reaction weights of the forward and reverse directions at the s-th stage, z_i are concentrations, and α_i^s, β_i^s are stoichiometric coefficients (nonnegative integers). For convenience we rewrite (14.5) in vector form:

$$b_s^+ \mathbf{z}^{\alpha^s} - b_s^- \mathbf{z}^{\beta^s} = \nu_s w, \qquad s = 1, \ldots, n, \quad z_1 + \ldots + z_n = 1, \tag{14.6}$$

where

$$\mathbf{z}^{\alpha^s} = z_1^{\alpha_1^s} \ldots z_n^{\alpha_n^s},$$

with

$$\boldsymbol{\alpha}^s = (\alpha_1^s, \ldots, \alpha_n^s) \quad \text{and} \quad \boldsymbol{\beta}^s = (\beta_1^s, \ldots, \beta_n^s)$$

being multi-indices of equal lengths

$$||\alpha^s|| = \alpha_1^s + \ldots + \alpha_n^s = ||\beta^s||.$$

Moreover, the rank of the matrix

$$\Gamma = ((\alpha_i^s - \beta_i^s))_{i=1,\ldots,n}^{s=1,\ldots,n}$$

is equal to $n-1$. Introducing the determinants

$$\Delta_j = \det((\alpha_i^s - \beta_i^s))_{i=1,\ldots,n-1}^{s=1,\ldots,[j],\ldots,n} \tag{14.7}$$

where the symbol $[j]$ indicates that the number j is being omitted, we see that the ν_j are proportional to $(-1)^{j-1}\Delta_j$, so that the components of $(\nu_1, \ldots, \nu_n)$ have no factors in common. We note that in order to define this system of determinants, we could have deleted any column from Γ instead of the last one, since the sum of the columns of Γ is equal to 0, and the vector $(\nu_1, \ldots, \nu_n)$ would then be the same up to a sign.

Our purpose is to construct of the resultant for the system (14.6), with respect to w, and to determine its coefficients.

LEMMA 14.1. *Consider the system*

$$\mathbf{z}^{\gamma^s} = C_s, \qquad s = 1, \ldots, n, \tag{14.8}$$

where

$$\gamma^s = (\gamma_1^s, \ldots, \gamma_n^s),$$

the γ_k^s are integers (not necessary positive) and

$$\Delta = \det((\gamma_k^j))_{k,j=1,\ldots,n} \neq 0, \quad C_1 \neq 0, \ \ldots, \ C_n \neq 0.$$

Then: a) the number of roots $\mathbf{z}$ of the system (14.8) with nonzero coordinates i.e.

$$\mathbf{z} \in (\mathbf{C} \setminus \{0\})^n$$

is equal to $|\Delta|$, b) all such roots are simple, c) they all have the form

$$z_j = \mathbf{C}^{\mathbf{\Delta}^j/\Delta} e^{2\pi i <\mathbf{m}, \mathbf{\Delta}^j/\Delta>}, \quad j = 1, \ldots, n, \tag{14.9}$$

where

$$\mathbf{\Delta}^j = (\Delta_1^j, \ldots, \Delta_n^j)$$

is the vector formed from the algebraic complements Δ_k^j to the elements γ_k^j in the matrix $((\gamma_k^j))$; $\mathbf{m} = (m_1, \ldots, m_n)$ is a vector of integers; $<a, b>$ denotes a scalar product of the vectors **a** *and* **b** *i.e.*

$$<a, b> = a_1 b_1 + \ldots + a_n b_n,$$

$$\mathbf{C}^{\mathbf{\Delta}^j/\Delta} = C_1^{\Delta_1^j/\Delta} \ldots C_1^{\Delta_n^j/\Delta}.$$

Proof. a). We make use of a result of Bernstein [**20**] to the following effect: The number of roots of the system (14.8) lying in $(\mathbf{C} \setminus \{0\})^n$ is equal to the mixed volume of the system of sets S_j, where (in our case) S_j is the segment $[0, \boldsymbol{\gamma}_j]$ in $\mathbf{R}^n$. This mixed volume is equal to the volume V_n of the set $S_1 + \ldots + S_n$, since $V_n(S_{i_1} + \ldots + S_{i_k}) = 0$ for $k < n$. The set $S_1 + \ldots + S_n$ consists of all vectors $\mathbf{x} = (x_1, \ldots, x_n)$ of the form

$$x_1 = <\boldsymbol{\gamma}^1, \mathbf{t}>, \ \ldots, \ x_n = <\boldsymbol{\gamma}^n, \mathbf{t}>, \quad \mathbf{t} \in \mathbf{R}^n, \ 0 \leq t_j \leq 1, \ j = 1, \ldots, n.$$

Thus

$$V_n(S_1 + \ldots + S_n) = \int_{S_1+\ldots+S_n} dx_1 \ldots dx_n =$$

$$= |\Delta| \int_{J_n} dt_1 \ldots dt_n = |\Delta|,$$

where J_n is the unit cube in $\mathbf{R}^n$. It is also easy to check that the system (14.8) does not belong to the class of systems for which the number of roots is less than the mixed volume.

b). Let $\mathbf{z}$ be a root of (14.8) lying in $(\mathbf{C} \setminus \{0\})^n$. Then the Jacobian of (14.8) is given by

$$J = \begin{vmatrix} \gamma_1^1 z_1^{-1} \mathbf{z}^{\gamma^1} & \ldots & \gamma_n^1 z_n^{-1} \mathbf{z}^{\gamma^1} \\ \ldots & \ldots & \ldots \\ \gamma_1^n z_1^{-1} \mathbf{z}^{\gamma^n} & \ldots & \gamma_n^n z_n^{-1} \mathbf{z}^{\gamma^n} \end{vmatrix} = \mathbf{z}^{\gamma^1+\ldots+\gamma^n-\mathbf{I}} \Delta \neq 0,$$

where $\mathbf{I} = (1, \ldots, 1)$, and we conclude that $\mathbf{z}$ is a simple root.

c). Let

$$C_j = r_j e^{it_j} \quad \text{and} \quad z_j = \rho_j e^{i\varphi_j}, \quad j = 1, \ldots, n.$$

The system (14.8) gives rise to the following two linear systems of equations

$$< \ln \boldsymbol{\rho}, \gamma^1 >= \ln r_1, \qquad < \gamma^1, \mathbf{t} >= t_1 + 2m_1\pi,$$

$$\ldots \qquad \ldots$$

$$< \ln \boldsymbol{\rho}, \gamma^n >= \ln r_n, \qquad < \gamma^n, \mathbf{t} >= t_n + 2m_n\pi.$$

Solving them we get

$$\rho_j = \mathbf{r}^{\boldsymbol{\Delta}^j/\Delta}, \quad \varphi_j =< \mathbf{t}, \boldsymbol{\Delta}^j/\Delta > +2\pi < \mathbf{m}, \boldsymbol{\Delta}^j/\Delta >,$$

from which assertion c) follows. □

We remark that for $w = 0$ the system (14.6) has no roots in $(\mathbf{C} \setminus \{0\})^n$. Indeed, suppose that $\mathbf{z}$ is a root lying in $(\mathbf{C} \setminus \{0\})^n$ and $\Delta_1 \neq 0$ i.e. $\nu_1 \neq 0$. Then for $\mathbf{z}$ we have the system

$$\mathbf{z}^{\alpha^1-\beta^1} = b_1^+/b_1^- = C_1,$$

$$\ldots$$

$$\mathbf{z}^{\alpha^n-\beta^n} = b_n^+/b_n^- = C_n,$$

or

$$\begin{aligned} z_1^{\alpha_1^2-\beta_1^2} \ldots z_{n-1}^{\alpha_{n-1}^2-\beta_{n-1}^2} &= C_2 z_n^{\beta_n^2-\alpha_n^2}, \\ &\ldots\ldots \\ z_1^{\alpha_1^n-\beta_1^n} \ldots z_{n-1}^{\alpha_{n-1}^n-\beta_{n-1}^n} &= C_n z_n^{\beta_n^n-\alpha_n^n}. \end{aligned} \tag{14.10}$$

Moreover,

$$\mathbf{z}^{\alpha^1-\beta^1} = \frac{b_1^+}{b_1^-},$$

so applying Lemma 14.1 to (14.10) we obtain

$$z_1 = \mathbf{C}^{\boldsymbol{\Delta}^1/\Delta} z_n e^{2\pi i < \mathbf{m}, \boldsymbol{\Delta}^1/\Delta_1 >},$$

$$\ldots\ldots$$

$$z_{n-1} = \mathbf{C}^{\boldsymbol{\Delta}^{n-1}/\Delta} z_n e^{2\pi i < \mathbf{m}, \boldsymbol{\Delta}^{n-1}/\Delta_1 >}.$$

Here we used the identity

$$\sum_{j=2}^{n}(\beta_n^j - \alpha_n^j)\Delta_j^s = \Delta_1, \quad s = 1, \dots, n-1.$$

Now substituting these solutions into the first equation $\mathbf{z}^{\alpha^1-\beta^1} = b_1^+/b_1^-$ we have

$$z_n^{\|\alpha^1\|-\|\beta^1\|}\mathbf{C}^{<\Delta^1/\Delta,\alpha^1-\beta^1>}e^{2\pi i} = b_1^+/b_1^-.$$

From here we arrive at a contradiction, since

$$z_n^{\|\alpha^1\|-\|\beta^1\|} = 1,$$

and only the constants $b_2^+, \dots, b_n^+, b_2^-, \dots, b_n^-$ are contained in $\mathbf{C}$. Therefore the last equality is impossible if the constants $b_s^\pm$ are to be arbitrary.

From now we shall assume that for $w = 0$ the system (14.6) has no roots in $\mathbf{C}^n$.

LEMMA 14.2. *If $w = 0$ is not a root of system (14.6) and if $\nu_j \neq 0$, then the system*

$$\begin{array}{rcl} b_1^+\mathbf{z}^{\alpha^1} - b_1^-\mathbf{z}^{\beta^1} & = & \nu_1 w, \\ & \cdots & \\ b_{j-1}^+\mathbf{z}^{\alpha^{j-1}} - b_{j-1}^-\mathbf{z}^{\beta^{j-1}} & = & \nu_{j-1} w, \\ b_{j+1}^+\mathbf{z}^{\alpha^{j+1}} - b_{j+1}^-\mathbf{z}^{\beta^{j+1}} & = & \nu_{j+1} w, \\ & \cdots & \\ b_n^+\mathbf{z}^{\alpha^n} - b_n^-\mathbf{z}^{\beta^n} & = & \nu_n w, \\ z_1 + \dots + z_n & = & 1 \end{array} \tag{14.11}$$

has only isolated roots in $\mathbf{C}^n$, and it has no roots in the hyperplane at infinity $\mathbf{CP}^n \setminus \mathbf{C}^n$ (for almost all $b_j^\pm$).

Proof. We can assume that $\nu_1 \neq 0$. For the proof of Lemma 14.2 it is sufficient, by Lemma 8.1, to show that the system

$$\begin{array}{rcl} b_2^+\mathbf{z}^{\alpha^2} - b_2^-\mathbf{z}^{\beta^2} & = & 0, \\ & \cdots & \\ b_n^+\mathbf{z}^{\alpha^n} - b_n^-\mathbf{z}^{\beta^n} & = & 0, \\ z_1 + \dots + z_n & = & 0 \end{array} \tag{14.12}$$

has only the root $\mathbf{z} = (0, \dots, 0)$. First we show that (14.12) has no roots in $(\mathbf{C} \setminus \{0\})^n$. Indeed, if $\mathbf{z}$ were such a root, then from the system

$$\mathbf{z}^{\alpha^2-\beta^2} = b_2^+ b_2^- = C_2,$$

$$\cdots\cdots$$

$$\mathbf{z}^{\alpha^n-\beta^n} = b_n^+ b_n^- = C_n,$$

we would by Lemma 14.1 get

$$z_1 = z_n \mathbf{C}^{\Delta^1/\Delta_1} e^{2\pi i <\mathbf{m},\Delta^1/\Delta_1>},$$

$$\cdots\cdots$$

$$z_{n-1} = z_n \mathbf{C}^{\Delta^{n-1}/\Delta_1} e^{2\pi i <\mathbf{m},\Delta^{n-1}/\Delta_1>}.$$

Substituting these solutions into the last equation of (14.12) we would have

$$z_n\left(1+\mathbf{C}^{\Delta^1/\Delta_1}e^{2\pi i<\mathbf{m},\Delta^1/\Delta_1>}+\ldots+\mathbf{C}^{\Delta^{n-1}/\Delta_1}e^{2\pi i<\mathbf{m},\Delta^{n-1}/\Delta_1>}\right)=0.$$

Since here the coefficient of z_n is not identically equal to zero, this means that $z_n = 0$. But this would imply that $z_j = 0,\ j = 1,\ldots,n$, and we reach a contradiction.

Suppose now that (14.12) has a root $\mathbf{z}$ such that $z_1 \neq 0,\ldots,z_k \neq 0$ and $z_{k+1} = 0,\ldots,z_n = 0$. Inserting this root into (14.12) we get (after re-labelling the variables)

$$\begin{aligned} b_2^+\mathbf{z}^{\alpha^2}-b_2^-\mathbf{z}^{\beta^2} &= 0,\\ &\cdots\\ b_s^+\mathbf{z}^{\alpha^s}-b_s^-\mathbf{z}^{\beta^s} &= 0,\\ z_1+\ldots+z_n &= 0. \end{aligned} \tag{14.13}$$

We see that the form of system (14.12) is retained, since if $\mathbf{z}^{\beta^2} = 0$ then $\mathbf{z}^{\alpha^2} = 0$ too, otherwise some other coordinate, except $z_{k+1},\ldots,z_n$, would also have to be zero. Let us prove that $s = k$ in (14.13).

a). Let $s < k$ and consider the system

$$\begin{aligned} b_1^+\mathbf{z}^{\alpha^1}-b_1^-\mathbf{z}^{\beta^1} &= 0,\\ &\cdots\\ b_s^+\mathbf{z}^{\alpha^s}-b_s^-\mathbf{z}^{\beta^s} &= 0,\\ z_1+\ldots+z_n &= 1, \end{aligned} \tag{14.14}$$

where $z_{k+1} = \ldots = z_n = 0$. We claim that this system has roots, which then means that the system (14.6) with $w = 0$ also has roots. Indeed, setting $z_1 = 0,\ldots,z_{j-1} = 0, z_j = 1, z_{j+1} = \ldots = z_k = 0$ in (14.14) we find that (14.14) must contain a monomial $z_1^{j_1},\ldots,z_k^{j_k}$, unless we just obtain the trivial system. Therefore at least one equation in (14.14) has the form

$$b_j^+z_k^{||\alpha^j||}-b_j^-z_{k-1}^{||\beta^j||}=0,$$

from which we get $z_k = cz_{k-1}$. Substituting this solution into (14.14) we are left with

$$\begin{aligned} \overline{b}_1^+\mathbf{z}^{\overline{\alpha}^1}-\overline{b}_1^-\mathbf{z}^{\overline{\beta}^1} &= 0,\\ &\cdots\\ \overline{b}_{s-1}^+\mathbf{z}^{\overline{\alpha}^{s-1}}-\overline{b}_{s-1}^-\mathbf{z}^{\overline{\beta}^{s-1}} &= 0,\\ z_1+\ldots+z_{k-1}(1+c) &= 1. \end{aligned}$$

This system has the same form as (14.14), and we apply the same method to it. Finally we obtain the system

$$c_1z_1^{\alpha}-c_2z_2^{\alpha}=0,\quad c_3z_1+c_4z_2=1, \tag{14.15}$$

where c_1, c_2 are products of $b_s^{\pm}$, and c_3, c_4 are sums of such products. From (14.15) we have $z_2 = c_5z_1$ and $z_1(c_3 + c_4c_5) = 1$. Moreover, $c_3 + c_4c_5 \not\equiv 0$ since it is obvious that $c_3+c_4c_5$ is a non-constant polynomial in $b_s^{\pm}$. We thus found the claimed solution to the system (14.14), and thereby produced a contradiction.

b). Let now $s > k$. In this case $\Delta_1 = 0$, since this determinant must contain a $s \times (n-k-1)$ minor consisting of zeros, and $s+n-k-1 > n-1$.

Thus, system (14.13) takes the form

$$\begin{aligned} b_2^+\mathbf{z}^{\alpha^2} - b_2^-\mathbf{z}^{\beta^2} &= 0, \\ &\;\;\cdots \\ b_k^+\mathbf{z}^{\alpha^k} - b_k^-\mathbf{z}^{\beta^k} &= 0, \\ z_1 + \ldots + z_k &= 0. \end{aligned} \tag{14.16}$$

This means that the matrix

$$((\alpha_j^s - \beta_j^s))_{j=1,\dots n}^{s=2,\dots n}$$

is equal to

$$\mathbf{A} = \begin{pmatrix} \alpha_1^2 - \beta_1^2 & \ldots & \alpha_k^2 - \beta_k^2 & 0 & \ldots & 0 \\ \ldots & \ldots & \ldots & \ldots & \ldots & \ldots \\ \alpha_1^k - \beta_1^k & \ldots & \alpha_k^k - \beta_k^k & 0 & \ldots & 0 \\ \alpha_1^{k+1} - \beta_1^{k+1} & \ldots & \alpha_k^{k+1} - \beta_k^{k+1} & \ldots & \ldots & \alpha_n^{k+1} - \beta_n^{k+1} \\ \ldots & \ldots & \ldots & \ldots & \ldots & \ldots \\ \alpha_1^n - \beta_1^n & \ldots & \alpha_k^n - \beta_k^n & \ldots & \ldots & \alpha_n^n - \beta_n^n \end{pmatrix}.$$

Since

$$\Delta_1 = \det((\alpha_j^s - \beta_j^s))_{j=1,\dots,n-1}^{s=2,\dots,n} \neq 0,$$

it follows by the Laplace formula that one of the determinants of order $k-1$ in the left corner, say

$$\det((\alpha_j^s - \beta_j^s))_{j=1,\dots,k-1}^{s=2,\dots,k},$$

must be non-zero. To (14.16) we may then apply the same reasoning that we used in the beginning of Lemma 14.2. The proof of Lemma 14.2 is thereby completed. □

For $\nu_1 \neq 0$ we can, in view of Lemma 14.2, compute the resultant of (14.6) with respect to w by the formula

$$\frac{R(w)}{Q(w)} = \frac{B_0 + B_1 w + \ldots + B_N w^N}{Q(w)} =$$

$$= \prod_{j=1}^{M} (b_1^+ \mathbf{z}_{(j)}^{\alpha^1}(w) - b_1^- \mathbf{z}_{(j)}^{\beta^1}(w) - \nu_1 w), \tag{14.17}$$

where $\mathbf{z}_{(j)}(w)$ are roots of the system

$$\begin{aligned} b_2^+\mathbf{z}^{\alpha^2} - b_2^-\mathbf{z}^{\beta^2} &= \nu_2 w, \\ &\;\;\cdots \\ b_n^+\mathbf{z}^{\alpha^n} - b_n^-\mathbf{z}^{\beta^n} &= \nu_n w, \\ z_1 + \ldots + z_n &= 1 \end{aligned} \tag{14.18}$$

for fixed w.

Since $w = 0$ is not a root of system (14.6), we have $B_0 \not\equiv 0$. Our first purpose is now to find those terms in B_0 which may vanish for $b_1^+, \dots, b_n^+, b_1^-, \dots, b_n^- \neq 0$. We shall write that $B_0 \sim B$ if $B_0/B \neq 0$, for $b_1^\pm, \dots, b_n^\pm \neq 0$.

THEOREM 14.1. *If the system (14.6) (for $w = 0$) does not have any roots, then the constant term B_0 satisfies*

$$B_0 \sim ((b_1^+)^{\nu_1} \dots (b_n^+)^{\nu_n} - (b_1^-)^{\nu_1} \dots (b_n^-)^{\nu_n})^p. \tag{14.19}$$

(we are assuming that none of the rational fractions $B_0, \dots, B_N$ is reducible).

Proof :

$$B_0 \sim \prod_{j=1}^{M} \left(b_1^+ \mathbf{z}_{(j)}^{\alpha^1} - b_1^- \mathbf{z}_{(j)}^{\beta^1} \right), \tag{14.20}$$

where $\mathbf{z}_{(j)}$ are roots of system (14.18) for $w = 0$ (i.e. $\mathbf{z}_{(j)} = \mathbf{z}_{(j)}(0)$).

LEMMA 14.3. *Suppose $\nu_{k+1} = \dots = \nu_n = 0$, $\nu_1 \neq 0$, $\dots$, $\nu_k \neq 0$. Then in formula (14.20) it suffices to take into account only those roots $\mathbf{z}_{(j)}$ that have at least k non-zero coordinates.*

Proof. Consider a root $\mathbf{z}$ for which $z_1 \neq 0, \dots, z_{k-1} \neq 0$ and $z_k = \dots = z_n = 0$. There are three possible cases: a) $\mathbf{z}^{\alpha^1} \neq 0$, $\mathbf{z}^{\beta^1} = 0$, b) $\mathbf{z}^{\alpha^1} = 0$, $\mathbf{z}^{\beta^1} \neq 0$ and c) $\mathbf{z}^{\alpha^1} \neq 0$, $\mathbf{z}^{\beta^1} \neq 0$. Let us show that case c) does not occur. Indeed, the system (14.18) will, for $w = 0$ and $z_{k+1} = \dots = z_n = 0$, contain exactly k equations and hence the matrix $\mathbf{\Gamma}$ will have the form

$$\mathbf{\Gamma} = \begin{pmatrix} \alpha_1^1 - \beta_1^1 & \dots & \alpha_{k-1}^1 - \beta_{k-1}^1 & 0 & \dots & 0 \\ \dots & \dots & \dots & \dots & \dots & \dots \\ \alpha_1^{k-1} - \beta_1^{k-1} & \dots & \alpha_{k-1}^{k-1} - \beta_{k-1}^{k-1} & 0 & \dots & 0 \\ \dots & \dots & \dots & \dots & \dots & \dots \\ \alpha_1^n - \beta_1^n & \dots & \alpha_{k-1}^n - \beta_{k-1}^n & \dots & \dots & \alpha_n^n - \beta_n^n \end{pmatrix}.$$

Therefore if the last column is deleted at least $n - k + 1$ of the determinants of order $n - 1$ are equal to zero, and hence besides $\nu_{k+1} = \dots = \nu_n = 0$ one more ν_j is equal to zero, which is impossible (as before we use the relation $||\boldsymbol{\alpha}^j|| = ||\boldsymbol{\beta}^j||$). □

Let us return to the proof of Theorem 14.1. First we consider the roots $\mathbf{z}_{(j)}$ for which none of the coordinates vanishes. Finding these roots by means of Lemmas 14.1 and 14.2, we get

$$\mathbf{z}_1 = z_n \mathbf{C}^{\mathbf{\Delta}^1/\Delta_1} e^{2\pi i<\mathbf{m},\mathbf{\Delta}^1/\Delta>},$$

$$\dots\dots$$

$$\mathbf{z}_{n-1} = z_n \mathbf{C}^{\mathbf{\Delta}^{n-1}/\Delta_1} e^{2\pi i<\mathbf{m},\mathbf{\Delta}^{n-1}/\Delta>},$$

where

$$\mathbf{C} = (C_2, \dots, C_n): \quad C_2 = b_1^+/b_2^-, \ \dots, \ C_n = b_n^+/b_n^-.$$

Hence

$$\prod_j (b_1^+ \mathbf{z}_{(j)}^{\alpha^1} - b_1^- \mathbf{z}_{(j)}^{\beta^1}) \sim \prod_j \left(\frac{b_1^+}{b_1^-} \mathbf{z}_{(j)}^{\alpha^1 - \beta^1} - 1 \right) \sim$$

$$\prod_j \left(\frac{b_1^+}{b_1^-} \left(\frac{b_2^+}{b_2^-} \right)^{-\Delta_2/\Delta_1} \dots \left(\frac{b_n^+}{b_n^-} \right)^{(-1)^{n-1}\Delta_n/\Delta_1} e^{2\pi i(-m_2\Delta_2 + \dots + m_n\Delta_n(-1)^{n-1})/\Delta_1} - 1 \right) =$$

$$= \prod_j \left(\frac{b_1^+}{b_1^-} \left(\frac{b_2^+}{b_2^-}\right)^{\nu_2/\nu_1} \cdots \left(\frac{b_n^+}{b_n^-}\right)^{\nu_n/\nu_1} e^{2\pi i(m_2\nu_2+\ldots+m_n\nu_n)/\nu_1} - 1 \right).$$

Assuming $\nu_1 > 0$ we get that the set of roots $\mathbf{z}_{(j)}$ may be divided into ν_1 classes, according to the remainders after division of $m_2\nu_2 + \ldots + m_n\nu_n$ by ν_1. These remainders are all distinct, because the numbers $\nu_1, \ldots, \nu_n$ being relatively prime, one can find $m_1, \ldots, m_n$ such that $m_1\nu_1 + \ldots + m_n\nu_n = 1$, i.e.

$$\frac{m_2\nu_2 + \ldots + m_n\nu_n}{\nu_1} = \frac{1}{\nu_1} - m_1.$$

Now it is sufficient just to multiply $m_1, \ldots, m_n$ by j, to obtain any remainder

$$\prod_{j=0}^{\nu_1-1} \left(\left(\frac{b_1^+}{b_1^-}\right) \left(\frac{b_2^+}{b_2^-}\right)^{\nu_2/\nu_1} \cdots \left(\frac{b_n^+}{b_n^-}\right)^{\nu_n/\nu_1} e^{2\pi i/\nu_1} - 1 \right) =$$

$$= (-1)^{\nu_1+1} \left(\prod_{j=1}^{n} \left(\frac{b_j^+}{b_j^-}\right)^{\nu_j} - 1 \right).$$

It follows that

$$\prod_j \left(b_1^+ \mathbf{z}_{(j)}^{\alpha^1} - b_1^- \mathbf{z}_{(j)}^{\beta^1} \right) \sim \left(\prod_j \left(\frac{b_j^+}{b_j^-}\right)^{\nu_j} - 1 \right)^p,$$

where the product is taken over all roots in $(\mathbf{C} \setminus \{0\})^n$.

Let now $\nu_1 = 0$. Consider the roots (if any) for which one coordinate vanishes, say $z_1 \neq 0, \ldots, z_{n-1} \neq 0$ and $z_n = 0$. If z_n is contained in $\mathbf{z}^{\alpha^1}$ or in $\mathbf{z}^{\beta^1}$ then this root gives no contribution to B_0. If z_n occurs neither in $\mathbf{z}^{\alpha^1}$ nor in $\mathbf{z}^{\beta^1}$ then, setting $\mathbf{z}$ equal to zero, we get a system of $(n-1)$ equations of the same form as (14.18) but with $(n-1)$ unknowns. Moreover, from the form of the matrix $\mathbf{\Gamma}$ it is clear that the determinants of order $(n-2)$ in this new system are proportional to $\nu_1, \ldots, \nu_{n-1}$. Applying the previous reasoning to this system, we then get

$$\prod_j \left(b_1^+ \mathbf{z}_{(j)}^{\alpha^1} - b_1^- \mathbf{z}_{(j)}^{\beta^1} \right) \sim \left(\prod_j \left(\frac{b_j^+}{b_j^-}\right)^{\nu_j} - 1 \right)^p,$$

etc. □

The original proof of the formula for the constant term B_0 in the kinetic polynomial was given by Lazman in [**111**] under some additional assumptions. In our proof these restrictions have been removed. Moreover, it enables us to determine the degree p of the cycle characteristic of B_0.

COROLLARY 14.1. *If $\nu_1 \neq 0, \ldots, \nu_n \neq 0$ then the exponent p in formula (14.19) is equal to*

$$\left|\frac{\Delta_1}{\nu_1}\right| = \ldots = \left|\frac{\Delta_n}{\nu_n}\right|.$$

Proof. Since $\nu_1 \neq 0, \ldots, \nu_n \neq 0$, then in B_0 we only have to account for the roots with $z_1 \neq 0, \ldots, z_n \neq 0$, and their number is equal to $|\Delta_1|$ by Lemma 14.1. □

How can we find p in the general case, for instance if $\nu_n = 0$? Apart from the roots in $(\mathbf{C} \setminus \{0\})^n$ we must take into account also the roots having one coordinate equal to zero (if such roots exist). Choose such a z_j, say z_n, which is not contained in $\mathbf{z}^{\alpha^1}$ and in $\mathbf{z}^{\beta^1}$. Then set w and z_n equal to zero in (14.18). If there remain exactly $(n-1)$ equations consisting of two monomials, then there exist roots with $z_n = 0$ and $z_1 \neq 0, \ldots, z_{n-1} \neq 0$. Assume that it was the last equation that disappeared. Then the number of such roots is equal to

$$\frac{|\Delta_1|}{|\alpha_n^n - \beta_n^n|}.$$

Hence the exponent p increases by

$$\frac{|\Delta_1|}{|\nu_1|(\alpha_n^n - \beta_n^n)}$$

etc.

COROLLARY 14.2. *If $p > 1$ and if $\nu_1, \ldots, \nu_n \neq 0$, then the factor*

$$B_C = ((b_1^+)^{\nu_1} \ldots (b_n^+)^{\nu_n} - (b_1^-)^{\nu_1} \ldots (b_n^-)^{\nu_n})$$

(B_C is the cycle characteristic of the rate of the catalytic reaction [**150**]*) is contained in the coefficients B_s of the resultant with an exponent at least equal to $p - s$, $s = 0, 1, \ldots, p-1$, but is not contained in the leading coefficient B_N.*

Proof. Let $\nu_1 \neq 0$. Then the roots of system (14.6) are not isolated or their number is less than usual, precisely if $B_N = 0$ and

$$1 + \left(\frac{b_2^+}{b_2^-}\right)^{s_2} e^{i\alpha_2} + \ldots + \left(\frac{b_n^+}{b_n^-}\right)^{s_n} e^{i\alpha_n} = 0.$$

This latter surface intersects the surface

$$b_1^{+\nu_1} \ldots b_n^{+\nu_n} = b_1^{-\nu_1} \ldots b_n^{-\nu_n}$$

transversally, and it follows that $B_N \not\equiv 0$ for $b_1^{+\nu_1} \ldots b_n^{+\nu_n} = b_1^{-\nu_1} \ldots b_n^{-\nu_n}$.

Furthermore,

$$R(w) = \prod_j (w_1(\mathbf{z}_{(j)}(w)) - \nu_1 w),$$

$$B_1 = R'(w)|_{w=0} = \sum_m (w_1'(\mathbf{z}_{(m)}(0)) - \nu_1) \prod_{j \neq m} w_1(\mathbf{z}_{(j)}(0)).$$

Since the set of roots $\mathbf{z}_{(j)}(0)$ is divided into classes Γ_α, and since

$$\prod_{j \in \Gamma_\alpha} w_1(\mathbf{z}_{(j)}(0)) \sim \left(\prod_i b_i^{+\nu_i} - \prod_i b_i^{-\nu_i}\right),$$

it follows that if we delete one root, then

$$\prod_{j \neq m} w_1(\mathbf{z}_{(j)}(0)) \sim \left(\prod_i b_i^{+\nu_i} - \prod_i b_i^{-\nu_i}\right)^{p-1}.$$

For the second derivatives the exponent decreases by two units, for the third by three and so on. (Here we may safely differentiate the product $R(w)$, for $\nu_1 \neq 0, \ldots \nu_n \neq 0$

we only have to consider roots with non-zero coordinates in this product, and such roots are simple by Lemma 14.1). □

LEMMA 14.4. *The function $R(w)$ given by formula (14.17) is a polynomial in w (i.e. $Q(w) = 1$).*

Proof. General arguments (see e.g. Ch. 2 and [**11**]) show that $R(w)$ is a rational function. In order to prove that $R(w)$ is a polynomial it is sufficient to show that for $k = 0, 1, \ldots, M_1$ the integrals

$$\sum_{j=1}^{M_1} \int_{\Gamma_j(w)} \frac{(b_1^+ \mathbf{z}^{\alpha^1} - b_1^- \mathbf{z}^{\beta^1} - \nu_1 w)^k d(b_2^+ \mathbf{z}^{\alpha^2} - b_2^- \mathbf{z}^{\beta^2}) \wedge \ldots \wedge d(z_1 + \ldots + z_n)}{(b_2^+ \mathbf{z}^{\alpha^2} - b_2^- \mathbf{z}^{\beta^2} - \nu_2 w) \ldots (z_1 + z_2 + \ldots z_n - 1)} \tag{14.21}$$

are polynomials in w. Here

$$\Gamma_j(w) = \{\mathbf{z} : \left|b_2^+ \mathbf{z}^{\alpha^2} - b_2^- \mathbf{z}^{\beta^2} - \nu_2 w\right| = \varepsilon_1, \ \ldots, |z_1 + \ldots + z_n - 1| = \varepsilon_n\}$$

is a cycle corresponding to the j-th root of system (14.18), and M_1 is the number of roots of system (14.18). Since these integrals are equal to the sums of the values of the function

$$\left(b_1^+ \mathbf{z}^{\alpha^1} - b_1^- \mathbf{z}^{\beta^1} - \nu_1 w\right)^k$$

in the roots of (14.18), we see from the Newton recursion formulas (2.1) that $R(w)$ is a polynomial in these sums.

We now introduce the notation

$$w_j(\mathbf{z}) = b_j^+ \mathbf{z}^{\alpha^j} - b_j^- \mathbf{z}^{\beta^j}, \quad j = 1, \ldots, n,$$
$$w_{n+1}(\mathbf{z}) = z_1 + \ldots + z_n - 1,$$

and we let Γ_j be a cycle of the form

$$\Gamma_j = \{\mathbf{z} : |w_2(\mathbf{z})| = \delta_2, \ \ldots, |w_{n+1}(\mathbf{z})| = \delta_{n+1}\} = \Gamma_j(0)$$

corresponding to the j-th root of system (14.18) with $w = 0$. For sufficiently small $|w|, \varepsilon$ and δ, the cycle Γ_j is homological to $\Gamma_j(w)$, since $w = 0$ is not a root of $R(w)$ (we can suppose that $|w_j| > |\nu_j w|$, $j = 2, \ldots, n$, on Γ_j). Therefore the integral (14.21) is equal to

$$\sum_{j=1}^{M_1} \int_{\Gamma_j} \frac{(w_1(\mathbf{z}) - \nu_1 w)^k dw_2 \wedge \ldots \wedge dw_{n+1}}{(w_2 - \nu_2 w) \ldots (w_n - \nu_n w) w_{n+1}}.$$

Expanding the integrand in a geometric series we get a series in w, and in order that this series should terminate it is necessary that integrals of the form

$$\sum_{j=1}^{M_1} \int_{\Gamma_j} \frac{w_1^{k-s_1} dw_2 \wedge \ldots \wedge dw_{n+1}}{w_2^{s_2} \ldots w_n^{s_n} w_{n+1}} \tag{14.22}$$

should vanish for sufficiently large $||\mathbf{s}|| = s_1 + \ldots + s_n$ $(0 \le s_1 \le k)$. Passing to homogeneous coordinates

$$z_1 = \frac{\xi_1}{\xi_0}, \ \ldots, \ z_n = \frac{\xi_n}{\xi_0}$$

we obtain the integrand

$$\varphi = \frac{w_1^{*k-s_1}(\boldsymbol{\xi})dw_2^* \wedge \ldots \wedge dw_{n+1}^*}{w_2^{*s_2} \ldots w_n^{*s_n} w_{n+1}^* \xi_0^l},$$

where

$$l = ||\boldsymbol{\alpha}^1||k + ||\boldsymbol{\alpha}^2|| + \ldots + ||\boldsymbol{\alpha}^n|| + 1 - \sum_{j=1}^{n} s_j ||\boldsymbol{\alpha}^j||.$$

For sufficiently large $||\mathbf{s}||$ we clearly have $l < 0$. Hence the integrand has no singularities on the hyperplane at infinity $\mathbf{CP}^n \setminus \mathbf{C}^n$ (recall that system (14.8) has no roots at infinity). It therefore follows from the theorem on the complete sum of residues [**80**], that the integrals (14.22) vanish for sufficiently large $||\mathbf{s}||$. □

Let us now write

$$R(w) = B_N w^N + \ldots + B_1 w + B_0 = B_N(w - w_1)\ldots(w - w_N),$$

with $B_0 \neq 0$. Then for sufficiently small $|w|$ we may talk about the logarithm $\ln R(w)$. Introducing the notation

$$d_k = \left. \frac{d^k \ln R(w)}{dw^k} \right|_{w=0},$$

we have $d_k = (k-1)! s_k$, where

$$s_k = \sum_j \frac{1}{w_{(j)}^k}.$$

The numbers $1/w_{(j)}^k$ are zeros of the polynomial

$$P(w) = w^N R\left(\frac{1}{w}\right) = B_0 w^N + \ldots + B_N,$$

and hence the s_k are the power sums of the zeros of $P(w)$. Therefore the Newton formula and the Waring formula (Theorems 2.1 and 2.2) are valid for them:

$$s_k B_0 + s_{k-1} B_1 + \ldots + s_1 B_{k-1} + k B_k = 0, \;\; k \leq N,$$

$$\frac{B_k}{B_0} = \frac{(-1)^k}{k!} \begin{vmatrix} s_1 & 1 & 0 & \ldots & 0 \\ s_2 & s_1 & 2 & \ldots & 0 \\ \ldots & \ldots & \ldots & \ldots & \ldots \\ s_k & s_{k-1} & & \ldots & s_1 \end{vmatrix}$$

Substituting the expressions for d_k into these formulas we get relations between B_k and d_k.

LEMMA 14.5. *The coefficients of $R(w)$ and d_k are connected by the following formulas:*

$$B_k = \frac{1}{k!} \sum_{j=1}^{k} B_{k-j} \frac{d_j}{(j-1)!}, \quad 1 \leq k \leq N,$$

$$\frac{B_k}{B_0} = \frac{1}{k!}\begin{vmatrix} d_1 & -1 & 0 & \dots & 0 \\ d_2 & d_1 & -2 & \dots & 0 \\ \dots & \dots & \dots & \dots & \dots \\ \frac{d_k}{(k-1)!} & \frac{d_{k-1}}{(k-2)!} & & \dots & d_1 \end{vmatrix}. \tag{14.23}$$

Observe that the determinant in (14.23) is a polynomial in $d_1, \dots, d_k$ with positive coefficients. From here it is obvious that in order to find the coefficients of $R(w)$ we must find d_k. We get

$$d_1 = -\sum_{l=1}^{N} \frac{1}{w_l},$$

$$d_2 = -\sum_{l=1}^{N} \frac{1}{w_l^2},$$

$$\dots\dots$$

$$d_s = -(s-1)!\sum_{l=1}^{N} \frac{1}{w_l^s}.$$

In particular, Lemmas 14.5 and 14.2 imply that

$$d_1 = \frac{B_1}{B_0}, \quad d_2 = \frac{2B_2B_0 - B_1^2}{B_0^2}.$$

It is clear from this that in order to find $B_1, \dots, B_N$ one must first find the corresponding derivatives of $\ln R(w)$ or, equivalently, the sums

$$\sum_{l=1}^{N} \frac{1}{w_l^s}.$$

We can also apply Theorem 2.3.

THEOREM 14.2. *Suppose that the system (14.6) satisfies the conditions of Theorem 14.1. Then*

$$-\frac{B_1}{B_0} = \sum_{l=1}^{N} \frac{1}{w_l} = \sum_{j=1}^{n} \nu_j \sum_{s=1}^{M_j} \frac{1}{w_s\left(\mathbf{z}_{(s)}^j\right)} \tag{14.24}$$

where M_j is the number of roots of system (14.11) and $\mathbf{z}_{(s)}^j$ are the roots of system (14.11) for $w = 0$.

Theorem 9.1 (and Corollary 9.1) imply formula (14.24).

Thus, knowing all the roots of the subsystems (14.11) we can calculate one of the coefficients of the resultant $R(w)$ (namely B_1/B_0). Let us next show how to find the second derivative $d^2 \ln R/dw^2$ for $w = 0$:

$$\left.\frac{d^2 \ln R}{dw^2}\right|_{w=0} = \sum_{j=1}^{M_1}\left(\sum_{1<s\le l} 2\int_{\Gamma_j} \ln w_1 \frac{\nu_s\nu_l \mathbf{dw}[1]}{w_s w_l \mathbf{w}[1]} - \right.$$

$$\left. -2\sum_{s>1}\int_{\Gamma_j} \frac{\nu_1\nu_s \mathbf{dw}[1]}{w_s \mathbf{w}[1]} - \int_{\Gamma_j} \frac{\nu_1^2 \mathbf{dw}[1]}{w_1^2 \mathbf{w}[1]}\right) =$$

$$= \sum_{j=1}^{M_1} \left(\sum_{s \neq l} \int_{\Gamma_j} \frac{(-1)^s \nu_s \nu_l \mathbf{dw}[s]}{w_l \mathbf{w}} + \sum_{s=1}^{n} \int_{\Gamma_j} \frac{(-1)^s \nu_s^2 \mathbf{dw}[s]}{w_s \mathbf{w}} \right).$$

Here we used the formula

$$2\frac{\ln w_1}{w_s^3} dw_s = -d\left(\frac{\ln w_1}{w_s^2}\right) + \frac{dw_1}{w_1 w_s^2}.$$

Now, using Theorem 5.3 to pass to the global residue Γ_j^l, we get

$$\left.\frac{d^2 \ln R}{dw^2}\right|_{w=0} = \sum_{s,l=1}^{n} \sum_{j=1}^{M_l} \int_{\Gamma_j^l} \frac{(-1)^{s+l-1} \nu_s \nu_l \mathbf{dw}[s]}{w_l w}. \tag{14.25}$$

THEOREM 14.3. *If all the roots of system (14.11), corresponding to non-zero ν_l, are simple for $w = 0$, then*

$$-2\frac{B_2}{B_0} + \left(\frac{B_1}{B_0}\right)^2 = \sum_{l=1}^{N} \frac{1}{w_l^2} = \sum_{s,l=1}^{n} \sum_{j=1}^{M_l} \left.\frac{(-1)^{s+l} \nu_s \nu_l J_s(\mathbf{z})}{J_l(\mathbf{z}) w_l^2(\mathbf{z})}\right|_{\mathbf{z}=\mathbf{z}_{(j)}^l}, \tag{14.26}$$

where J_s is the Jacobian of the system (14.11) for $j = s$.

Proof. In formula (14.24) we make the transformation

$$\frac{\mathbf{dw}[s]}{w_l \mathbf{w}} = \frac{J_s J_l \mathbf{dz}}{J_l w_l \mathbf{w}} = \frac{J_s \mathbf{dw}[l]}{J_l w_l \mathbf{w}}$$

(since the roots are simple we have $J_l \neq 0$ on Γ_j^l). This gives us the logarithmic residue of the function $J_s/(J_l w_l)$.

In general, to find

$$\left.\frac{d^s \ln R}{dw^s}\right|_{w=0}$$

we must calculate integrals of the form

$$\sum_{j=1}^{M_1} \int_{\Gamma_j} \frac{\mathbf{dw}[1]}{w_1^{s_1} w_2^{s_2} \ldots w_n^{s_n} w_{n+1}}.$$

They can be rather easily computed if all roots of all subsystems are simple. In this case, introducing new variables in neighborhoods of the roots $w_2, \ldots, w_{n+1}$, we get that

$$\int_{\Gamma_j} \frac{\mathbf{dw}[1]}{w_1^{s_1} \ldots w_{n+1}} = c \frac{\partial^{s_2 + \ldots + s_n - n + 1}}{\partial w_2^{s_2 - 1} \ldots \partial w_n^{s_n - 1}} \cdot \frac{1}{w_1^{s_1}}(\mathbf{z}_{(j)}),$$

and these derivatives can be found by using the implicit function theorem. Thanks to Theorem 5.3 one can decrease the order of derivation by one (for the case of multiple roots see [**11**], where such integrals are computed). In just the same way as (14.24), formula (14.26) can be derived directly from Theorem 9.1.

In the proofs of Theorems 14.2 and 14.3 obtained by Kytmanov methods from multidimensional residue theory play an essential role. Another proof of these assertions was given by Lazman (see [**40**]), using traditional methods of mathematical analysis and the Euler–Jacobi formula (see [**14**]).

Theorems 14.1–14.3 thus enable us to write down explicitly the first coefficients B_2, B_1, B_0 of the kinetic polynomial. For finding the other coefficients one can use the algorithm described above. Examples indicate that the expressions for the coefficients $B_1, \ldots, B_N$ are rather unwieldy, and here the tools from computer algebra can be useful. The explicit form of $B_1, \ldots, B_N$ can simplify the asymptotic analysis of the dependence of w on different parameters. The simple structure of B_0 allows one to deduce a number of substantial results [**38, 39, 109, 111, 112, 113, 154, 155, 156**].

Originally the form of the constant term in the kinetic polynomial was obtained [**111**], and the algorithm based on multidimensional logarithmic residues for finding the other coefficients was given later. By using the modified transformation formulas for the Grothendieck residue given in [**98**], we shall now provide simpler formulas for B_j, $j \le 1$, and also give various applications of the new algorithm.

Consider the system (14.5), and write as before

$$w_j = b_j^+ \mathbf{z}^{\alpha^j} - b_j^- \mathbf{z}^{\beta^j}, \; j = 1, \ldots, n, \; w_{n+1} = z_1 + \ldots + z_n - 1. \tag{14.27}$$

By Lemma 14.1 and the Hilbert Nullstellensatz (or by the Macaulay theorem) there exists a set of homogeneous polynomials $a_{jk}(\mathbf{z})$, $j, k = 1, n$, such that

$$z_j^{L+1} = \sum_{k=1}^{n} a_{jk} w_k(\mathbf{z}), \quad j = 1, \ldots, n. \tag{14.28}$$

Moreover, the degree L can be chosen so that $L \le r_1 + \ldots + r_n - n$ (the polynomials $a_{jk}(\mathbf{z})$ can be found by using e.g. the method of indeterminate coefficients).

Denote by $\mathbf{A}$ the matrix

$$((a_{jk}(\mathbf{z})))_{j,k=1}^{n}$$

and by J_s the Jacobian of the subsystem obtained from (14.5) by deleting the s-th equation (w is fixed).

In our earlier formulas for the coefficients of the resultant, the coefficients were calculated in terms of the roots of subsystems. Here we shall give formulas for computing these coefficients directly from the coefficients of the initial system. The difference comes from the fact that the resultant is now calculated with respect to the linear equation of (14.5).

THEOREM 14.4. *If system (14.5) has no roots for $w = 0$, then*

$$d_1 = \frac{B_1}{B_0} = \sum_{s=1}^{n} (-1)^{n+s-1} \nu_s \mathfrak{M} \left[\frac{(z_1 + \ldots + z_n)^{r_s - 1} J_s \det \mathbf{A}}{z_1^L \ldots z_n^L} \right], \tag{14.29}$$

where $\mathfrak{M}$ is the linear functional assigning to a Laurent polynomial

$$Q = \frac{P(z_1, \ldots, z_n)}{z_1^{L_1} \ldots z_n^{L_n}}$$

its constant term. (In other words, in the polynomial $P(z_1, \ldots, z_n)$ we must take the coefficient of the monomial $z_1^{L_1} \ldots z_n^{L_n}$.)

The proof is carried out with the use of multidimensional residues theory (see [**98**] and also Ch. 2, Sec. 9). First we give a formula for computing d_2.

Theorem 14.5. *Under the conditions of Theorem 14.4 we have*

$$d_2 = 2\frac{B_2}{B_0} - \left(\frac{B_1}{B_0}\right)^2 =$$

$$= \sum_{k=1}^{n} \sum_{s,l=1}^{n} (-1)^{n+s-1} \nu_s \nu_l \mathfrak{M}\left[\frac{(z_1 + \ldots + z_n)^{r_s + r_l - 1} a_{kl} J_s \det \mathbf{A}}{z_1^L \ldots z_k^{2L+1} \ldots z_n^L}\right]. \quad (14.30)$$

For the calculation of the other coefficients of the resultant we obtain the following formulas.

Theorem 14.6 (Kytmanov [98]). *Under the conditions of Theorem 14.4 we have*

$$d_s = \sum_{l=1}^{n} \sum_{\substack{||\alpha||=s \\ \alpha_1=\ldots=\alpha_{l-1}=0,\ \alpha_l>0}} (-1)^{n+l-1} \nu_s \frac{s!}{\alpha!} \boldsymbol{\nu}^\alpha \sum_{j_1,\ldots,j_{s-1}=1}^{n} C_{j_1,\ldots,j_{s-1}} \mathfrak{M}[Q(\mathbf{z})], \quad (14.31)$$

where

$$Q(\mathbf{z}) = \frac{(z_1 + \ldots + z_n)^{r_l\alpha_l + \ldots + r_n\alpha_n - 1} a_{j_1 l} \ldots a_{j_{s-1} n} J_l \det \mathbf{A}}{z_{j_1}^{L+1} \ldots z_{j_{s-1}}^{L+1} z_1^L \ldots z_n^L},$$

$$\boldsymbol{\alpha} = (\alpha_1, \ldots, \alpha_l, \ldots, \alpha_n), \quad \alpha! = \alpha_1! \ldots \alpha_n!, \quad \boldsymbol{\nu}^\alpha = \nu_1^{\alpha_1} \ldots \nu_n^{\alpha_n},$$

and if the set $j_1, \ldots, j_{s-1}$ *contains* β_1 *ones,* β_2 *twos and so on, then*

$$C_{j_1 \ldots j_{s-1}} = \beta_1! \ldots \beta_n!.$$

An alternative version of formula (14.31)is given by:

$$d_s = -\sum_{||\alpha||=s} \frac{s! \boldsymbol{\nu}^\alpha}{\alpha!(r_1\alpha_1 + \ldots + r_n\alpha_n)} \sum_{j_1 \ldots j_s = 1}^{n} C_{j_1 \ldots j_s} \mathfrak{M}[Q_1(\mathbf{z})], \quad (14.32)$$

where

$$Q_1(\mathbf{z}) = \frac{(z_1 + \ldots + z_n)^{r_l\alpha_l + \ldots + r_n\alpha_n} a_{j_1 1} \ldots a_{j_s n} J_{n+1} \det \mathbf{A}}{z_{j_1}^{L+1} \ldots z_{j_s}^{L+1} z_1^L \ldots z_n^L}.$$

Formula (14.32) was obtained by considering the resultant of the system (14.6) with respect to the last equation (i.e. the balance equation). It is somewhat simpler than the expression (14.31) since there is no summation over l. However, the degree of the numerator is greater. In particular, starting from (14.32) we can write:

$$d_1 = -\sum_{i=1}^{n} \frac{\nu_l}{r_l} \sum_{k=1}^{n} \mathfrak{M}\left[\frac{(z_1 + \ldots + z_n)^{r_l} a_{kl} J_{n+1} \det \mathbf{A}}{z_k^{L+1} z_1^l \ldots z_n^L}\right],$$

$$d_2 = -\sum_{l \leq s=1}^{n} \frac{\nu_l \nu_s}{r_l + r_s} \sum_{k,p=1}^{n} C_{kp} \mathfrak{M}\left[\frac{(z_1 + \ldots + z_n)^{r_l + r_s} a_{kl} a_{ps} J_{n+1} \det \mathbf{A}}{z_k^{L+1} z_p^{L+1} z_1^l \ldots z_n^L}\right].$$

Formulas (14.29)–(14.32) are nothing but Corollaries of Theorem 9.1. We now give some applications of these formulas.

EXAMPLE 14.1. The linear transformation scheme. Assuming that the system (14.5) is linear, we can write it in the form

$$\begin{aligned} b_j z_j - b_{-j} z_{j+1} &= w, \qquad j = 1, \dots, n, \\ z_1 + \dots + z_n &= 1 \qquad (z_{n+1} = z_1). \end{aligned}$$

Since $r_j = 1$ we have $L = 0$ so that

$$z_j = a_{jl}(b_1 z_1 - b_{-1} z_2) + \dots + a_{jn}(b_n z_n - b_{-n} z_1), \; j = 1, \dots, n,$$

with $a_{jk}(\mathbf{z}) \equiv \text{const}$.

Writing $\mathbf{z} = \mathbf{A}\mathbf{w}$ we then have $\mathbf{w} = \mathbf{A}^{-1}\mathbf{z}$, where

$$\mathbf{A}^{-1} = \begin{pmatrix} b_1 & -b_{-1} & 0 & \dots & 0 \\ 0 & b_2 & -b_{-2} & \dots & 0 \\ \dots & \dots & \dots & \dots & \dots \\ -b_{-n} & 0 & 0 & \dots & b_n \end{pmatrix}.$$

From here we get

$$\det \mathbf{A} = 1/\det \mathbf{A}^{-1} = 1/(b_1 \dots b_n - b_{-1} \dots b_{-n}),$$

and further

$$J_s = (-1)^{n+s} \overline{J}_s,$$

$$\overline{J}_s = b_1 \dots [s] \dots b_n + b_1 \dots [s] \dots b_{n-1} b_{-n} + \dots + b_{-1} \dots [s] \dots b_{-n}.$$

As usual, $[s]$ means that the term with number s should be omitted. This yields

$$\frac{B_1}{B_0} = \sum_{s=1}^{n} (-1)^{n+s-1} \nu_s J_s \det \mathbf{A} = -\sum_{s=1}^{n} \frac{\overline{J}_s}{b_1 \dots b_n - b_{-1} \dots b_{-n}},$$

and hence

$$R(w) = \left(\sum_{s=1}^{n} \overline{J}_s \right) w - (b_1 \dots b_n - b_{-1} \dots b_{-n}),$$

in complete accordance with (14.3).

EXAMPLE 14.2. A two stage nonlinear mechanism. The system

$$\begin{aligned} b_1 z_1^2 - b_{-1} z_2^2 &= w, \\ b_2 z_2 - b_{-2} z_1 &= 2w, \\ z_1 + z_2 &= 1 \end{aligned}$$

corresponds to the tranformations scheme

$$2Z_1 \rightleftarrows 2Z_2, \qquad Z_2 \rightleftarrows Z_1.$$

It is easy to check that for the system

$$w_1 = b_1 z_1^2 - b_{-1} z_2^2, \qquad w_2 = b_2 z_2 - b_{-2} z_1$$

the entries of the matrix $\mathbf{A}$ have the form

$$a_{11} = \frac{b_2^2}{B}, \quad a_{12} = \frac{b_{-1}(b_{-2} z_1 + b_2 z_2)}{B},$$

$$a_{21} = \frac{b_{-2}^2}{B}, \quad a_{22} = \frac{b_1(b_{-2} z_1 + b_2 z_2)}{B},$$

$$B = b_1 b_2^2 - b_{-1} b_{-2}^2,$$

Furthermore,

$$\det \mathbf{A} = \frac{b_{-2} z_1 + b_2 z_2}{B},$$

$$J_1 = -(b_2 + b_{-2}), \qquad J_2 = 2(b_1 z_1 + b_{-1} z_2).$$

Using Theorems 14.4 and 14.5 we therefore get

$$\frac{B_1}{B_0} = -\frac{1}{B}(4b_{-1}b_{-2} + 4b_1 b_2 + (b_2 + b_{-2})^2),$$

$$2\frac{B_2}{B_0} - \left(\frac{B_1}{B_0}\right)^2 = -\frac{1}{B^2}((b_2 + b_{-2})^4 + 16(b_2 + b_{-2})^2(b_1 b_2 + b_{-1}b_{-2}) +$$

$$+ 32 b_1 b_2 b_{-1} b_{-2} + 8b_1^2 b_2^2 + 8b_{-1}^2 b_{-2}^2 + 8b_1 b_2^2 b_{-1} + 8b_1 b_{-1} b_{-2}^2).$$

In conclusion we have

$$R(w) = 4(b_1 - b_{-1})w^2 - (4b_{-1}b_{-2} + 4b_1 b_2 + (b_2 + b_{-2})^2)w + B.$$

(See also Example 9.2).

EXAMPLE 14.3. Consider the system

$$\begin{aligned} b_1 z_1^p - b_{-1} z_2^p &= w, \\ b_2 z_2 - b_{-2} z_1 &= pw, \\ z_1 + z_2 &= 1 \end{aligned}$$

corresponding to the transformation scheme

$$pZ_1 \rightleftarrows pZ_2, \qquad Z_2 \rightleftarrows Z_1.$$

In this case we find

$$a_{11} = \frac{b_2^p}{B}, \quad a_{12} = \frac{b_{-1}((b_2 z_2^{p-1}) + \ldots + (b_{-2} z_1)^{p-1})}{B},$$

$$a_{21} = \frac{b_{-2}^p}{B}, \quad a_{22} = \frac{b_1((b_2 z_2^{p-1}) + \ldots + (b_{-2} z_1)^{p-1})}{B},$$

where $B = b_1 b_2^p - b_{-1} b_{-2}^p$. We also obtain

$$\det \mathbf{A} = \frac{1}{B}((b_2 z_2)^{p-1} + \ldots + (b_{-2} z_1)^{p-1}),$$

$$J_1 = -(b_2 + b_{-2}), \qquad J_2 = p b_1 z_1^{p-1} + p b_{-1} z_2^{p-1},$$

and hence by Theorem 14.4

$$\frac{B_1}{B_0} = -\frac{1}{B}\left((b_2 + b_{-2})^p + p^2\left(b_1 b_2^{p-1} + b_{-1} b_{-2}^{p-1}\right)\right).$$

EXAMPLE 14.4. Three stage mechanism. The system

$$\begin{aligned} b_1 z_1^2 - b_{-1} z_2^2 &= w, \\ b_2 z_1 - b_{-2} z_3 &= 2w, \\ b_3 z_2 z_3 - b_{-3} z_1^2 &= 2w, \\ z_1 + z_2 + z_3 &= 1 \end{aligned}$$

corresponds to the transformation scheme

$$2Z_1 \rightleftarrows 2Z_2, \quad Z_1 \rightleftarrows Z_3, \quad Z_2 + Z_3 \rightleftarrows 2Z_1.$$

To find the entries of the matrix $\mathbf{A}$ we solve a linear system of equations. This gives us

$$\begin{aligned}
a_{11} &= \frac{b_2 b_3^2 b_{-2} z_3}{B}, \\
a_{12} &= \frac{B z_1^2 + b_2 b_3 b_{-1} b_{-2} b_{-3} z_1 z_2 + b_3 b_{-1} b_{-2}^2 b_{-3} z_2 z_3}{b_2 B}, \\
a_{13} &= \frac{\left(b_2^2 b_3 b_{-1} z_2 + b_{-1} b_{-2}^2 b_{-3} z_3\right) b_{-2}}{b_2 B}, \\
a_{21} &= \frac{-B z_2 + b_1 b_3 b_{-2}^2 b_{-3} z_3}{b_{-1} B}, \\
a_{22} &= \frac{(b_2 z_1 z_2 + b_{-2} z_2 z_3) b_1^2 b_3^2}{b_{-1} B}, \\
a_{23} &= \frac{(b_{-1} b_{-3} z_2 + b_1 b_3 z_3) b_1 b_{-2}^2}{b_{-1} B}, \\
a_{31} &= \frac{b_2^4 b_3^2 z_3}{b_{-2}^2 B}, \\
a_{32} &= \frac{1}{b_2^2 B}\left(-b_{-2} B z_3^2 + b_2^3 b_3 b_{-1} b_{-3} z_1 z_3 - \right. \\
&\quad \left. - b_2 B z_1 z_3 + b_2^2 b_3 b_{-1} b_{-2} b_{-3} z_2 z_3\right), \\
a_{33} &= \frac{\left(b_2^2 b_3 z_2 + b_{-2}^2 b_{-3} z_3\right) b_1^2 b_{-1}}{b_{-2}^2 B},
\end{aligned}$$

where

$$B = b_1 b_2^2 b_3^2 - b_{-1} b_{-2}^2 b_{-3}^2.$$

The determinant of the matrix $\mathbf{A}$ has the form

$$\det \mathbf{A} = \left(b_{-1} z_2 \left(b_2^4 b_3 z_1^2 z_3 + b_2^2 b_{-2}^2 b_{-3} z_1^2 z_3 + z_1 z_2 z_3 b_2^2 b_3 b_{-2} + \right.\right.$$
$$\left. + z_2 z_3^2 b_2^2 b_3 b_{-2}^2 + z_3^3 b_{-2}^4 b_{-3} + z_1 z_3^2 b_2 b_{-2}^3 b_{-3}\right) + z_1^2 z_3^2 b_1 b_2^2 b_3 b_{-2}^2 +$$
$$\left. + z_3^2 (z_1 b_2 + z_3 b_{-2}) b_1 b_3 b_{-2}^3\right) \frac{1}{b_2 b_{-1} b_{-2}^2 B},$$

and the Jacobians J_1, J_2, J_3 are equal to

$$\begin{aligned}
J_1 &= 2 b_{-2} b_{-3} z_1 - b_2 b_3 z_2 + z_3 b_3 (b_2 + b_{-2}), \\
J_2 &= 2 b_1 b_3 z_1 z_3 - 2 z_1 z_2 (2 b_{-1} b_{-3} + b_1 b_3) - 2 b_3 b_{-1} z_2^2, \\
J_3 &= 2 b_1 b_{-2} z_1 + 2 b_{-1} b_2 z_2.
\end{aligned}$$

Applying Theorem 14.4 we get

$$\frac{B_1}{B_0} = \left(b_2 b_3 b_{-1} b_{-2}^3 b_3 + b_2^2 b_3^2 b_{-1} (b_2 + b_{-2})^2 + 2 b_{-1} b_{-2}^4 b_{-3}^2 - \right.$$
$$- b_1 b_2^2 b_3^2 b_{-2}^2 + b_2 b_3 b_{-1} b_{-2}^2 b_{-3} (b_2 + b_{-2}) + b_2^2 b_3 b_1 b_{-2}^2 b_{-3} -$$

$$-4b_1b_2^2b_3^2b_{-1}b_{-2}+b_{-1}^2b_{-2}^3b_{-3}^2+4b_1b_3b_{-1}b_{-2}^3b_{-3}+4b_1b_{-1}b_{-2}^4b_{-3}+$$
$$+8b_1b_2^2b_3b_{-1}b_{-2}^2+4b_1b_2b_3b_{-1}b_{-2}^3+4b_2^2b_{-1}^2b_{-2}^2b_3)\,\frac{1}{b_{-1}b_2^2B}.$$

EXAMPLE 14.5. Schemes without interactions between different intermediates. Consider the system of equations

$$\begin{aligned} b_1z_1^{\alpha_1}-b_{-1}z_2^{\alpha_1} &= \nu_1 w,\\ &\;\;\cdots \\ b_nz_n^{\alpha_n}-b_{-n}z_1^{\alpha_n} &= \nu_n w,\\ z_1+\ldots+z_n &= 1, \end{aligned} \tag{14.33}$$

corresponding to a single-route transformation scheme with no interactions between different intermediates. It is known [**152**] that this rather wide class of nonlinear mechanisms has the following quasi-thermodynamic properties: The steady state is unique and stable for fixed values of balances. As will now be shown this result can be obtained within the framework of the approach developed here [**36**].

For (14.33) we have the matrix

$$\Gamma=\begin{pmatrix} \alpha_1 & -\alpha_1 & 0 & \ldots & 0\\ 0 & \alpha_2 & -\alpha_2 & \ldots & 0\\ \ldots & \ldots & \ldots & \ldots & \ldots\\ -\alpha_n & 0 & 0 & \ldots & \alpha_n \end{pmatrix},$$

and the algebraic complement Δ_j to the j-th entry of the last column is equal to

$$\Delta_j=\prod_{i\neq j}\alpha_i.$$

Therefore we can take them as ν_j.

It is obvious that for $w=0$ the system (14.33) has no roots, so the theory developed here applies to it. We consider first the equation determining the coefficients of the resultant $R(w)$.

We assume that

$$R(w)=B_0+B_1+\ldots+B_Nw^N,$$

with $B_0\neq 0$, and we write $L+1=\alpha_1\ldots\alpha_n$. We must find polynomials $a_{jk}(z)$ such that

$$z_j^{L+1}=a_{j1}(z)(b_1z_1^{\alpha_1}-b_{-1}z_2^{\alpha_1})+\ldots+a_{jn}(z)(b_nz_n^{\alpha_n}-b_{-n}z_1^{\alpha_n}),\quad j=1,\ldots,n.$$

These polynomials can be chosen of the form

$$a_{jk}(z)=c_{jk}((b_kz_k^{\alpha_k})^{\nu_k-1}+(b_kz_k^{\alpha_k})^{\nu_k-2}(b_{-k}z_{k+1}^{\alpha_k})+\ldots+(b_{-k}z_{k+1}^{\alpha_k})^{\nu_k-1}),$$

with the convention that $k+1=1$ for $k=n$. Indeed, we then have

$$a_{j1}w_1+\ldots+a_{jn}w_n=c_{j1}(b_1^{\nu_1}z_1^{L+1}-b_{-1}^{\nu_1}z_2^{L+1})+\ldots+$$
$$+c_{jn}(b_n^{\nu_n}z_n^{L+1}-b_{-n}^{\nu_n}z_1^{L+1})=z_j^{L+1},$$

where $w_k=b_kz_k^{\alpha_k}-b_{-k}z_{k+1}^{\alpha_k}$. So if we write $z_j^{L+1}=\omega_j$ and

$$b_j^{\nu_j}\omega_j-b_{-j}^{\nu_j}\omega_{j+1}=\omega_j'$$

we find that the matrix $\mathbf{C} = ((c_{jk}))$ satisfies the condition

$$\mathbf{C}\boldsymbol{\omega}' = \boldsymbol{\omega},$$

Then $\boldsymbol{\omega}' = \mathbf{C}^{-1}\boldsymbol{\omega}$, and hence

$$\mathbf{C}^{-1} = \begin{pmatrix} b_1^{\nu_1} & -b_{-1}^{\nu_1} & 0 & \dots & 0 \\ 0 & b_2^{\nu_2} & -b_{-2}^{\nu_2} & \dots & 0 \\ \dots & \dots & \dots & \dots & \dots \\ -b_{-n}^{\nu_n} & 0 & 0 & \dots & b_n^{\nu_n} \end{pmatrix}.$$

From this we obtain the expression

$$(\det \mathbf{C})^{-1} = \det \mathbf{C}^{-1} = b_1^{\nu_1} \dots b_n^{\nu_n} - b_{-1}^{\nu_1} \dots b_{-n}^{\nu_n} = B$$

for the cycle characteristic.

LEMMA 14.6. *The adjoint of the matrix*

$$\mathbf{T} = \begin{pmatrix} a_1 & -b_1 & 0 & \dots & 0 \\ 0 & a_2 & -b_2 & \dots & 0 \\ \dots & \dots & \dots & \dots & \dots \\ -b_n & 0 & 0 & \dots & a_n \end{pmatrix}$$

is given by

$$\mathbf{T}^* = \begin{pmatrix} a_2 \dots a_n & b_1 a_3 \dots a_n & \dots & b_1 \dots b_{n-1} \\ b_2 \dots b_n & a_1 a_3 \dots a_n & \dots & a_1 b_2 \dots b_{n-1} \\ a_2 b_3 \dots b_n & b_1 b_3 \dots b_n & \dots & a_1 a_2 \dots b_{n-1} \\ \dots & \dots & \dots & \dots \\ a_2 a_3 \dots b_n & b_1 a_3 \dots a_{n-1} b_n & \dots & a_1 a_2 \dots a_{n-1} \end{pmatrix}.$$

Knowing $\mathbf{C}^{-1}$ we can now easily find the matrix $\mathbf{C}$ with the help of Lemma 14.6. Using the form of a_{jk} we find that

$$\det \mathbf{A} = \frac{1}{B} \prod_{j=1}^{n} (b_j^{\nu_j - 1} z_j^{L-\alpha_j+1} + \dots + b_{-j}^{\nu_j - 1} z_{j+1}^{L-\alpha_j+1}).$$

To find the Jacobians J_s we first note that

$$J_s = (-1)^{n+s} \nu_s \overline{J}_s,$$

where

$$\overline{J}_s = \begin{vmatrix} b_1 z_1^{\alpha_1 - 1} & -b_{-1} z_2^{\alpha_1 - 1} & 0 & \dots & 0 \\ 0 & b_2 z_2^{\alpha_2 - 1} & -b_{-2} z_3^{\alpha_2 - 1} & \dots & 0 \\ \dots & \dots & \dots & \dots & \dots \\ 1 & 1 & 1 & 1 & 1 \\ \dots & \dots & \dots & \dots & \dots \\ -b_{-n} z_1^{\alpha_n - 1} & 0 & 0 & \dots & b_n z_n^{\alpha_n - 1} \end{vmatrix} \begin{matrix} \\ \\ \\ \leftarrow s^{th} \ line \\ \\ \\ \end{matrix}.$$

LEMMA 14.7. *If*

$$\overline{J}_s = \begin{vmatrix} a_1 & -b_1 & 0 & \dots & 0 \\ 0 & a_2 & -b_2 & \dots & 0 \\ \dots & \dots & \dots & \dots & \dots \\ 1 & 1 & 1 & 1 & 1 \\ \dots & \dots & \dots & \dots & \dots \\ -b_n & 0 & 0 & \dots & a_n \end{vmatrix} \begin{matrix} \\ \\ \\ \leftarrow s^{th}\ \mathit{line} \\ \\ \\ \end{matrix},$$

then

$$\overline{J}_s = b_1 \dots [s] \dots b_n + b_1 \dots [s] a_{s+1} b_{s+2} \dots b_n + \dots + a_1 a_2 \dots [s] \dots a_n.$$

Notice that the $\overline{J}_s$ are polynomials with positive coefficients. From here we get that

$$\frac{B_1}{B_0} =$$

$$= -\frac{1}{B} \sum_{s=1}^{n} \nu_s^2 \mathfrak{M} \left[\frac{(z_1 + \dots + z_n)^{\alpha_s - 1} \overline{J}_s \prod_{j=1}^{n} \left(b_j^{\nu_1 - 1} z_j^{L+1-\alpha_j} + \dots + b_{-j}^{\nu_1 - 1} z_{j+1}^{L+1-\alpha_j} \right)}{z_1^L \dots z_n^L} \right].$$

Moreover, theorem 14.6 (formula (14.32)) implies that

$$d_k = -\frac{C_k}{B_k}, \qquad C_k > 0.$$

We claim that system (14.33) has only one root with positive coordinates $z_1, \dots, z_n$. To prove this we need a result from [**132**] (see also Theorem 11.4):

Let $\mathbf{f} = (f_1, \dots, f_n)$ be a system of polynomials in $\mathbf{R}^n$ and D a bounded domain with piecewise smooth boundary ∂D. The degree $\Omega(f, D)$ of the mapping $\mathbf{f}$ in D is then equal to

$$\Omega(f, D) = \frac{1}{2^n} \sum_{\Gamma_0} \operatorname{sign} \frac{\partial(f_1, \dots, f_n)}{\partial(x_1, \dots, x_n)},$$

where

$$\Gamma_0 = \partial D \cap \{x : \ |f_2| = \varepsilon, \dots, |f_n| = \varepsilon\}.$$

Recall that if the Jacobian $\partial \mathbf{f}/\partial \mathbf{x}$ preserves the sign in $\overline{D}$ then Ω is the number of roots of the system $\mathbf{f}$ in D.

LEMMA 14.8. *The matrix*

$$\Delta_n = \begin{vmatrix} a_1 & -b_1 & 0 & \dots & 0 & -\nu_1 \\ 0 & a_2 & -b_2 & \dots & 0 & -\nu_2 \\ \dots & \dots & \dots & \dots & \dots & \\ -b_n & 0 & 0 & \dots & a_n & -\nu_n \\ 1 & 1 & 1 & 1 & 1 & 0 \end{vmatrix}$$

satisfies the relation

$$\Delta_n = \nu_1 \overline{J}_1 + \dots + \nu_n \overline{J}_n,$$

where $\overline{J}_s$ *are the determinants from Lemma 14.6.*

Since the Jacobian of system (14.33) is equal to

$$J = \begin{vmatrix} b_1\alpha_1 z_1^{\alpha_1-1} & -b_{-1}\alpha_1 z_2^{\alpha_1-1} & 0 & \dots & 0 & -\nu_1 \\ 0 & b_2\alpha_1 z_2^{\alpha_2-1} & -b_{-2}\alpha_2 z_3^{\alpha_2-1} & \dots & 0 & -\nu_2 \\ \dots & \dots & \dots & \dots & \dots & \dots \\ -b_{-n}\alpha_n z_1^{\alpha_n-1} & 0 & 0 & \dots & b_n\alpha_n z_n^{\alpha_n-1} & -\nu_n \\ 1 & 1 & 1 & 1 & 1 & 0 \end{vmatrix},$$

we see from Lemmas 14.7 and 14.8 that $J > 0$ if $z_j = x_j$ and either $x_1 > 0, \dots, x_n > 0$, or $x_1 = 0, x_2 > 0, \dots, x_n > 0$, or $x_1 > 0,\ x_{n-1} > 0,\ x_n = 0$.

Consider now the cube

$$D_R^n = \{(\mathbf{x}, w):\ 0 < x_1 < R, \dots, 0 < x_n < R, -R < w < R\} \subset \mathbf{R}^{n+1}$$

and the set

$$\Gamma_R^n = \{\mathbf{x} \in \partial D_R^n:\ |w_2| = \varepsilon, \dots, |w_{n+1}| = \varepsilon\}.$$

Then Γ_R^n lies entirely in the sides $\{x_1 = 0\}, \dots, \{x_n = 0\}$, since the hyperplanes

$$x_1 + \dots + x_n = 1 \pm \varepsilon$$

do not intersect the sides $\{x_1 = R\}, \dots, \{x_n = R\}$, and $|w_2| = \varepsilon$ does not meet $\{w = \pm R\}$ (for sufficiently large R).

By the Tarkhanov Theorem 11.4 the number of roots of (14.33) with positive coordinates $x_1, \dots, x_n$ is equal to the number of points in Γ_R^n divided by 2^{n+1}.

So we must consider Γ_R^n on the sides $x_1 = 0, \dots, x_n = 0$. If $x_3 = 0$ then $\Gamma_R^n \cap \{x_3 = 0\} = \emptyset$ since the equations

$$\begin{aligned} b_2 x_2^{\alpha_2} &= \nu_2 w \pm \varepsilon, \\ -b_3 x_4^{\alpha_3} &= \nu_3 w \pm \varepsilon \end{aligned}$$

have no solutions. Similarly we have $\{x_j = 0\} \cap \Gamma_R^n = \emptyset$, for $j \geq 3$. If $x_1 = 0$ then we obtain

$$b_2 x_2^{\alpha_2} - b_{-2} x_3^{\alpha_3} = \nu_2 w \pm \varepsilon,$$

$$b_n x_n^{\alpha_n} = \nu_n w \pm \varepsilon, \tag{14.34}$$

$$x_2 + \dots + x_n = 1 \pm \varepsilon.$$

This system has practically the same form as (14.33) (if we assume $b_{-n} = 0$), but in $(n-1)$ variables. Applying our previous arguments we reduce it to a system of the form

$$\begin{aligned} b_n x_n^{\alpha_n} &= \nu_n w \pm \delta, \\ x_n &= 1 \pm \delta. \end{aligned} \tag{14.35}$$

The system (14.35) has four roots, and hence (by induction over n) the system (14.34) has 2^n roots. The case

$$\{x_2 = 0\} \cap \Gamma_R^n \neq \emptyset$$

is handled analogously. we thus find that Γ_R^n consists of 2^{n+1} points, and hence the system (14.33) has only one positive solution $x_1 > 0, \dots, x_n > 0$. The following theorem is thereby proved.

THEOREM 14.7. *The system (14.33) has only one root in the set*

$$x_1 > 0, \dots, x_n > 0, \quad w \in \mathbf{R}^1.$$

EXAMPLE 14.6. For $n = 2$ the system (14.33) takes the form

$$b_1 z_1^\alpha - b_{-1} z_2^\alpha = \beta w,$$
$$b_2 z_2^\beta - b_{-2} z_1^\beta = \alpha w,$$
$$z_1 + z_2 = 1.$$

We then have

$$J_1 = -\beta(b_2 z_2^{\beta-1} + b_{-2} z_1^{\beta-1}),$$
$$J_2 = \alpha(b_1 z_1^{\alpha-1} + b_{-1} z_2^{\alpha-1}),$$
$$B = b_1^\beta b_2^\alpha - b_{-1}^\beta b_{-2}^\alpha, \qquad L + 1 = \alpha\beta,$$
$$\det A =$$
$$= \frac{1}{B}((b_1 z_1^\alpha)^{\beta-1} + (b_1 z_1^\alpha)^{\beta-2}(b_{-1} z_2^\alpha) + \dots + (b_{-1} z_2^\alpha)^{\beta-1})((b_2 z_2^\beta)^{\alpha-1} + \dots + (b_{-2} z_1^\beta)^{\alpha-1}),$$

$$C^{-1} = \begin{pmatrix} b_1^\beta & -b_{-1}^\beta \\ -b_{-2}^\alpha & b_2^\alpha \end{pmatrix}, \qquad C = \begin{pmatrix} b_2^\alpha & b_{-1}^\beta \\ b_{-2}^\alpha & b_1^\beta \end{pmatrix}.$$

Furthermore,

$$a_{11}(z) = \frac{b_2^\alpha}{B}\left((b_1 z_1^\alpha)^{\beta-1} + \dots + (b_{-1} z_2^\alpha)^{\beta-1}\right),$$
$$a_{12}(z) = \frac{b_{-1}^\beta}{B}\left((b_2 z_2^\beta)^{\alpha-1} + \dots + (b_{-2} z_1^\beta)^{\alpha-1}\right),$$
$$a_{21}(z) = \frac{b_{-2}^\alpha}{B}\left((b_1 z_1^\alpha)^{\beta-1} + \dots + (b_{-1} z_2^\alpha)^{\beta-1}\right),$$
$$a_{22}(z) = \frac{b_1^\beta}{B}\left((b_2 z_2^\beta)^{\alpha-1} + \dots + (b_{-2} z_1^\beta)^{\alpha-1}\right).$$

It follows that

$$\frac{B_1}{B_0} =$$
$$= -\frac{1}{B}\mathfrak{M}\left[\left((\beta^2(z_1 + z_2)^{\alpha-1}(b_2 z_2^{\beta-1} + b_{-2} z_1^{\beta-1}) + \right.\right.$$
$$+\alpha^2(z_1 + z_2)^{\beta-1}(b_1 z_1^{\alpha-1} + b_{-1} z_2^{\alpha-1})) \times$$
$$\times \left((b_1 z_1^\alpha)^{\beta-1} + \dots + (b_{-1} z_2^\alpha)^{\beta-1}\right)\left((b_2 z_2^\beta)^{\alpha-1} + \dots + \right.$$
$$\left.\left. + (b_{-2} z_1^\beta)^{\alpha-1}\right) \Big/ \left(z_1^{\alpha\beta-1} z_2^{\alpha\beta-1}\right)\right] =$$
$$= -\frac{1}{B}\left(\beta^2 \sum_{k=0}^{\beta-1} \sum_{\beta s = \alpha k}^{\alpha(k+1)} C_{\beta s - \alpha k}^\alpha b_1^k b_{-1}^{\beta-k-1} b_2^s b_{-2}^{\alpha-s} + \right.$$
$$\left. + \alpha^2 \sum_{s=1}^{\alpha} \sum_{\alpha k = \beta(s-1)}^{\beta s} C_{\beta s - \alpha k}^\beta b_1^k b_{-1}^{\beta-k} b_2^{s-1} b_{-2}^{\alpha-s}\right).$$

EXAMPLE 14.7. Consider the system

$$\begin{aligned} b_1 z_1^\alpha - b_{-1} z_2^\alpha &= w, \\ &\cdots \\ b_{n-1} z_{n-1}^\alpha - b_{-(n-1)} z_n^\alpha &= w, \\ b_n z_n - b_{-n} z_1 &= \alpha w, \\ z_1 + \ldots + z_n &= 1, \end{aligned}$$

corresponding to a mechanism without interactions between different intermediates, in which one stage is linear and the others are nonlinear and of equal order. In this case we get

$$L + 1 = \alpha, \; B = b_1 \ldots b_{n-1} b_n^\alpha - b_{-1} \ldots b_{-(n-1)} b_{-n}^\alpha,$$

$$\det A = \frac{1}{B}\left((b_n z_n)^{\alpha-1} + (b_n z_n)^{\alpha-2}(b_{-n} z_1) + \ldots + (b_{-n} z_1)^{\alpha-1}\right),$$

$$\begin{aligned} J_1 &= (-1)^{n+1}\alpha^{n-2}(b_{-2}\ldots b_{-n} z_3^{\alpha-1}\ldots z_n^{\alpha-1} + \\ &\quad b_2 \ldots b_{-n} z_2^{\alpha-1}[3]\ldots z_n^{\alpha-1} + \ldots + b_1 \ldots b_n z_2^{\alpha-1}\ldots z_{n-1}^{\alpha-1}), \\ &\cdots \\ J_n &= \alpha^{n-1}(b_1 \ldots b_{-(n-1)} z_2^{\alpha-1}\ldots z_n^{\alpha-1} + \\ &\quad b_1 \ldots z_1^{\alpha-1}[2]\ldots z_n^{\alpha-1} + \ldots + b_1 \ldots b_{n-1} z_1^{\alpha-1}\ldots z_{n-1}^{\alpha-1}). \end{aligned}$$

We thus have

$$\begin{aligned} &\frac{B_1}{B_0} = \\ &= -\frac{1}{B}\mathfrak{M}\left[\left((z_1 + \ldots + z_n)^{\alpha-1}(\alpha^{n-2}J_1 + \ldots + \alpha^n J_n\right)\times \right. \\ &\left.\times\left((b_n z_n)^{\alpha-1} + \ldots + (b_{-n} z_1)^{\alpha-1}\right)\right)/\left(z_1^{\alpha-1}\ldots z_n^{\alpha-1}\right)\Big] = \\ &= -\frac{\alpha^{n-2}}{B}\left(\alpha^2(b_1 \ldots b_{n-1} b_n^{\alpha-1} + b_{-1}\ldots b_{-(n-1)} b_{-n}^{\alpha-1} + \ldots + \right. \\ &\left. + b_2 \ldots b_{n-1}(b_n + b_{-n})^\alpha\right). \end{aligned}$$

As before we restrict ourselves to finding the two first coefficients of the kinetic polynomial.

Formulas (14.30)–(14.32) permit us to write the other coefficients in symbolic form. Experience shows however that the expressions one obtains are rather cumbersome. But there does exist a finite algorithm for finding all the coefficients B_j $(j > 1)$ of the kinetic polynomial in symbolic form. The methods and software of computer algebra can become decisive for this approach. The basic steps of the algorithm are the determination of the expressions for $a_{jk}(\mathbf{z})$ in (14.28) and the reduction of similar terms in the product of polynomials appearing in (14.29)–(14.32). All of these operations can result in rather unwieldy analytic transformations for B_j (only the structure of B_0 is simple), and therefore along with the construction of the kinetic polynomial it is necessary also to develop methods to simplify it, in view for instance of the sensitivity of its coefficients to variations of the parameters $b_s^\pm$.

The results of this section (Theorem 14.7) show that, by using our new approach, we have succeeded not only to simplify the algorithm for finding the kinetic polynomial but also to obtain some meaningful conclusions concerning quasi-thermodynamic properties of systems without interacting substances.

It seems promising to apply this approach also on many-route schemes. Furthermore, it is quite reasonable to ask for a way of writing the equations of chemical kinetics in terms of the observed reaction rates also in the nonstationary case. Here the initial system of equations

$$\mathbf{x} = \mathbf{\Gamma}^T \mathbf{w}(\mathbf{x})$$

can be written (after an elimination procedure), for example, in the form $F(\dot{w}) = R(w)$, where F is a linear differential operator and $R(w)$ is the kinetic polynomial.

15. Construction of the kinetic polynomial in the general case

The general case differs from the single-route mechanism (see Sec. 14) in that the stationary equation (14.1) (analogous to (14.2)) is of the form

$$w_s(\mathbf{z}) = \sum_{i=1}^{p} \nu_s^{(i)} w^{(i)}, \quad s = 1, \dots, n, \qquad \mathbf{Az} = \mathbf{a}, \tag{15.1}$$

where $\nu_s^{(i)}$ is the stoichiometric number at the s-th stage in the i-th route, $w^{(i)}$ is the reaction rate of the i-th route, p is the number of routes. In general there can be several linear conservation laws $\mathbf{Az} = \mathbf{a}$. The many-routeness of the reaction means that for the stoichiometric matrix $\mathbf{\Gamma}$ there exists p vectors $\boldsymbol{\nu}^{(i)}$ of stoichiometric numbers corresponding to linear dependencies among the rows, i.e. $\boldsymbol{\nu}^{(i)}\mathbf{\Gamma} = 0$. It is these last relations that allow one to write the stationary equations $\mathbf{\Gamma}^T\mathbf{w} = 0$ in the form (15.1).

In the linear case all necessary quantities can be written explicitly. For this purpose the language of graph theory is the most natural [**152**]. To each chemical transformation scheme one associates its reaction graph. For example, a two-route linear scheme can be represented by the following graphs

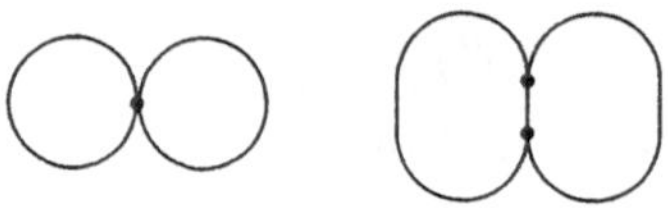

(15.2)

in which the two cycles correspond to the two routes. The difference between the two graphs (15.2) is that the first has one common vertex (a substance), while the second has two common vertices (an entire stage). The general form of the stationary kinetic equations for a complex catalytic reaction with a many-route linear mechanism was presented and investigated in the papers [**70, 145, 146, 147, 148, 153**]. The equations look like

$$w^{(s)} = \frac{\sum_{i=1}^{p} B_C^{(i)} P_i}{\sum_x D_x}, \tag{15.3}$$

where $B_C^{(i)}$ is the cycle characteristic for the i-th route, and P_i is a conjugation parameter of the i-th cycle. The expressions $B_C^{(i)}$ are analogous to B_0 for each cycle, i.e.

$$B_C^{(i)} = \prod_j b_j^{\nu_j^{(i)}} - \prod_j b_{-j}^{\nu_j^{(i)}}.$$

The conjugation characteristic P_i corresponds to structural singularities of the reaction graph.

From our experience with the construction of the kinetic polynomial for a single-route mechanism we expect that also in the general case the structural singularities of the initial system should appear in the final result of the elimination of variables from the stationary equations.

The problem of constructing the kinetic polynomial for a many-route nonlinear reaction can be formulated as an elimination problem for the system (15.1) with respect to

$$w^{(i)}, \quad i = 1, \dots, p.$$

After elimination from (15.1) we obtain one kinetic polynomial equation

$$R_i(w^{(i)}) = 0 \tag{15.4}$$

for the determination of each $w^{(i)}$. This problem fits quite well into the general scheme of elimination that we developed in Ch. 2. Here we try to trace the specific character of the structure of the reaction graph in the final result, and also to indicate those cases where the methods we used to construct the kinetic polynomial for a single-route mechanism are applicable in a more general setting.

In Sec. 9 (system (9.12), Theorem 9.3, Example 9.5) we already encountered a system of nonlinear algebraic equations corresponding to a certain many-route scheme. Such schemes arise when only one rate w for some route and only one linear conservation law are singles out in the initial stationary equations

$$\mathbf{\Gamma}^T(\mathbf{w}(\mathbf{z})) = 0.$$

In this way the problem of constructing the kinetic polynomial can be viewed as several elimination problems for each route.

In system (9.12) it is the homogeneity assumption on the polynomials

$$P_j, \quad j = 1, \dots, n+1,$$

that restricts the generality. The scheme for elimination of variables developed in Ch. 2 can be extended also to the general case but it becomes quite awkward.

We now consider some particular cases of two-route mechanisms, for which we can apply the construction methods for the kinetic polynomial of a single-route transformation scheme. The elimination problem for the system (15.1) can be formulated as follows: Eliminate all variables z_j from (15.1), i.e. find a system of polynomials in $w^{(k)}$ for which the zeros are the projections of the roots of the system (15.1) on the axes $w^{(k)}$. This procedure can indeed be carried out by using the classical method of reducing the system to "triangular" form. But in general the number of equations may increase and the system for the variables $w^{(k)}$ can contain a greater number of equations than the number of variables $w^{(k)}$. For general systems it is easy to construct such examples. Let us show that this situation is common also in the case of system (15.1).

EXAMPLE 15.1. We consider the following system of equations for a two-route mechanism

$$\begin{aligned} b_1 z_1^2 - b_{-1} z_2^2 &= w^{(1)}, \\ b_2 z_2^2 - b_{-2} z_1^2 &= w^{(1)}, \\ b_3 z_2^3 - b_{-3} z_3^3 &= w^{(2)}, \\ b_4 z_3^3 - b_{-4} z_2^3 &= w^{(2)}, \\ z_1 + z_2 + z_3 &= 1. \end{aligned}$$

Suppose

$$b_{-j} = 0 \text{ and } b_j = 1, \; j = 1,2,3,4.$$

Then we get

$$\begin{aligned} z_1^2 &= w^{(1)}, \\ z_2^2 &= w^{(1)}, \\ z_2^3 &= w^{(2)}, \\ z_3^3 &= w^{(2)}, \\ z_1 + z_2 + z_3 &= 1. \end{aligned} \tag{15.5}$$

The following vectors are the roots of system (15.5) $(z_1, z_2, z_3, w^{(1)} w^{(2)})$:

1) $\left(\frac{1}{3}, \frac{1}{3}, \frac{1}{3}, \frac{1}{9}, \frac{1}{27}\right)$,

2) $\left(\frac{1}{2} - i\frac{\sqrt{3}}{6}, \frac{1}{2} - i\frac{\sqrt{3}}{6}, i\frac{\sqrt{3}}{3}, \frac{1}{6} - i\frac{\sqrt{3}}{6}, -i\frac{\sqrt{3}}{9}\right)$,

3) $\left(\frac{1}{2} + i\frac{\sqrt{3}}{6}, \frac{1}{2} + i\frac{\sqrt{3}}{6}, -i\frac{\sqrt{3}}{3}, \frac{1}{6} + i\frac{\sqrt{3}}{6}, i\frac{\sqrt{3}}{9}\right)$,

4) $(-1, 1, 1, 1, 1)$,

5) $\left(\frac{1}{2} + i\frac{\sqrt{3}}{2}, -\frac{1}{2} - i\frac{\sqrt{3}}{2}, 1, -\frac{1}{2} + i\frac{\sqrt{3}}{2}, 1\right)$,

6) $\left(\frac{1}{2} - i\frac{\sqrt{3}}{2}, -\frac{1}{2} + i\frac{\sqrt{3}}{2}, 1, -\frac{1}{2} - i\frac{\sqrt{3}}{2}, 1\right)$.

Assume that the coordinates $w^{(1)}$ and $w^{(2)}$ of these roots are the roots of the system

$$\begin{aligned} P_1(w^{(1)}, w^{(2)}) &= 0, \\ P_2(w^{(1)}, w^{(2)}) &= 0, \end{aligned} \tag{15.6}$$

and moreover, that (15.6) has no other roots (including infinite roots). Then by the Bézout theorem the degrees of P_1 and P_2 can be either 1 and 6, or 2 and 3. Let us show that there exists no polynomial P_1 of degree two, whose zero set contains the six points $(w^{(1)}, w^{(2)})$. Indeed, assume

$$P_1 = a_1 w^{(1)2} + a_2 w^{(1)} w^{(2)} + a_3 w^{(2)2} + a_4 w^{(1)} + a_5 w^{(2)} + a_6.$$

Substituting the six pairs $(w^{(1)}, w^{(2)})$ into the equation $P_1 = 0$, we get six linear equations with respect to a_j. It is not difficult to check that the determinant of this system is not equal to zero, and hence the only solution is

$$a_j = 0, \quad j = 1, \dots, 6.$$

Therefore, after the elimination of all variables z_j from the system (15.5) the number of equations in $w^{(1)}, w^{(2)}$ will be greater than 2.

Now we consider some particular cases of two-route mechanisms.

1. One route is arbitrary and the second is linear:

$$\begin{aligned}
b_1 \mathbf{z}^{\alpha^1} - b_{-1} \mathbf{z}^{\beta^1} &= \nu_1 w^{(1)}, \\
&\cdots \\
b_k \mathbf{z}^{\alpha^k} - b_{-k} \mathbf{z}^{\beta^k} &= \nu_k w^{(1)}, \\
b_{k+1} z_k - b_{-(k+1)} z_{k+1} &= w^{(2)}, \\
&\cdots \\
b_{n+1} z_n - b_{-(n+1)} z_k &= w^{(2)}, \\
z_1 + \dots + z_n &= 1,
\end{aligned} \tag{15.7}$$

where

$$\alpha^j = (\alpha_1^j, \dots, \alpha_k^j), \quad \beta^j = (\beta_1^j, \dots, \beta_k^j).$$

Eliminating all variables except z_k from the system

$$\begin{aligned}
b_{k+1} z_k - b_{-(k+1)} z_{k+1} &= w^{(2)}, \\
&\cdots \\
b_n z_n - b_{-(n+1)} z_k &= w^{(2)},
\end{aligned}$$

we have

$$\begin{aligned}
z_{k+1} &= c_{k+1} z_k, \\
&\cdots \\
z_n &= c_n z_k, \\
w^{(2)} &= c_0 z_k,
\end{aligned}$$

where all constants $c_j \not\equiv 0$, and are expressed through b_j and b_{-j}, $j \geq k+1$.

Inserting these expressions for z_j into the last equation of system (15.7) arrive at

$$\begin{aligned}
b_1 \mathbf{z}^{\alpha^1} - b_{-1} \mathbf{z}^{\beta^1} &= \nu_1 w^{(1)}, \\
&\cdots \\
b_k \mathbf{z}^{\alpha^k} - b_{-k} \mathbf{z}^{\beta^k} &= \nu_k w^{(1)}, \\
z_1 + \dots + z_{k-1} + c z_k &= 1,
\end{aligned} \tag{15.8}$$

where $c = 1 + c_{k+1} + \ldots + c_n$. In order to exactly get the system of a single-route mechanism, we must make the change $cz_k \to z_k$. We then have

$$b_1'\mathbf{z}^{\alpha^1} - b_{-1}'\mathbf{z}^{\beta^1} = \nu_1 w,$$

$$\cdots\cdots$$

$$b_k'\mathbf{z}^{\alpha^k} - b_{-k}'\mathbf{z}^{\beta^k} = \nu_k w,$$

$$z_1 + \ldots + z_k = 1,$$

where

$$b_j' = b_j\, c^{-\alpha_k^j}, b_{-j}' = b_{-j}\, c^{-\beta_k^j}.$$

It is interesting that this transformation does not influence the form of the constant term of the kinetic polynomial (in w). Indeed, the constant term is given by

$$a_0 \sim \left(b_1'^{\nu_1} \ldots b_k'^{\nu_k} - b_{-1}'^{\nu_1} \ldots b_{-k}'^{\nu_k}\right)^s =$$

$$= \left(b_1^{\nu_1} \ldots b_k^{\nu_k} c^{-\sum \nu_j \alpha_k^j} - b_{-1}^{\nu_1} \ldots b_{-k}^{\nu_k} c^{-\sum \nu_j \beta_k^j}\right) \sim$$

$$\sim \left(b_1^{\nu_1} \ldots b_k^{\nu_k} - b_{-1}^{\nu_1} \ldots b_{-k}^{\nu_k} c^{\sum \nu_j (\alpha_k^j - \beta_k^j)}\right),$$

but since $\mathbf{\Gamma}$ is a singular matrix

$$\sum \nu_j \left(\alpha_k^j - \beta_k^j\right) \sim \det \mathbf{\Gamma} = \det\left(\left(\alpha_j^i - \beta_j^i\right)\right)_{i,j=1}^k = 0,$$

and hence

$$a_0 \sim \left(b_1^{\nu_1} \ldots b_k^{\nu_k} - b_{-1}^{\nu_1} \ldots b_{-k}^{\nu_k}\right).$$

It is clear that a similar procedure can be applied to a many-route mechanism if all its stages except one are linear.

2. Consider now the system

$$\begin{aligned}
b_1\mathbf{z}^{\alpha^1} - b_{-1}\mathbf{z}^{\beta^1} &= \nu_1 w^{(1)}, \\
&\cdots \\
b_k\mathbf{z}^{\alpha^k} - b_{-k}\mathbf{z}^{\beta^k} &= \nu_k w^{(1)}, \\
b_{k+1}\mathbf{t}^{\alpha^{k+1}} - b_{-(k+1)}\mathbf{t}^{\beta^{k+1}} &= \nu_{k+1} w^{(2)}, \\
&\cdots \\
b_{n+1}\mathbf{t}^{\alpha^{n+1}} - b_{-(n+1)}\mathbf{t}^{\beta^{n+1}} &= \nu_{n+1} w^{(2)}, \\
z_1 + \ldots + z_n &= 1,
\end{aligned} \tag{15.9}$$

where $\mathbf{z} = (z_1, \ldots, z_k)$, $\mathbf{t} = (z_k, \ldots, z_n)$. Assuming that all monomials $\mathbf{t}^{\alpha^j}$ and $\mathbf{t}^{\beta^j}$, $j \geq k+1$, are of the same degree, and eliminating $\mathbf{t}$ from the last equations, we get the system

$$\begin{aligned} b_1\mathbf{z}^{\alpha^1} - b_{-1}\mathbf{z}^{\beta^1} &= \nu_1 w^{(1)}, \\ &\dots \\ b_k\mathbf{z}^{\alpha^k} - b_{-k}\mathbf{z}^{\beta^k} &= \nu_k w^{(1)}, \\ P_{k+1}(\mathbf{z}) &= 0, \\ &\dots \\ P_{n+1}(\mathbf{z}) &= 0, \\ z_1 + \ldots + z_n &= 1, \end{aligned} \qquad (15.10)$$

and

$$b_{k+1}\mathbf{t}^{\alpha^{k+1}} - b_{-(k+1)}\mathbf{t}^{\beta^{k+1}} = \nu_{k+1} w^{(2)}, \qquad (15.11)$$

(if $\nu_{k+1} \neq 0$). The system (15.10) together with equation (15.11) is equivalent to system (15.9). But for (15.10) the Theorem 9.3 is valid ($P_{k+1}, \ldots, P_{n+1}$ are homogeneous). Therefore, using Theorem 9.3 we can find the coefficients of the logarithmic derivative of the resultant $R(w^{(1)})$ of system (15.9).

Thus, if in a two-route mechanism all the equations corresponding to one route are of the same degrees, then Theorem 9.3 can be applied to the system (see also Example 9.5). The same goes also for many-route mechanism.

In the general the picture is not yet as clear as it is with single-route mechanisms. It can be conjectured that the structure of the initial reaction graph entirely determines the structure of the kinetic polynomials for each route. In addition, the most transparent results are obtained for mechanisms for which the cycles in the reaction graph have only common vertices (as for example in the first scheme of (15.2)). The experience from the analysis of linear schemes shows that the expressions for the reaction rate essentially depend on the character of the "linkage" among the cycles. For "soft linkage" (common vertices) the cycle characteristics are summed with certain positive weights, but for "rigid linkage" there appears one more conjugation characteristic of cycles. In terms of systems of equations these two cases correspond to initial systems having subsystems with intersections in one or several variables.

Now we show that the approaches that we have developed in Ch. 2 and 3 can be used in the more general case when the kinetic equations are not algebraic. For example, nonisothermal systems are transcendental with respect to the temperature variable. Moreover, the multidimensional logarithmic residue permits us to give a generalized interpretation of the kinetic polynomial.

Recall that the kinetic polynomial P is the polynomial in the rates $w(\mathbf{z})$ corresponding to the mass-action law. For more complicated kinetic laws at the individual stages, or in the nonisothermal case we have no longer a polynomial, and here the elimination procedure does end up. But then the integral form for representing the kinetic equations can be useful. We assume that it is written in the form

$$f_i(\mathbf{z}) = 0, \qquad i = 1, \ldots, n, \qquad (15.12)$$

where f_j are quite arbitrary (but holomorphic) functions, and $\mathbf{z} = (z_1, \ldots, z_n)$, with n being the number of independent variables. Introducing the additional variable w

we consider the system

$$F_i(\mathbf{z},w)=0, \qquad i=1,\dots,n+1, \tag{15.13}$$

instead of (15.12).

In (15.13) we may now try to eliminate all variables except w, i.e. reduce (15.13) to a single equation

$$\Phi(w)=0. \tag{15.14}$$

For the function Φ we then have the following relations (by a Theorem due to Yuzhakov ([**11**, p.55], or Sec. 9)):

$$\ln\Phi=\sum_j \ln F_{n+1}(\mathbf{z}_{(j)}(w),w)= \tag{15.15}$$

$$=\frac{1}{(2\pi i)^n}\sum_j\int_{\Gamma_j}\ln F_{n+1}(\mathbf{z},w)\frac{dF(z,w)[n+1]}{F_1\dots F_n},$$

where the surfaces Γ_j are given by

$$\Gamma_j=\{\mathbf{z}: \ |F_1(\mathbf{z},0)|=\varepsilon_1,\dots,|F_n(\mathbf{z},0)|=\varepsilon_n\},$$

$\mathbf{z}_{(j)}(w)$ are the roots (for fixed w) of the system (15.13) without the last equation, and

$$dF[k]=dF_1\wedge\dots\wedge dF_{k-1}\wedge dF_{k+1}\wedge\dots\wedge dF_n.$$

It can be shown that (15.15) implies the formulas (as in Theorem 9.1)

$$\frac{\Phi'(w)}{\Phi(w)}=\frac{1}{(2\pi i)^n}\sum_j\int_{\Gamma_j}\sum_{k=1}^{n+1}(-1)^{n+k-1}\frac{F_k(\mathbf{z},w)dF[k]}{F_1\dots F_{n+1}} \tag{15.16}$$

and

$$\frac{1}{k!}\frac{d^k\ln\Phi(w)}{dw^k}\bigg|_{w=0}= \tag{15.17}$$

$$=\frac{1}{(2\pi i)^n}\sum_j\sum_{l=1}^{n+1}\sum_{\substack{||\alpha||=k\\ \alpha_1=\dots=\alpha_{l-1}=0\\ \alpha_l>0}}\frac{(-1)^{n+l-1}}{\alpha!}\int_{\Gamma_j}\left(\frac{F_l'}{F_l}\right)^{(\alpha_l-1)}\left(\frac{dF}{F}[l]\right)\Bigg|_{w=0}$$

where (α) denotes a derivative of order α in w.

Formulas (15.17) give expressions for the coefficients in the power series expansion in w of the logarithm of the resultant Φ. The coefficients in the series expansion of the resultant Φ itself can then be found by means of the Newton recursion formulas, starting from (15.17).

If the F_i are polynomials then the integrals in (15.16), (15.17) may be calculated in closed form and the resultant Φ is a polynomial in w. If the F_i are more complicated functions then we must resort to approximate computations of these integrals.

A more general situation, in comparison with the mass-action-law is the kinetics of Marselen de Donde. In this case the kinetic law of an elementary stage is given by

$$w_j^+ = \exp\left(\sum \alpha_{ij}\mu_i(\mathbf{z})\right),$$

where $\mu_i(\mathbf{z})$ is the chemical potential of the i-th substance. For ideal kinetics $\mu_i = \mu_i^0 + \ln z_i$ and w_j^+ has the usual polynomial form

$$w_j^+ = k_j \prod_i z_i^{\alpha_{ij}}.$$

If the chemical potentials contain correction terms, so that $\mu_i = \mu_i^0 + \ln z_i + f_i(z_i)$, then

$$w_j^+ = k_j^+ \prod_i z_i^{\alpha_{ij}} e^{\alpha_{ij} f_i(z_i)}.$$

Introducing the notation $x_i = z_i e^{f_i(z_i)}$, we can represent the kinetic law of the individual stage in the typical form for the mass-action-law

$$w_j^+ = k_j^+ \prod_i x_i^{\alpha_{ij}}.$$

From here on the procedure for obtaining the kinetic polynomial remains without changes. When μ_i is a more complicated function of $\mathbf{z}$, for example if f_i depends also on the other concentrations, the algorithm for obtaining the resultant $\Phi(w)$ becomes substantially more complicated. The kinetic polynomial for the simplest case of Marselen de Donde kinetics is of the same form as the polynomial $P(w)$ obtained for the same reaction mechanism with ideal kinetics.

It is interesting that the structure of $P(w)$ in the analyzed non-ideal case only depends on the stoichiometrics. The form of the correction term $f_i(\mathbf{z})$ in no way influences the final result. According to (15.12)–(15.17) the standard reduction of the chemical kinetic equations (CKE) to the kinetic polynomial (KP) can be schematically represented by the diagram

$$\mathrm{CKE}(\mathbf{z}) \Rightarrow \mathrm{CKE}(\mathbf{z}, w) \Rightarrow \Phi(w) \Rightarrow \exp \int \frac{\Phi'}{\Phi} dw \Rightarrow \mathrm{KP}(w), \tag{15.18}$$

where the expression for Φ'/Φ is given by formula (15.16).

In the standard procedure (15.18) the last passage from the integral representation of the kinetic equations to the kinetic polynomial is relatively simple to carry out in the case of the mass-action law. An analogue of the kinetic polynomial for more complicated kinetic laws can be obtained, if the integrals in (15.16) can be calculated in closed form. In the general case the elimination procedure terminates at the stage were $\ln \Phi$ is obtained:

$$\ln \Phi(w) = \int \frac{\Phi'(w)}{\Phi(w)} dw. \tag{15.19}$$

The representation (15.19) can be interpreted as an integral representation of the kinetic equations. The left hand side in the identity (15.19) depends in general on the measured value w in a complex fashion, while on the right hand side we have a function which is calculated by the formulas (15.16) and (15.17), on the basis of theoretical considerations regarding the reaction mechanism, such as the kinetic law of each individual stage, the steady state equation, the conservation laws, etc. The integral formula (15.19) has a well defined physical and chemical sense. If the integrand $\Phi' dw/\Phi$ represents the sense of the local reaction order (which is quite natural) then relation (15.19) shows that the integral of the local reaction order is the global brutto-order $\ln \Phi(w)$ of the reaction. As shown by formulas (15.16) the local reaction order can be calculated from theoretical representations which are put into the initial kinetic model. In this context we remark that some of the kinetic details will be reflected in the final result $\Phi(w)$. They are cancelled in the process of elimination of the concentrations z_i.

We consider the non-isothermal case separately. The equations of non-isothermal kinetics can be written in the form

$$f_i(\mathbf{z}, T) = 0, \qquad i = 1, \dots, n, \tag{15.20}$$

$$G(\mathbf{z}, T) = 0. \tag{15.21}$$

where the equations (15.20) correspond to the stationary concentration conditions, and (15.21) expresses the thermal balance. In simplest case T occurs as an exponential function of the rate constants of the isolated stages in the detailed reaction mechanism, and therefore the system (15.20) and (15.21) is already non-algebraic.

Proceeding by analogy with the case of linear dependence in $\mathbf{z}$ (monomolecular reactions), we find from (15.20) a relation $\mathbf{z} = \varphi(T)$. After substituting φ into (15.21) we get the temperature equation

$$G(\varphi(T), T) = 0. \tag{15.22}$$

However, in the general case such an elimination method for $\mathbf{z}$ becomes complicated, since for nonlinear transformation mechanisms the function $\varphi(T)$ in (15.22) can be many-valued.

For a complex reaction the thermal discharge is determined by the brutto-reaction, and then the equations of mass and thermal balances can be written in the form

$$F_i(\mathbf{z}, w, T) = 0, \qquad i = 1, \dots, n+1, \tag{15.23}$$

$$G(w, T) = 0. \tag{15.24}$$

Regarding T as a parameter in (15.23), we can, by analogy with the isothermal case, construct the resultant $\Phi(w, T)$, and together with (15.24) get equations for the steady states w, T:

$$\begin{aligned} \Phi(w, T) &= 0, \\ G(w, T) &= 0. \end{aligned} \tag{15.25}$$

Systems of type (15.25) are obtained in the quasi-stationary case with respect to all or some of the concentrations $\mathbf{z}$. Hence the method of elimination of unknowns that we have developed in Ch. 2, 3 can be readily adapted to the non-isothermal case. Thus, the concept of the kinetic polynomial, originally developed for single-route reactions with ideal kinetics, can be extended to more general types of kinetics and also to the non-isothermal case. For the interpretation of complicated experimentally observed data of kinetic dependencies the integral version (15.19) of the chemical kinetic equations can be useful.

CHAPTER 4

COMPUTER REALIZATIONS

16. Analytic manipulations on the computer

16.1. SPECIAL FEATURES, FIELD OF APPLICATIONS

Over the last three decades a new part of computational mathematics has been formed. It is called computer algebra. By definition [28] this is the part of computer science which deals with development, analysis, realization and applications of algebraic algorithms. Computer algebra has the following special features:

1) The algorithms work with data represented in symbolic form.

2) The solution is carried out on the basis of an algebraization of the given problem.

3) There exist algebraic analogues of the analytic operations.

In the systems for analytic calculations (CA) that have been developed, solutions to most of the classical algebraic problems are realized: Calculation and simplification of polynomial expressions, determination of the greatest common divisor of polynomials, matrix algebra, etc. [28, 122]. Although the word "algebra" is part of the title of this discipline, the analytic transformations that are performed include most of the operations from mathematical analysis: The determination of limits, differentiation, integration (and integrability checking), the solving of differential equations, expansion in series [67].

The use of CA requires (as a rule) a larger computer capacity than is needed for numerical analysis. So a pure replacement of "approximate" numerical methods by "exact" analytic ones is usually not motivated. Computer algebra has its own specific class of problems, some of which are related to mathematical modelling.

Firstly, there are problems where numerical results are uninformative or represented by a lot of data (for example, the problem of computing the number of real roots of a polynomial). Moreover, the round-off errors in numerical methods can lead to a false solution (for example, in calculating the rank of a matrix).

Secondly, there are problems connected with unwieldy mathematical models and the development of numerical methods of solution (for example, generation of differential equations of chemical kinetics [69]). Here analytic manipulations can be used to handle the computer programs [83].

Thirdly, there exist promising systems, where a combination of symbolic and numerical methods is used. For example, in [58] the problem of decomposing the parameter space for a linearized dynamical model into suitable domains was solved via analytic solutions of the linear algebra problems followed by an application of numerical methods.

The basic mathematical problems which can be solved with the help of CA are:

Various (canonical) simplifications of expressions, i.e. the construction of a simpler object being equivalent to the given one, or the determination of a unique representation for equivalent objects (for example, the simplification of expressions with radicals);

Analysis of algebraic structures (for example, groups); Integration of functions and summation of series (this problem is considered in detail in [**62**]); Solution procedures in algebra, in particular, the method of elimination of quantors. This type of problems (Sec. 17) has stimulated a lot of research in computer algebra: The determination of resultants of polynomials in several variables, the localization of roots of polynomials, the calculations of greatest common divisors (GCD) for polynomials, and of the sequence of polynomial remainders.

The capabilities of modern systems of CA have been widely exploited in different branches of science. One of the most developed systems — MACSYMA — has for instance been applied in acoustics, econometrics, elementary particle physics, robotechnology, thermodynamics, genetics, control theory, theory of aerials, deformation theory, sky mechanics, fluid mechanics, solid physics, theory of relativity, chemistry and chemical technology [**123**]. The application of CA systems to physics have been particularly successful [**76**]. On the other hand (as was pointed out in [**28**]) "the striking feature of the most important applications is that they are based on simple and rather straightforward computational methods" (for instance, the arithmetic of polynomials, calculation of determinants of matrices with symbolic elements).

A typical example is the evolution of computer algebra applications in chemical kinetics. In 1967 Silvestri and Zahner [**131**] simply used the Cramer rule to derive the kinetic equation for catalytic reactions with a linear mechanism. (Cristiansen, Way and Preter were pioneers in applying determinants to chemical kinetics). The more recent PL-program for deriving the equations of fermentative reactions is based on an analogous transfer [**74**]. At the same time ideas based on graph theoretical methods were developed: First for fermentative [**93**], and later for heterogeneous catalytic reactions [**150, 152**]. It was exactly this approach that was used by Semenov [**71**] to create the system ARAP for analysis of the kinetics of complex reactions with linear mechanisms. In contrast to previous developments he used more complicated and effective mathematical tools, namely nontrivial representations of graphs [**56**]. This system was realized with the help of the specialized language YARMO for analytic transformations. Problems which were closely related to computer algebra were solved through the creation of a system for automatic kinetic calculations [**121**]. Here the values of the target function and its derivatives, which are necessary for solving data processing problems of kinetic experiments, are determined on the basis of the concept of numerical graphs. The expressions are represented in the form of a graph with the nodes being algebraic operators, while the edges represent variables and the expressions are decomposed into blocks of the type

$$y = \sum_{i=1}^{n} a_i x_i, \ dx_1/dx_2, \ \exp(x)$$

etc.

Another direction for computer algebra applications in chemical kinetics is the analysis of the dynamics of complex reactions. Procedures for analyzing conditions for (non)-uniqueness of steady states and their stability, were realized in a program of Ivanova and Tarnopolsky [**91**] which calculates the characteristic equation in symbolic form, and which permits to distinguish structural fragments of the reaction graph, thereby predicting the type of dynamic behavior. Clarke created a system (in the language APL) for qualitative analysis of the dynamics and stability of reaction systems [**60**]. This allows one (on the basis of the theory in [**60**]) to determine the basic dynamical features of the given reaction: In particular it generates all reactions which are possible for a given collection of substances. It also checks the validity, uniqueness and global attraction of stationary manifolds, as well as the stability of the steady states in terms of properties of the reaction graph etc.

Thus the level of a computer algebra application is largely determined by the theoretical level of the branch where it is applied. On the other hand, the methods of computer algebra also allow one to formulate applied problems in a new way.

16.2. THE SYSTEM MAPLE

There exist a lot of computer algebra systems. In Russia alone more than 40 different systems have been elaborated [**83**]. The best known Russian systems are ANALITIK, Algebra 0.5, SIRIUS, and SRM. Their capabilities and applications are described in [**67, 76**]. Widespread western systems such as ALTMAN, FORMAS, MACSYMA, muMATH, REDUCE, SAC etc. are classified in [**28**]. Lately, the systems Axiom, Maple [**54, 55**] and Mathematica [**144**] have gained wide recognition.

In the Russian edition of our book we used the system REDUCE [**84**]. Earlier in [**108**] we used a system from [**126**]. For the English edition we have chosen the system Maple. This system was given to us by Sergei Znamenskii (Krasnoyarsk EmNet Center, DOS - 5203 - 429023 - 1). We use the Manual of the authors [**54, 55**].

Maple is a mathematical manipulation language. A basic feature of such a language is the ability to perform simplification of expressions and other transformations involving unevaluated elements.

In Maple, statements are normally evaluated as far as possible in the current environment. For example, the statement

`a := 1;` assigns the value `1` to the name `a`. If this statement is later followed by the statement

`x := a+b;` then the value `1+b` is assigned to the name `x`.

As each complete statement is entered by the user, it is evaluated and the results are printed on the output device. The printing of expressions is normally presented in a two-dimensional, multi-line format designed to be as readable as possible on standard computer terminals.

There is an on-line *help* facility in the Maple system. To start using it, enter the command ? and some information about the help facility will be displayed at the terminal. For example,

```
?index
?gcd.
```

A Maple session is typically initiated by the command **maple**. When a session is initiated, the maple leaf logo is displayed and a prompt character, such as '>', appears.

The capabilities of Maple include:

algebraic transformations, simplification, ordering of polynomial and rational functions,

a developed apparatus of substitutions,

calculation with symbolic matrices,

integer and real arithmetic of arbitrary exactness,

introduction of new functions and extension of syntactical program,

analytic differentiation and integration,

factorization of polynomials.

Maple supports univariate and multivariate polynomials in both expanded and unexpanded form.

```
> (x+1)^4*(x+2)^2;
```

$$(x+1)^4(x+2)^2,$$

```
> expand(");
```

$$x^6+8x^5+26x^4+44x^3+41x^2+20x+4,$$

```
> factor(");
```

$$(x+1)^4(x+2)^2.$$

The double-quote symbol '"' refers to the latest expression. (Similarly the '""' command refers to the second last expression and so on.)

Notice that every Maple statement ends with a semicolon. The exponential operator is '^', multiplication is denoted by '*', and division by '/'. The assignment operator is ':=' as in Pascal.

Maple does not automatically transform rational expressions into a standard form

```
> ((x*y/2-y^2/3)*(x-y)*(3*x+y))/(x^3-x^2*y-x*y+y^2);
```

$$\frac{((xy/2-y^2/3)(x-y)(3x+y)}{(x^3-x^2y-xy+y^2)}.$$

However, facilities are available for simplification on demand

```
> normal(");
```

$$1/6\frac{(9x^2-3xy-2y^2)y}{x^2-y}.$$

There are explicit functions for greatest common divisor and least common multiple computations with polynomials. Ending a command with ':' causes Maple to compute the result without printing it

```
> gcd(x^3 + 1, x^2 + 3*x + 2);
```

$$x+1,$$

```
> p := 55*x^2*y + 44*x*y - 15*x*y^2 - 12*y^2:
> q := 77*x^2*y - 22*x^2 - 21*x*y^2 + 6*x*y:
> gcd(p, q);
```

$$11x - 3y,$$

```
> lcm(15*(x-5)*y, 9*(x^2 - 10*x + 25));
```

$$45x^2y - 450xy + 1125y.$$

Maple knows how to factor polynomials over various domains, including the integers, finite fields and algebraic number fields

```
> factor(x^4-2);
```

$$x^4 - 2,$$

```
> factor(x^4-2, sqrt(2));
```

$$(x^2 - 2^{1/2})(x^2 - 2^{1/2}).$$

The **degree** and **ldegree** functions are used to determine the highest degree and lowest degree of the polynomial a in the intermediate(s) x, which is most commonly a single name but may be any intermediate or list or set of such.

The polynomial a may have negative integer exponents in x. Thus **degree** and **ldegree** functions may return a negative or positive integer. If a is not a polynomial in x in this sense, then **FAIL** is returned.

The polynomial a must be in collected form in order for **degree/ldegree** to return an accurate result. Applying the function **collect** or **expand** to the polynomial before calling **degree** avoids this problem.

If x is a set of intermediates, the **total** degree/ldegree is computed. If x is a list of intermediates, then the **vector** degree/ldegree is computed. Examples:

```
> degree(2/x^2+5+7*x+x^3); ⟶ 3
> ldegree(2/x^2+5+7*x+x^3); ⟶ -2
> degree(x*sin(x), x); ⟶ FAIL
> degree(x*sin(x), sin(x)); ⟶ 1
> degree(x*y^3+x^2, [x,y]); ⟶ 2
> degree(x*y^3+x^2, {x,y}); ⟶ 4.
```

Of course, Maple can also do arithmetic. However, unlike your calculator, rational arithmetic is exact

```
> 1/2 + 1/3 + 2/7;
```

$$\frac{47}{42},$$

and calculations are performed using as many digits as necessary. Constants in Maple may be approximated by floating-point numbers with the user control over the number of digits carried via the global variable **Digits** (its default value is 10). The **evalf** function causes evaluation to floating-point number. Floating-point arithmetic takes place automatically when floating-point numbers are present

```
> 1/2.0 + 1/3.0 +2/7.0
```

$$1.119047619.$$

Trigonometric functions use radians

```
> h := tan(3*Pi/7):
> evalf(h);
```

$$4.381286277$$

```
> Digits := 40:
> evalf(h);
```

$$4.381286267534823072404689085032695444160$$

```
> Digits := 10:
```

We will not consider examples from calculus since we will use only polynomial computer algebra.

Maple is a programming language as well as a mathematical system. As such it supports a variety of data structures such as arrays and tables, as well as mathematical objects like polynomials and series. Sets and lists are two other basic structures. They are constructed from sequences, i.e. expressions separated by commas. Sequences can also be generated using the **seq** command

```
> seq(i^2, i=1..10);
```

$$1, 4, 9, 16, 25, 36, 49, 64, 81, 100.$$

Sets are represented using braces and set operators are **union**, **intersect**, **minus** (for set difference) and **member** (for set membership). Note that the order of elements in a set is arbitrary and the Maple system chooses a specific ordering based on the internal addresses of the expressions.

Lists, unlike sets, retain the user-specified order and multiplicity. They are represented using square brackets. Selection of elements from sets and lists (and more generally, selection of operands from any expression) is accomplished using the op function: **op(i,expr)** yields i-th operand. Also, **nops(expr)** yields the number of operand in **expr**.

Maple has the standard **array** data structure found in most programming languages. Arrays are created using the **array** command which defines the dimensions of the array. The syntax for selecting an element is similar to other programming languages.

Maple has a **solve** function which can solve many kinds of equations, including single equations involving elementary transcendental functions and systems of linear or polynomial equations. Maple attempts to find all solutions for a polynomial equation

```
> poly := 2*x^5 - 3*x^4 + 38*x^3 - 57*x^2 - 300*x + 450:
> solve(poly = 0,x);
```

$$3/2, 6^{1/2}, -6^{1/2}, 5I, -5I.$$

Maple includes algorithms for solving systems of polynomial equations

```
> eqns := {x^2 + y^2 = 1, x^2 + x =y^2}:
```

```
> solve(eqns, {x, y});
```

$$\{y = 0, x = -1\}, \{y = 0, x = -1\},$$

$$\{x = 1/2, y = 1/2\, 3^{1/2}\}, \{y = -1/2\, 3^{1/2}, x = 1/2\}.$$

An important component is the Maple programming language which may be used to write procedures. The following procedure computes the Fibonacci numbers. The purpose of `option remember` is to tell the system to store computed values as it proceeds

```
> F := proc(n) option remember; F(n) := F(n-1) + F(n-2) end:
> F(0) := 0:
> F(1) := 1:
> F(101);
```

$$573147844013817084101,$$

```
> seq(F(i), i=1..10);
```

$$1, 1, 2, 3, 5, 8, 13, 21, 34, 55.$$

One can access the Maple library by using the command `interface` with `verboseproc= 2`.

The linear algebra package `linalg` allows standard matrix manipulations as well as many other functions. To use a function in the `linalg` package one could write `linalg[functionname]`. To avoid using the long names for `linalg` functions we first tell Maple we want to use the linear algebra package

```
> with(linalg):
Warning:  new definition for norm
Warning:  new definition for trace.
```

The `with` function sets up definitions of the functions in a package such as `linalg`. After the `with` we can use `det` instead of `linalg[det]`. Had we ended the `with` command with ';' instead of ':' we would have seen a complete list of all the functions in the `linalg` package, which is too long to be included realiz here. Naming conflicts are always reported to ensure the user is informed.

Matrices are represented as two dimensional arrays. The elements of a matrix can be specified row-by-row in the matrix function and they are displayed in a two-dimensional format. Note the use of the doubly nested lists

```
> a := matrix([[x,y,z],[y,x,y],[z,y,x]]):
> det(a);
```

$$x^3 - 2xy^2 + 2zy^2 - z^2x$$

```
> factor(");
```

$$(x - z)(x^2 + zx - 2y^2).$$

17. Basic problems in computer algebra of polynomials

17.1. APPLIED PROBLEMS AND POLYNOMIAL ALGEBRA

In many parts of mathematics and its applications to the natural sciences, there arise problems connected with the study of systems of polynomial equations and inequalities. This is for example the case in algebra, algebraic geometry, qualitative theory of differential equations, optimization theory, approximation theory, perturbation theory, analysis and synthesis of multidimensional electric chains, digital filtration, and kinetics of complex reactions.

Let a dynamic system be described by the equation

$$\dot{\mathbf{x}} = \mathbf{f}(\mathbf{x}, \mathbf{p}), \tag{17.1}$$

where $\mathbf{f}(\mathbf{x}, \mathbf{p})$ are polynomials in the coordinates $\mathbf{x}$ and the parameters $\mathbf{p}$. In order to make a qualitative analysis of this system, we must first of all analyze the steady states (s.s.), i.e. the solutions to the system

$$\mathbf{f}(\mathbf{x}, \mathbf{p}) = 0 \tag{17.2}$$

of algebraic equations.

The following problems then have to be handled:

1) Estimation of the total number of s.s. in whole, or the number of s.s.in a certain domain (for example, in a real polyhedron);

2) calculation of the s.s.;

3) classification of different types of coordinate dependencies in terms of the parameters, and calculation of bifurcation boundaries in the parameter space;

4) analysis of the asymptotics of s.s. (if the system (17.2) has small parameters).

Then we should carry out an analysis of the stability of the s.s. (in the Lyapunov seance). Among the problems involved here we mention the following:

5) Calculation of stability conditions for s.s., and localization of domains with different types of s.s. (nodes, saddles, foci etc.) in the parameter space;

6) investigation of cycle bifurcations (Hopf bifurcation);

7) classification of bifurcation diagrams.

Let us look at the algebraic meaning of these problems. The estimation of the number of s.s. is reduced to estimating the number of solution of system (17.2). If $\mathbf{f}$ are polynomials in $\mathbf{x}$ then it is always possible to eliminate unknowns from this system and to arrive at one (or more) equation in one unknown $P(x_1) = 0$. A natural requirement is that the elimination method should preserve the multiplicity of the solutions. It is exactly such methods that are presented in this book. Other methods with these properties have been considered in [**28, 104, 105, 106, 107**]. In the typical case the degree of the polynomial $P(x_1)$ is equal to the number of solutions of (17.2) (see, e.g. [**11**]). Yet another method for solving this problem makes use of the resultant from [**21**]: The number of solutions of a system of polynomial equations

$$P_1(x_1, \dots, x_n) = 0, \ \dots, \ P_n(x_1, \dots, x_n) = 0$$

without zero coordinates x_i is equal to the Minkowski mixed volume of the convex hulls of the supports of the polynomials P_i multiplied by $n!$ We remark that the first of these methods gives an algorithm for determining all solutions simultaneously with estimates of their number. The second method, based on the geometry of Newton polyhedra, is an effective method for obtaining the bifurcation conditions of the s.s. [**114**].

The determination of the number of s.s. in a given domain of real space can, after elimination of unknowns, in principle be reduced to the problem of determining the number of solutions of the single equation $P(x_1) = 0$ in a domain defined by conditions of the type $\Phi(x_1, \dots, x_n) = 0$, $\psi(\mathbf{x}) \geq 0$ or > 0. Several elimination procedures, such as the method of Gröbner bases which we shall consider below, typically allows one to solve for $x_2, \dots, x_n$ in the equations $\Phi(x_1, \dots, x_n) = 0$. Therefore our problem can be reduced to counting the roots of the polynomial $P(x_1)$ in a domain defined by polynomial inequalities of the form $\psi(x_1) \geq 0$ or > 0. An algorithm for solving this problem was obtained in [**19**]. In its simplest form this is the classical problem of determining the number of roots of a polynomial on a segment of the real axis, which can be settled by the Sturm method or its modifications (for example, by means of inners [**92**]).

In general case there arises the problem of investigating the solvability of a system of polynomial equations and inequalities. The solution to this problem for real fields is given by the Tarski theorem [**133**].

A subset in $\mathbf{R}^n$ is called semialgebraic if it is a finite union of finite intersections of sets, given by algebraic equations or inequalities.

THEOREM 17.1 (Tarski, Seidenberg [**129, 133**]). *If a set A is semialgebraic set in $\mathbf{R}^{n+m} = \mathbf{R}^n \oplus \mathbf{R}^m$, then its projection A' to $\mathbf{R}^m$ is also semialgebraic.*

A proof of this theorem, including a simple constructive solution procedure, is given in [**88**] (see also [**61**]). It is based on the following elementary lemma:

Let $p_1, \dots, p_s$ be polynomials in one variable of degree not greater than m. Suppose the degree of the polynomial p_s is exactly equal to $m > 0$ and that none of the polynomials vanishing identically. If $g_1, \dots, g_s$ are the remainders obtained after division of p_s by $p_1, \dots, p_{s-1}$ and p'_s, then the set $\mathrm{SGN}(p_1, \dots, p_{s-1}, p'_s, g_1, \dots, g_s)$ determines the set $\mathrm{SGN}(p_1, \dots, p_s)$. ($\mathrm{SGN}(p_1, \dots, p_s)$ is a certain set which encodes the information on the zeros and sign changes of the polynomials $p_1, \dots, p_s$, see [**88**]).

The solvability theory has already found practical applications e.g. in the theory of multidimensional systems occurring in (radio) electronics, [**25**]. Here we content ourselves with an example.

Consider the polynomials

$$f_1(x) = x^2 + \alpha_1 x + \alpha_0, \quad f_2(x) = x^2 + \beta_1 x + \beta_0.$$

We want to find the domains in the $\{\alpha_0, \alpha_1, \beta_0, \beta_1\}$ space, corresponding to different configurations of the real roots of the system $f_1(x) = 0$, $f_2(x) = 0$. By repeatedly applying the lemma we obtain conditions for $\alpha_0, \alpha_1, \beta_0, \beta_1$ corresponding to 9 cases (• is a root of $f_1(x) = 0$ and ∘ is a root of $f_2(x) = 0$):

1) ——————— $\alpha_0 > \alpha_1^2/4, \quad \beta_0 > \beta_1^2/4;$

2)  $\beta_0 > \beta_1^2/4, \quad \alpha_0 < \alpha_1^2/4;$

3) ———•——•——○——○——— $\beta_0 < \beta_1^2/4, \quad \alpha_1 > \beta_1 + \sqrt{D_\beta}.$
$D_\beta = \beta_1^2 - 4\beta_0,$
$\beta_0 + (\alpha_1 - \beta_1)(\beta_1 + \sqrt{D_\beta})/2 < \alpha_0 < \alpha_1^2/4;$

4) ———•——○——•——○——— $\beta_0 < \beta_1^2/4, \quad \alpha_1 > \beta_1,$
$\beta_0 + (\alpha_1 - \beta_1)(\beta_1 - \sqrt{D_\beta})/2 < \alpha_0 < \beta_0 +$
$+(\alpha_1 - \beta_1)(\beta_1 + \sqrt{D_\beta})/2;$

5) ———•——○——○——•——— $\beta_0 < \beta_1^2/4, \quad \alpha_1 > \beta_1,$
$\alpha_0 < \beta_0 + (\alpha_1 - \beta_1)(\beta_1 - \sqrt{D_\beta})/2$
or
$\alpha_1 < \beta_1,$
$\alpha_0 < \beta_0 + (\alpha_1 - \beta_1)(\beta_1 + \sqrt{D_\beta})/2;$

6) ———○——•——○——•——— $\beta_0 < \beta_1^2/4,$
$\beta_0 - \sqrt{D_\beta} < \alpha_1 < \beta_1$
$\beta_0 - (\beta_1 - \alpha_1)(\beta_1 + \sqrt{D_\beta})/2 < \alpha_0 < \beta_0 +$
$+(\beta_1 - \alpha_1)(-\beta_1 + \sqrt{D_\beta})/2$
or
$\alpha_1 < \beta_1 - \sqrt{D_\beta},$
$\beta_0 - (\beta_1 - \alpha_1)(\beta_1 + \sqrt{D_\beta})/2 < \alpha_0 < \beta_0 +$
$+(\beta_1 - \alpha_1)(-\beta_1 + \sqrt{D_\beta})/2;$

7) ———○——○——•——•——— $\beta_0 < \beta_1^2/4, \quad \alpha_1 < \beta_1 - \sqrt{D_\beta},$
$\beta_0 - (\beta_1 - \alpha_1)(\sqrt{D_\beta} - \beta_1)/2 < \alpha_0 < \alpha_1'/4;$

8) ———○———○——— $\beta_0 < \beta_1^2/4, \quad \alpha_0 > \alpha_1^2/4;$

9) ———○——•——•——○——— $\beta_0 < \beta_1^2/4, \quad \beta_1 < \alpha_1 < \beta_1 + \sqrt{D_\beta},$
$\beta_0 + (\alpha_1 - \beta_1)(\beta_1 + \sqrt{D_\beta})/2 < \alpha_0 < \alpha_1^2/4$
or
$\beta_1 - \sqrt{D_\beta} < \alpha_1 < \beta_1,$
$\beta_0 + (\beta_1 - \alpha_1)(-\beta_1 + \sqrt{D_\beta})/2 < \alpha_0 < \alpha_1^2/4.$

It is essential that the boundaries of the domains 1)-9) are found without actually computing the roots of $f_1(x)$, $f_2(x)$. The simplicity of this particular example allows here us to explicitly find the inequalities for the parameters $\alpha_0, \alpha_1, \beta_0, \beta_1$. In general case the result of the reduction procedure is formulated in terms of zeros of certain polynomials, for which the explicit boundaries can still be indicated.

The scheme of finding the conditions 1)-9) is the following:

1) 'descent', i.e. successively applying the lemma we decrease the degree in x of the polynomials down to 0;

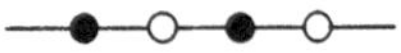

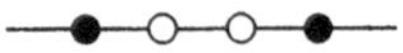

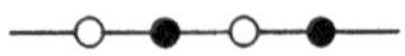

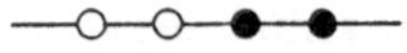

2) 'ascent', i.e. for the obtained expressions in the parameters of f_1, f_2 we generate all possible cases (> 0, $= 0$, < 0) and lift up. We recover the location of the roots of f_1, f_2 for these cases,

3) finally we gather the data corresponding to each case and analyze the obtained inequalities.

In the process of the computation the discriminants and resultants of the polynomials are automatically determined (in fact by the Euclid algorithm). In spite of the fact that every step of the algorithm is elementary, it requires inspection of a lot of cases. Here it would be useful to have general criteria like the well-known criteria based on 'the number of sign changes' (of Sturm, Descartes, etc.). We present one result in this direction known for us [**68**].

THEOREM 17.2 (Dubovitskii). *Let $f(t)$, $g(t)$ be polynomials (it is assumed that f has no multiple zeros and that g is relatively prime to f) and let s be a number such that* $\deg f^{2s+1} \geq \deg(gf)'$. *If for the pair of polynomials $(f^{2s+1}, f'g)$ we construct the Sturm series $X_0 = f^{2s+1}$, $X_1 = f'g$, X_2, ..., X_n, then the number of sign changes at the point $t = a$, minus the number of signs changes at the point $t = b$ is equal to $n_{ab}(f|g > 0) - n_{ab}(f|g < 0)$, where $n_{ab}(f|g \lesseqgtr 0)$ is the number of roots of $f(t) = 0$ into the domain $g(t) \lesseqgtr 0$, $a \leq t \leq b$.*

Then $n_{ab}(f|g > 0) = (n_{ab}(f) + n_{ab}(f|g))/2$ where $n_{ab}(f|g) = n_{ab}(f|g > 0) - n_{ab}(f|g < 0)$.

The problem of calculating the s.s. for algebraic systems, can after elimination of unknowns be reduced to finding the zeros of a single one-variable polynomial. This is a classical problem of computational mathematics and various algorithms for solving it are to be found in any package of scientific programs (see e.g. [**159**], where the solution of a system of two algebraic equations are calculated by means of resultant). We must remark, however, that situations similar to the well-known example of Forsight may occur, where the calculation of the roots leads to 'catastrophes'. Another difficulty is the high degrees of the obtained polynomials (the standard algorithms work if the degree is not more than 30–36).

The classification of different types of coordinate dependencies with respect to the parameter λ in the models, and the calculation of bifurcation boundaries require an analysis of bifurcations of high codimensions. The method for solving this problem in the case when the system is reduced to one equation $F(x, \lambda, p) = 0$ (see e.g. [**17, 138**]) requires a compatibility check of the conditions

$$F = 0, \ \partial^{(i+j)} F / \partial x^i \partial x^j = 0$$

in the given domain. If F is a polynomial then the methods of solvability theory can be used here. The same can be said about the method of Newton polyhedra [**114**]: It is to analyze the compatibility of the obtained system of conditions.

The analysis of the asymptotics of s.s. (which is of interest for applications) requires in general the use of methods from the theory of multi-valued solutions to non-linear equations [**139**], which are based on Newton diagrams. A computer construction of the Newton diagram was realized in [**18**]. In the multidimensional case a necessary step in the construction of asymptotics is the elimination of unknowns ([**139**] particularly describes the classical procedure of Kronecker elimination).

The problems of stability 5)–7) can also in principle be approached by algebraic methods. As pointed out in [**13**] the problem of Lyapunov stability is algebraically solvable up to codimension 2 inclusively, and in some cases the algebraic investigation of the local problem can be extended to situations of higher codimensions.

One more aspect of the analysis of dynamical systems where polynomial algebra can be used is that of singular perturbed systems. In the method of integral varieties [**77**] the initial qualitative analysis problem is split into two parts: 1) The study of the form and singularities of the integral variety; 2) the study of a non-rigid system of lower dimension on the integral variety. In the case of polynomial systems the problem 1) is tackled with methods from elimination theory and solvability theory. Here it is important that the analytic classification of the dynamics of the system can be carried sufficiently far (see e.g. [**130**]).

We can thus distinguish two major groups of algorithms which are necessary for the qualitative analysis of systems of algebraic and differential equations: 1) The algorithms of elimination theory; 2) the algorithms of solvability theory.

17.2. METHODS AND ALGORITHMS

The problems of solvability theory are the one of the most general and important for applications. Among the subproblems that are encountered here we have the elimination of unknowns, the analysis of polynomial remainder sequences, the calculation of resultants and discriminants, the factorization of polynomials, the localization of real roots, and calculations with algebraic numbers. These problems are undoubtedly of great independent significance.

17.2.1. *The Resultant of Two Polynomials*

The resultant R of the polynomials

$$\begin{aligned} f(x) &= f_n x^n + \ldots + f_0, \\ g(x) &= g_n x^n + \ldots + g_0, \end{aligned} \tag{17.3}$$

is the determinant of the $(m+n)\times(m+n)$ matrix

$$R(f,g) = \begin{vmatrix} f_n & f_{n-1} & \ldots & f_0 & 0 & \ldots & 0 \\ \ldots & \ldots & \ldots & \ldots & \ldots & \ldots & \ldots \\ 0 & 0 & & f_n & f_{n-1} & \ldots & f_0 \\ g_m & g_{m-1} & \ldots & g_0 & 0 & \ldots & 0 \\ \ldots & \ldots & \ldots & \ldots & \ldots & \ldots & \ldots \\ 0 & 0 & & g_m & g_{m-1} & \ldots & g_0 \end{vmatrix} \begin{matrix} \left.\vphantom{\begin{matrix}a\\a\\a\end{matrix}}\right\} m \text{ lines} \\ \left.\vphantom{\begin{matrix}a\\a\\a\end{matrix}}\right\} n \text{ lines} \end{matrix} \tag{17.4}$$

(see Ch. 1). $R(f,g)=0$ if and only if $f(x)$ and $g(x)$ have a common zero or if $f_n=0$ and $g_m=0$ [**142**]. If the coefficients f_i, g_i are polynomials in the unknown variable y, then the calculation of the resultant permits one to eliminate x from the system $f(x)=0$, $g(x)=0$ and to reduce it to a single equation with respect to y. In the general case the calculation of the resultant is one of the steps in the classical Kronecker elimination procedure [**142**]. The resultants, with the discriminants $(R(f,f'))$ as a special case, are of great significance for the localization of polynomial roots in various given domains [**92**].

The simplest method for obtaining the resultant consists in computing the determinant (17.4) with symbolic coefficients. This is rather efficient (as the examples in Sec. 18 show) for small dimensions m and n. The efficiency here is determined on the one hand by the method of calculation (which must allow for zero entries in the determinant), and on the other hand by the complexity of the expressions for the coefficients f_i, g_i. But in order to obtain the resultant it is not necessary to actually calculate the determinant (17.4).

One of the first method,proposed by Collins [**64**],for calculating resultants, is based on sequences of polynomials remainders (see also [**28**]) and makes use of the following properties of a resultant:

$$R(f,g) = (-1)^{mn} R(g,f), \tag{17.5}$$

$$R(f,g) = f_n^{m-n+1} R(f, g//f) \tag{17.6}$$

(for $m \geq n$ $g//f$ is a remainder after division of g by f),

$$R(f,g) = g_m^{n-m+1}(-1)^{mn} R(f//g, g) \quad m \geq n, \tag{17.7}$$

$$R(f,g) = f_n^m, \quad \deg f(x) = 0.$$

For calculations with polynomials in several variables methods such as module arithmetic, Chinese remainder theorem and interpolation are used in [**64**]. The computational complexity of the algorithm is $\sim (n+m)^2 \log(n+m)$.

A modification of this polynomial approach leads to an algorithm with the complexity $\sim (n+m)\log^2(n+m)$ [**128**]. Other approaches are based on methods from linear algebra. In [**23**] an algorithm for finding the resultant is suggested, which requires the calculation of a determinant of order $\min(m,n)$:

$$R(f,g) = (-1)^{mn} \det \mathbf{C},$$

where for $m < n$ the entries in the matrix

$$\mathbf{C}(m \times m) = \begin{pmatrix} c_{n-m+1,n-m+2} & \cdots & c_{n-m+1,n+1} \\ \cdots & \cdots & \cdots \\ c_{n,n-m+2} & \cdots & c_{n,n+1} \end{pmatrix}$$

are given by the relations

$$c_{ij} = c_{i-1,j-1} - g_{i-n+m-1} c_{n,j-1}, \quad i = 0, \dots, n, \ j = 2, \dots, n+1;$$

$$c_{i,1} = f_i \quad (\text{for } i = 0, \dots, n), \quad c_{i-1,j-1} = 0, \quad i-1 < 0,$$

$$g_{i-n+m-1} = 0$$

for $i < n - m + 1$.

17.2.2. *Algorithms for Elimination of Unknowns*

1^0. The classical elimination scheme was considered in Ch. 1. This method is not suitable for computer realizations due to the fast growth of the dimension of the Sylvester matrix, whose determinant is calculated in the process of elimination, and also because of the risk of obtaining superfluous solutions (these correspond to the vanishing of f_n and g_m (see Sec. 17.2.1)).

2^0. One of the first realizations of the elimination algorithm [**118**] looks very much like the Gauss method for system of linear equations. The algorithm for eliminating the unknowns x_i from the polynomial equations

$$P_i(x_1, \dots, x_n) = 0, \quad i = 1, \dots, n,$$

has the form

$$\text{for } i := 1 \ \text{ step } 1 \ \text{ until } n-1 \ \text{ do}$$
$$\text{for } j := i+1 \ \text{ step } 1 \ \text{ until } n \ \text{ do,}$$

$$\left[\begin{array}{l} P^i = \text{ divisor } P_i \text{ and } P_j \text{ in variable } x_i \\ P_j = \text{ eliminant } P_i \text{ and } P_j \text{ in variable } x_i \end{array} \right].$$

The divisor and the eliminant for a pair of polynomials

$$R = \sum_{k=0}^{r} R_r(x_{i+1}, \ldots, x_n) x_i^k = 0, \quad R_r, S_s \not\equiv 0,$$

$$S = \sum_{k=0}^{s} S_k(x_{i+1}, \ldots, x_n) x_i^k = 0, \quad 0 < s \le r,$$

are obtained by a method similar to the algorithm for division of polynomials:

$$\longrightarrow \quad Q = S_s R - R_r S x_i^{r-s}$$
$$q = \text{ degree } Q \text{ by } x_i$$
$$q \ge s$$

no

$$r = q$$
$$R = Q$$

$$q > 0$$

no

$$r = s$$
$$R = S$$
$$s = q$$
$$S = Q$$

$$\text{divisor} = S$$
$$\text{eliminant} = Q$$

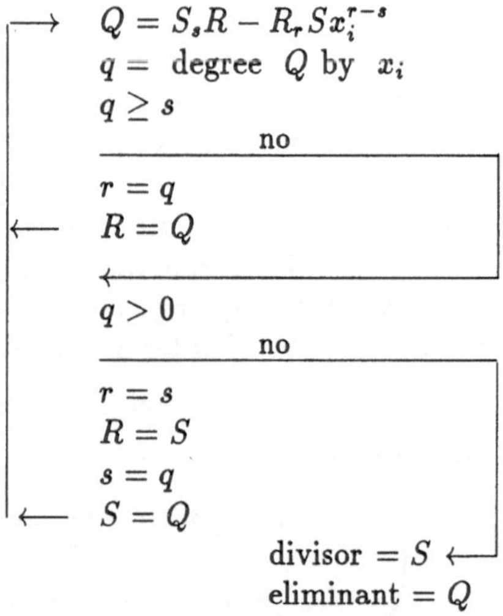

Any common zero of the polynomials P_j and P_i is also a common zero of the eliminant and of the divisor of P_j and P_i. This follows from the fact that the divisor is proportional to the remainder after division of P_i by P_j, while the eliminant is proportional to the resultant of P_i and P_j in the variable x_i.

In the reduced system every polynomial P_i is independent of x_j for $j < i$. Usually such a reduced system has the following form: The equations

$$P_i = 0, \quad i = 1, \ldots, n-1,$$

are linear in x_i and the equation $P_n = 0$ is a polynomial in x_n, and its degree reflects the number of solutions of the system. The simple algorithm [118] has, on the other hand, essential draw-backs: 1) Among the zeros of the polynomial P_n there may be superfluous solutions so therefore it is necessary to check the solutions by substitution into the equations; 2) if P_i and P_j have common zeros then their eliminant is identically equal to zero; 3) there is a fast growth in intermediate results (in floating point arithmetic this problem is not stable). This is why the author of the algorithm recommends it for small systems.

We now present a couple of examples.

1. The system

$$\begin{aligned} x^2 - y^2 &= 0, \\ x^2 - x + 2y^2 - y - 1 &= 0 \end{aligned} \tag{17.8}$$

is reduced to

$$\begin{aligned} x - 3y^2 + y + 1 &= 0, \\ 9y^4 - 6y^3 - 6y^2 + 2y + 1 &= 0. \end{aligned} \tag{17.9}$$

2. The system

$$\begin{aligned} x^2y + 3yz - 4 &= 0, \\ -3x^2z + 2y^2 + 1 &= 0, \\ 2yz^2 - z^2 - 1 &= 0 \end{aligned} \tag{17.10}$$

reduces to the form

$$\begin{aligned} x^2y + 3yz - 4 &= 0, \\ 2yz^2 - z^2 - 1 &= 0, \\ 36z^8 - 96z^7 + 42z^6 + 10z^4 + 6z^2 + 2 &= 0. \end{aligned} \tag{17.11}$$

3^0. In 1965 Buchberger suggested a method for simplification of polynomial ideals, which in honor of his scientific advisor be named the method of 'Gröbner bases' [**27, 143, 1**]. This method provides a tool for obtaining algorithmic solutions to problems such as finding exact solutions of algebraic equations, finding polynomial solutions to systems of homogeneous linear equations, multidimensional interpolation [**117**], and many other problems in the theory of polynomial ideals.

We briefly describe the basic concepts and the structure of the algorithm.

A polynomial g is reduced to h modulo the ideal F, written

$$\left(g \underset{F}{\rightarrow} h \right),$$

if there exists a polynomial $f \in F$, and a monomial u such that $h = g - buf$, where

$$cf(g,\ u \cdot \mathrm{lpp}(f)) \neq 0, \quad b = cf(g,\ u \cdot \mathrm{lpp}(f))/\mathrm{lc}(f).$$

Here $cf(g,\ r)$ is the coefficient of the monomial r in the polynomial g, $\mathrm{lpp}(f)$, is the leading monomial in $f(x_1, \dots, x_n)$, and $lc(f)$ is its coefficient. It is supposed that a linear ordering $<_T$ of the monomials $x_1^{i_1} \dots x_n^{i_n}$ has been fixed, for example by total degree

$$(1 <_T x_1 <_T x_2 <_T x_1^2 <_T x_1x_2 <_T \dots)$$

or by pure lexicographic ordering

$$(1 <_T x_1 <_T x_1^2 <_T \dots <_T x_1x_2 <_T \dots)$$

etc. The reduction can be considered as one of the steps in a generalized division algorithm.

The polynomial h is the normal form of the polynomial g modulo F (written $h = \mathrm{NF}(F, g)$) if there exists a sequence of reductions

$$g = k_0 \underset{F}{\rightarrow} k_1 \underset{F}{\rightarrow} \dots \underset{F}{\rightarrow} k_n = h$$

and if there is no polynomial h' such that $h \underset{F}{\to} h'$.

The algorithm for determining the normal form looks like

$$h := g,$$

$$\text{while } \{\exists f \in F, b, u : h \underset{F}{\to} \ \} \text{ do;}$$

$<$ choose $f \in F, b, u$ with leading product $u \operatorname{lpp}(f)$ with respect to the ordering $>_T$ such that

$$h \underset{F}{\to} ; \quad h := h - buf > .$$

A set F is called Gröbner basis if for arbitrary polynomials g, h_1, h_2 with $h_1 = \mathrm{NF}(F, g)$ and $h_2 = \mathrm{NF}(F, g)$, one always has $h_1 = h_2$.

The S-polynomial corresponding to f_1 and f_2 is defined to be the polynomial

$$\mathrm{SP}(f_1, f_2) = u_1 f_1 - \left(\frac{c_1}{c_2}\right) u_2 f_2,$$

where $c_i = \operatorname{lc}(f_i)$, and the product of the powers of u_i is such that the degree of $S_i u_i$ is equal to the least common multiple of the degrees of S_1 and S_2, where $S_i = \operatorname{lpp}(f_i)$.

The simplest algorithm for finding a Gröbner basis has the form:

$$G := F;$$

The set of pairs of polynomials $B = \{(f_1, f_2) | \ f_1, f_2 \in G, \ f_1 \neq f_2\}$:

while $B \neq \emptyset$ do;
$< (f_1, f_2)$ is a pair from B;
$B := B \setminus \{(f_1, f_2)\}$;
$h := \mathrm{SP}(f_1, f_2)$;
$h' := \mathrm{NF}(G, h)$;
if $h' \neq 0$ then $<$
$B := B\{(g, h') | g \in G\}$;
$G := G\{h'\} >>$.

A Gröbner basis F is said to be reduced if each polynomial $f \in F$ is represented in normal form modulo $F \setminus \{f\}$ and $\operatorname{lc}(f) = 1$. Every Gröbner basis can be put in reduced form by means of repeated reduction of all polynomials from the basis modulo the other polynomials, combined with normalization of the leading coefficients.

In [**27**] an improved version of the algorithm was presented (which is the one we have implemented in the system REDUCE (see Sec. 18.3)). This modified algorithm for calculating a Gröbner basis G for the ideal F has the form:

$R := F; \ P := 0, \ G := 0, \ B := 0$;
'Reduce everything' (R, P, G, B)
'New bases' (P, G, B)
while $B \neq 0$ do $<$
$(f_1, f_2) :=$ pair from B with minimal
least common multiple LCM(lpp(f_1),lpp(f_2)) with respect to $>_T$;
$B := B \setminus \{f_1, f_1\}$;
if (not 'Criterion 1' (f_1, f_2, G, B)
and not 'Criterion 2' (f_1, f_2)) then $<$
$h := \mathrm{NF}(G, \mathrm{SP}(f_1, f_2))$;

if $h \not\equiv 0$ then <
$G_0 := \{g \in G|$ lpp(h) is a divisor of lpp(g);
$R := G_0$; $P := \{h\}$; $G := G \setminus G_0$;
$B := B \setminus \{(f_1, f_2)|\ f_1 \in G_0$ or $f_2 \in G_0\}$;
'Reduce everything' (R, P, G, B);
'New bases' (P, G, B); >>
algorithm 'Reduce everything' (R, P, G, B);
while $R \neq 0$ do <
$h :=$ element from R; $R := R \setminus \{h\}$;
$h :=$NF(G, P, h);
if $h \neq 0$ then <
$G_0 := \{f \in G|$ lpp(h) is a divisor of lpp$(g)\}$;
$P_0 := \{p \in P|$ lpp(h) is a divisor of lpp$(p)\}$;
$G := G \setminus G_0$;
$P := P \setminus P_0$;
$R := R \cup G_0 \cup P_0$;
$B := B \setminus \{(f_1, f_2) \in B|\ f_1 \in G_0$ or $f_2 \in G_0\}$;
$P := P \cup \{h\}$ >>;
algorithm 'New basis' (P, G, B);
$G : G \cup P$;
$B := B \cap \{(g, p)|\ g \in G, p \in P, g \neq p\}$;
$H := G,\ K := 0$;
while $H \neq 0$ do <
$h :=$ element from H; $H := H \setminus \{h\}$;
$k :=$NF$(G \setminus \{h\}, h)$; $K := K\{k\}$;
$G := K$
algorithm 'Criterion 1' (f_1, f_2, G, B);
checking: exist $p \in G$ with $f_1 \neq p,\ f_2 \neq p$,
such that lpp(p) is a divisor of LCM(lpp(f_1), lpp(f_2)),
$(f_1, p) \in B,\ (p, f_2) \in B$;
algorithm 'Criterion 2' (f_1, f_2);
checking: LCM(lpp$(f_1$, lpp$(f_2))$) = lpp$(f_1) \cdot$ lpp(f_2).

The modification of the algorithm consists of the following steps: 1) The pair with minimal LCM of the leading products is chosen from the set B. Experiments have shown that this essentially accelerates the calculations, since it decreases the degrees of the polynomials with which the algorithm works; 2) the algorithm 'Reduce everything' and the corresponding part of basic algorithm uses the following fact: When a new polynomial is added to the basis all other polynomials can be reduced using this polynomial. The algorithm 'New basis' takes into account the set of pairs B, and performs the iterated reduction of the basis polynomials with respect to the others, which is required for the construction of a reduced Gröbner basis; 3) the criteria 1 and 2 in many cases (but not always) prevent superfluous calculations of normal forms that are already known to vanish.

The determination of a Gröbner basis makes it easy to settle questions of solvability and to find exact solutions of systems of algebraic equations. The solvability criterion for a system of equations is simply given by the condition: $1 \notin G$. The system of equations

$$\mathbf{F}(\mathbf{x}) = \mathbf{0}$$

has a finite number of solutions if and only if for all $i \in [1,n]$ a power of the form $x_i^{j_i}$ occurs as leading monomial of some polynomial from G. Finding a basis with respect to the pure lexicographical ordering reduces the system to 'triangular' form in the sense that G contains a polynomial in a single 'lowest' variable and the other x_i can be found successively from the corresponding basis polynomials (usually of low degrees) in x_i. For the above examples (17.8) and (17.9) from [**118**], the Gröbner basis calculated by the program described in Sec. 18.3 has the form

$$\begin{aligned} \mathrm{GB}(1) &= -3y^2 + y + x + 1, \\ \mathrm{GB}(2) &= \tfrac{1}{9}(9y^4 - 6y^3 - 6y^2 + 2y + 1) \end{aligned} \tag{17.12}$$

(with ordering $y >_T x$);

$$\begin{aligned} \mathrm{GB}(1) &= \frac{1}{2}(18z^6 - 48z^5 + 21z^4 + 5z^2 + 2y + 2), \\ \mathrm{GB}(2) &= \frac{1}{6}(18z^7 - 48z^6 + 3z^5 + 48z^4 + 2z^3 - 48z^2 + 19z + 6x^2), \\ \mathrm{GB}(3) &= \frac{1}{18}(18z^8 - 48z^7 + 21z^6 + 5z^4 + 4z^2 + 1) \end{aligned} \tag{17.13}$$

(with ordering $y >_T x >_T z$).

In the first example the algorithms from [**118**] and [**27**] both transform system (17.8) to the same reduced form, but in the second example the reduced forms are substantially different from each other: In case (17.11) x^2 and y are expressed as rational functions in z, whereas in case (17.13) polynomial dependencies are obtained for them. The solution to system (17.10) can be found by calculating the roots of the equation $\mathrm{GB}(3) = 0$ (see (17.13)). The coordinates x and y are then found from $\mathrm{GB}(1) = 0$ and $\mathrm{GB}(2) = 0$. In Sec. 19 we shall use Gröbner bases in an algorithm for computing the coefficients in kinetic polynomial.

Algorithms for Gröbner bases have been implemented in various computer algebra systems: The author of the Gröbner method realized it himself in the system SAC [**27**]. Pankratyev implemented it on REFAL [**116**], and Alekseev worked out a realization for solving problems from the qualitative theory of differential equations on FORTRAN [**12**]. Finally there is also a realization in the system SRM [**126**].

The complexity questions related to the method of Gröbner bases constitute a subject of active investigations. As pointed out in the book [**67**] of Davenport et al. one has in practice encountered both problems which have been very easily solved with the use of this algorithm, but also problems which one has not succeeded to solve even with the use of several megabytes of memory. The known estimates for needed memory capacity are not consoling, they depends exponentially on the number of variables [**83**]. During our work on the Russian edition of this book our computing means were restricted to the system REDUCE 3.2. for IBM PC, which does not include the algorithms of Gröbner bases. Our interest in polynomial computer algebra was so great that we ventured to implement this algorithm ourselves and to use it for our problems. Moreover, our limited personal computer resources forced us (after many brine and bums) to develop an effective and economized realization (in the sense of memory).

The results of these experiments are reflected in this book (e.g. in the work with 'dummy' polynomials described below). At a later stage we had the opportunity to work with the system REDUCE 3.3 which includes the package GROEBNER. Solving the examples from this book by means of this GROEBNER package we found that our program works somewhat slower. On the other hand, GROEBNER was unable (due to insufficient memory) to handle certain examples which were solved by our program. Therefore we cannot say that our activity was completely useless. Moreover, the package GROEBNER would not allow us to solve the problem of computing the coefficients in kinetic polynomials, since this, apart from finding the basis, also requires a calculation of a linear basis representation in terms of the polynomials of the initial systems. This is the reason why we have decided to keep this program in REDUCE code in Sec. 18.

During the work at the English text of this book we have also used the package Grobner in Maple [**55**], which of course is more efficient than our program.

4^0. Other algorithms for elimination of unknowns, and also for the more general problem of decomposing the algebraic varieties of all roots of polynomial systems into irreducible components, have been formulated in articles of Lazard [**104, 105, 106, 107**] and by the Grigoryev group [**57, 82**]. The connection between the algorithms from [**104, 105, 106**] and the method of Gröbner bases is analyzed in [**106**].

Elimination algorithms based on multidimensional complex analysis have been considered in this book (Ch. 2–3).

17.2.3. *The Algorithms of Solvability Theory*

Let there be given a formula of the first order

$$\exists X_{1,1}\ldots\exists X_{2,s_1}\forall X_{2,1}\ldots\forall X_{2,s_2}\exists X_{a,1}\ldots\exists X_{a,s_a}\left(\sum\right), \tag{17.14}$$

for real closed fields, where $\sum$ is a formula without quantifies containing k atomic subformulas of the form $(f_i \geq 0)$, $1 \leq i \leq k$, with f_i being polynomials in $X_{1,1},\ldots,X_{a,s_a}$. The first order solvability theory for real closed fields tells us that there exists an algorithm which checks the truth for every formula of the form (17.14), see [**81**]. In terms of this theory one can formulate many important and difficult mathematical problems, such as the solvability of systems of algebraic equations and inequalities, and problems in nonlinear optimization. The decision procedure was discovered by Tarski [**133**]. His procedure for eliminating quantifies associates to any given formula F a quantifier-free formula F' and the transfer $F \leftrightarrow F'$ is carried out via a sequence of axioms in the theory.

The algorithms in [**129, 133**] are characterized by exponential dependence on the number of polynomials including in formula, and on the maximal degree of these polynomials for a fixed number of variables. Collins [**63**] suggested an algorithm with a working time of order $L^{2^{O(n)}}$, where L is the size of initial formula, and n is the number of variables it involves. In [**19**] an elimination procedure with the same working time for a wide class of real closed fields is described. The main ingredience in the method of [**63**] is an algorithm for finding a 'cylindrical algebraic decomposition' of r-dimensional real space into a finite number of distinct connected sets (called cells), associated to a given finite set of polynomial of r variables. In these cells none of the polynomials changes its sign. Hence the sign of a polynomial in a cell can be found calculating its sign at any point inside the cell.

In [**63**] algorithms are used for constructing decompositions (not containing squares) adapted to the real roots. An improvement of the algorithm from [**63**] is described in [**65**].

In [**15**] it is shown that the elimination of quantifies can be used for obtaining a new version of the algorithms of cylindrical algebraic decomposition. The idea of [**15**] consists of formulating existence conditions for polynomials of given degree and with given root multiplicities.

In [**81**] an algorithm is constructed for solvability in first order theory for a wide class of real closed fields. Its working time is $L^{(O(n))^{5a-2}}$ where $a \leq n$ is a number quantifier alterations in formula (17.14). For small numbers a the working time of the algorithm in [**81**] is less than for previously known ones. The algorithm in [**81**] is a generalization of an algorithm in [**141**] for solving systems of polynomial inequalities. The latter finds the solutions (if any) to a system of inequalities

$$f_1 > 0, \ \dots, \ f_{k_1} > 0, \ f_{k_1+1} \geq 0, \ \dots, \ f_k \geq 0, \tag{17.15}$$

where $f_1, \ \dots, \ f_k$ are polynomials in $X_1, \ \dots, \ X_n$ with coefficients from ordered rings. The set V of all points satisfying (17.15) is represented as the union of a finite number of connected components V_i, each of which is a semialgebraic set. A finite set T is called a system of representatives for the set V if $V_i \cap T \neq \emptyset$ for all i. The algorithm in [**141**] associates to the system (17.15) a certain system of representatives T, with a number of elements bounded by $P\left(kd^{n^2}\right)$ (here $P(\cdot)$ is a polynomial, and $d > \deg_{X_1 \dots X_n}(f_i)$) in the time $\leq P\left(L^{\log^2 L}\right)$. For each point $(\xi_1, \ \dots, \ \xi_n) \in T$ the algorithm constructs an irreducible polynomial Φ, and also certain expressions

$$\xi_i = \xi_i(\theta) = \sum_j \alpha_j^{(i)} \theta^j,$$

where

$$\Phi(\theta) = 0, \quad 1 \leq i \leq n, \quad 0 \leq j < \deg(\Phi) \leq (kd)^n.$$

Furthermore,

$$\theta = \sum_{1 \leq i \leq n} \lambda_i \xi_i(\theta)$$

for suitable positive integers

$$1 \leq \lambda_i \leq \deg(\Phi).$$

The output of the algorithm is a family of polynomials of the form Φ together with the expressions $\xi_i(\theta)$. Moreover, every constructed Φ has a root θ_0, such that the point $(\xi_1(\theta_0), \ \dots, \ \xi_n(\theta_0))$ satisfies the system (17.15).

The scheme of the algorithm is the following: 1) Some estimate R of the solutions to the inequalities (17.15) is established; 2) By using system (17.15) a polynomial g is constructed

$$g = \left((R+1)^2 - X_0^2 - X_1^2 - \dots - X_{n+1}^2\right)^2 + g_1^2,$$

where

$$g_1 = \prod_{1\le i\le k+1} (f_i + \varepsilon) - \varepsilon^{k+1}, \qquad f_{k+1} = X_{n+1} f_1 \dots f_{k_1} - 1,$$

and an algorithm, which finds a system of representatives for the variety of zeros of the given polynomial, is applied to it. One considers an infinitesimal 'perturbation' of the initial polynomial, whose zero variety is a smooth hypersurface. The algorithm looks for points on this surface with some fixed directions of the gradient, solving a suitable system of algebraic equations by the methods of [**57, 82**]. Among the points of obtained system of representatives the algorithm then isolates the points that satisfy (17.15). Finally, by a method from [**19**] the compatibility of the system

$$\Phi_0(\theta) = 0, \ \tilde{f}_1(\theta) \ge 0, \ \dots, \ \tilde{f}_{k+1}(\theta) \ge 0$$

is checked, where $\tilde{f}_i(\theta)$ are obtained by substituting the expressions for $\xi_i \theta$ into (17.15).

18. Realization of the elimination method

18.1. CALCULATION OF THE RESULTANT OF TWO POLYNOMIALS

A simple `Maple`-procedure which computes the resultant of two polynomials $P1$ and $P2$ is given in Application F_1. (Here and below we use the version of Maple for personal computers IBM PC with operative system MS DOS). With the help of the function `degree` (see Sec. 16.2) the degrees N and M of the polynomials $P1$ and $P2$ are defined. In a working file called FOROUT the operator RES:=`matrix (...)` is generated, in which the Sylvester matrix (17.4) for 'dummy' of the polynomials $P1$ and $P2$ is written (in place of the proper coefficients of $P1$ and $P2$ indefinite coefficients $A(J)$ and $B(J)$ are written, and the commands `for...do...od` insert zeros in the place of zero coefficients of $P1$ and $P2$). This operator is entered into the program by the command `read`, and then the determinant `det(RES)` is calculated. The last two commands `for...do...od` finally plug in the true values of the coefficients of $P1$ and $P2$.

As an example we consider the determination of the kinetic polynomial for a three-stage parallel adsorption reaction mechanism

$$\begin{array}{ll} 1) & A_2 + 2Z \rightleftarrows 2AZ, \\ 2) & B + Z \rightleftarrows BZ, \\ 3) & AZ + BZ \rightleftarrows AB + 2Z, \end{array} \tag{18.1}$$

where Z, AZ, BZ are intermediate, A_2, B, are reagents and AB is the reaction product (the same symbols are used for the concentrations). The equations describing the stationary reaction rate

$$\begin{aligned} b_1 Z^2 - b_{-1} AZ^2 &= W, \\ b_2 Z - b_{-2} BZ &= 2W, \\ b_3 AZ \cdot BZ - b_{-3} Z^2 &= 2W, \\ Z + AZ + BZ &= 1 \end{aligned} \tag{18.2}$$

are reduced to the system [**110, 112**]

$$\begin{aligned} d_{01}Z^2 + d_{11}Z + d_{21} &= 0, \\ d_{02}Z^2 + d_{12}Z + d_{22} &= 0, \end{aligned} \tag{18.3}$$

where d_{ij} are polynomials in the kinetic parameters $b_{\pm i}$ (the so-called reaction weights) and the stationary reaction rate W, for which the expressions are given in Application F_1. The kinetic polynomial is obtained as the resultant of the polynomials (18.3) in the variable Z, and it has the form

$$KP(W) = K_0 + K_1W + K_2W^2 + K_3W^3 + K_4W^4. \tag{18.4}$$

The expressions for the coefficients K_i are given in Application F_1. (They were divided by the common factor b_{-2}^4).

To illustrate how our next program works, we consider an example of elimination of unknowns from the equations describing the quasi-stationary reaction mode with a two-route mechanism obtained from (18.1) by adding the stage

$$4)\ B + AZ \rightleftarrows Z + AB. \tag{18.5}$$

Its system of equations is given by

$$\begin{aligned} 2W_1 - W_3 - W_4 &= 0, \\ W_2 - W_3 &= 0, \end{aligned} \tag{18.6}$$

where

$$W_1 = b_1z^2 - b_{-1}x^2, \qquad W_2 = b_2z - b_{-2}y,$$
$$W_3 = b_3xy - b_{-3}z, \qquad W_4 = b_4x - b_{-4}z,$$
$$z + 1 - x - y,$$

with x, y, z denoting the concentrations of AZ, BZ, Z. Application F_2 performs the calculation of the resultant of the polynomial (18.6) in the variable y, by means of a standard routine called `resultant(...)`. As a result of the elimination the equations (18.6) are reduced to a polynomial of degree 4 in x.

For obtaining the kinetic polynomial corresponding to the model (18.6) we must transform it to the form

$$\begin{gathered} (b_{-3}(b_{-2} + b_4)^2 - b_3(b_{-4} + b_4 - b_2)(b_2 + b_{-2} - b_{-4}))x^2 + \\ +(-(b_2 + b_{-2} - b_{-4})(b_2b_4 + b_{-2}b_{-4} + b_{-2}b_4) + b_3(W + b_{-4} - b_2)\times \\ \times(b_2 + b_{-2} - b_{-4}) - 2b_{-3}(b_{-2} + W)(b_{-2} + b_4))x + \\ +(b_2 + b_{-2} - b_{-4}(W(b_2 + b_{-2}) + b_{-2}b_{-4}) + b_{-3}(b_{-2} + W)^2 = 0, \\ (2b_1(b_{-2} + b_4)^2 - 2b_{-1}(b_2 + b_{-2} - b_{-4})^2)x^2 + \\ +(-4b_1(b_{-2} + W)(b_{-2} + b_4))x + 2b_1(b_{-2} + W)^2 - \\ -W(b_2 + b_{-2} - b_{-4})^2 = 0, \end{gathered} \tag{18.7}$$

where

$$W = 2W_1 = W_3 + W_4$$

is the formation rate of the reaction product (18.1)+(18.5). The resultant of the polynomials (18.7) in the variable x is the kinetic polynomial for this reaction. It has the form (18.4), and the expression for $K_0, \ldots, K_4$ are given in Application F_3

(where BN1= $b_{-1}1, \ldots$, BN4= b_{-4}, respectively). One reason for the awkwardness is the presence of the factor

$$(b_2 + b_{-2} - b_{-4})^4$$

in the resultant of the polynomials (18.7).

In Sec. 14 we proved that the constant term of the kinetic polynomial for a single-route reaction mechanism must contain the factor

$$\left(\prod_{i=1}^{n} b_i^{\nu_i} - \prod_{i=1}^{n} b_{-i}^{\nu_i}\right).$$

As an example, for the mechanism (18.1) this factor is equal to

$$(b_1 b_2^2 b_3^2 - b_{-1} b_{-2}^2 b_{-3}^2).$$

In the case of a multi-route mechanism the situation is more complicated. For instance, for the double-route mechanism (18.1)+(18.5) the expression for $K(0)$ becomes

$$\begin{aligned} K(0) = 4b_{-2}^2(b_{-1}(b_1 b_2^2 b_3^2 - b_{-1} b_{-2}^2 b_{-3}^2) + 2b_{-1} b_1 (b_{-2} + b_3 + b_4) \times \\ \times (b_4 b_{-2} b_{-3} - b_{-4} b_2 b_3) + 2b_{-1}(b_2 + b_{-2} - b_{-4})(b_4 b_3 b_2 b_1 - \\ - b_{-4} b_{-3} b_{-2} b_{-1}) + (b_1 b_4^2 - b_{-1} b_{-4}^2)(b_{-1}(b_2 + b_{-2} - b_{-4})^2 - \\ - b_1(b_{-2} + b_3 + b_4)^2). \end{aligned} \qquad (18.8)$$

Notice that along with the terms

$$((b_1 b_2^2 b_3^2 - b_{-1} b_{-2}^2 b_{-3}^2) \text{ and } (b_1 b_4^2 - b_{-1} b_{-4}^2))$$

corresponding to the two basic routes, the expression (18.8) contains two more summands, corresponding to linear combinations of the routes (one of them corresponds to the brutto-equation of the reaction $0 \rightleftarrows 0$). But taking into account the thermodynamic restrictions on the kinetic parameters (see [**150**]) which in this case have the form

$$\frac{k_4}{k_{-4}} = \frac{k_2 k_3}{k_{-2} k_{-3}},$$

we get $b_4 b_{-2} b_{-3} = b_{-4} b_2 b_3$ for arbitrary concentrations A_2, B, AB. Hence the constant term will have the form

$$K(0) = K(0)' \left(k_1 k_2^2 k_3^2 [A_2][B]^2 - k_{-1} k_{-2}^2 k_{-3}^2 [AB]^2\right),$$

just as for the single-route reaction mechanism.

18.2. CONDITIONS FOR NON-UNIQUENESS OF STEADY-STATES

For applications, particularly to chemical kinetics, it is important to find criteria for non-uniqueness of solutions to systems of algebraic equations in a given domain, and also to calculate the bifurcation boundaries.

In the situation where the elimination of unknowns has been done and a polynomial in one variable has been obtained, the problem of finding the bifurcation boundaries is reduced to calculating conditions for multiple zeros of the polynomial.

For example, the conditions for a double zero are: $P(x) = 0$ and $P'(x) = 0$; for a triple zero: $P(x) = 0$, $P'(x) = 0$ and $P''(x) = 0$ etc. The condition for multiple zeros corresponds to the vanishing of the discriminant of the polynomial [**97, 142**]. The late is proportional to the resultant of $P(x)$ and $P'(x)$, so for finding the bifurcation boundaries it is sufficient to calculate the resultant of the polynomial and its derivative and to solve the obtained equation with respect to the appropriate parameters. Let us consider this procedure on the example of the kinetic model for the oxidation reaction: CO in CO_2. We assume that its mechanism is (18.1)+(18.5), where $k_{-3} = k_{-4} = 0$ and the substances A_2, B and AB correspond to O_2, CO, and CO_2 respectively. The result of the elimination of unknowns in this case is given in Application F_2. The bifurcation boundaries for this model were calculated in [**52**] with respect to the coordinates $\{p_{O_2}, p_{CO}\}$ (p_i are the partial pressures), whereby the equations for the boundaries were obtained in parametric form. We determine these boundaries by computing the resultant of the polynomial from Application F_2 and its derivatives in x. The values b_i, b_{-i} are given by the expressions

$$b_1 = k_1 p_{O_2}, \quad b_{-1} = k_{-1}, \quad b_2 = k_2 p_{CO}, \quad k_{-2} = b_{-2}, \quad b_3 = k_3,$$

$$b_4 = k_4 p_{CO}; \quad \text{with the values of } k_{\pm i} \text{ from [52]:}$$

$$k_1 = 202 \cdot 10^6, \quad k_{-1} = 16 \cdot 10^{13} \exp(-50000/(RT)), \quad k_2 - 45 \cdot 10^6,$$

$$k_{-2} = 10^{13} \exp(-35500/(RT)), \quad k_3 = 4 \cdot 10^4 \exp(-11000/(RT)),$$

$$k_4 = 45 \cdot 10^6, \quad \text{temperature } T = 450 \text{ K}, R = 1,987.$$

The obtained expressions have the form

$$D(p_{CO}, p_{O_2}) = a_{12} p_{CO}^{12} + \ldots + a_1 p_{CO} + a_0, \tag{18.9}$$

where $a_0, \ldots, a_{12}$ are polynomials in p_{CO} (see Application F_4). In the calculations we factored out the leading coefficient. To avoid loss of exactness we used the routine ON BIG FLOAT.

The roots of the equation $D(p_{CO}, p_{O_2}) = 0$ were calculated with the help of the system Eureka, using its function 'poly'. It finds all zeros of a given polynomial for IBM PC, and makes it possible to find the roots of nonlinear equations and systems, and to solve problems in nonlinear programming etc. Among the 12 zeros of the polynomial (18.9) only two (9 and 10) correspond to the boundary of plurality domain for zeros $0 < x < 1$ with physical sense. The other zeros determine bifurcation points outside the physical domain. In Table 1 we give, for some values of p_{O_2}, the calculated bifurcation values p_{CO}^-, p_{CO}^+ having physical sense, and the zeros of the polynomial from Application F_2 corresponding to these points (the columns $[ZO]^-$, $[ZO]^+$). As can be seen we obtained a fairly good approximation to the boundary of double zeros (two of the zeros agree up to 4-6 significant digits). We could not calculate the discriminant for this example in symbolic form on the personal computer because of limited memory. It was instead calculated on a EC 1055 computer with the help of the system SRM [**126**].

Let us compute, in symbolic form, the conditions of plurality of s.s. in the case of the mechanism (18.1) with $k_{-1} = k_{-3} = 0$. The s.s. belonging to the domain $0 < AZ < 1$ are in this case determined by the equation

$$f_0 = a_3 x^3 + a_2 x^2 + a_1 x + a_0 = 0, \tag{18.10}$$

where

$$a_3 = 2, \quad a_2 = 4\varepsilon c + b - 2, \quad a_1 = 2\varepsilon^2 c^2 + \varepsilon(b(b+c) - 4c),$$

TABLE 1. The boundaries of a domain of plurality of s.s. on the plane of $(p_{O_2} - p_{CO})$

$p_{O_2} \cdot 10^8$	$p_{CO}^{-} \cdot 10^8$	$[ZO]^-$	$p_{CO}^{+} \cdot 10^8$	$[ZO]^+$
5	0.43885000	0.00338500	1.0292460	0.20768334
		0.00338500		0.20768334
		0.64457984		0.00014564
		1.3423528		1.5781903
10	0.61383330	0.00332790	1.5070777	0.22191424
		0.00332790		0.22191424
		0.69038373		0.00012298
		1.2966631		1.5352948
50	1.2959491	0.00325413	3.3133996	0.26778907
		0.00325413		0.26778907
		0.76778193		0.00009482
		1.2194124		1.4580296
100	1.7622513	0.00323683	4.5007380	0.27992977
		0.00323683		0.27992977
		0.79115760		0.00008824
		1.1960713		1.4337558
500	3.4754019	0.00321401	8.6919485	0.29995268
		0.00321401		0.29995268
		0.82897289		0.00007905
		1.1583032		1.3936507

$$a_0 = -2\varepsilon^2 c^2, \quad x = AZ, \quad \varepsilon = b_1/b_3, \quad b + b_2/b_1, \quad c = b_{-2}/b_1.$$

In [**149**] the conditions for plurality were obtained in terms a_i. Let us find the Sturm system for the polynomial (18.10) in the segment [0,1]. By [**92**] this amounts to finding the sequence of central determinants (or inners) of the matrix

$$R = \begin{pmatrix} a_3 & a_2 & a_1 & a_0 & 0 \\ 0 & a_3 & a_2 & a_1 & a_0 \\ 0 & 0 & 3a_3 & 2a_2 & a_1 \\ 0 & 3a_3 & 2a_2 & a_1 & 0 \\ 3a_3 & 2a_2 & a_1 & 0 & 0 \end{pmatrix}. \tag{18.11}$$

The last term of Sturm system is $f_3 = \det R$. The preceding terms are given by the inners of the matrix R in which the last column is replaced by the sum of the columns situated to the right of it, and this column multiplied by x^i (see (18.11)). Then

$$f_2(0) = \det \begin{pmatrix} a_3 & a_2 & a_1 \\ 0 & 3a_3 & a_1 \\ 3a_3 & 2a_2 & 0 \end{pmatrix},$$

$$f_2(1) = \det \begin{pmatrix} a_3 & a_2 & a_1 + a_0 \\ 0 & 3a_3 & 2a_2 + a_1 \\ 3a_3 & 2a_2 & a_1 \end{pmatrix}, \tag{18.12}$$

$$f_1(0) = a_1, \qquad f_1(1) = 3a_3 + 2a_2 + a_1.$$

The values f_3, $f_2(0)$, $f_2(1)$ were obtained with the use of a system from [**126**]. The analysis of the obtained Sturm system gives the following necessary and sufficient conditions for plurality of solutions of (18.10) in [0,1]:

$$\delta < 1/8, \tag{18.13a}$$

$$b < b_* = 2(2 - \delta^{1/3})(1 + \delta^{1/3})^2, \tag{18.13b}$$

$$\begin{gathered} \delta b(14 - b - 4\delta)/(8\delta^2 + 2\delta^2(3b + 12) + \delta(b^2 - 2b + 24) + \\ +2(2 - b)^2) < c' < b(2 - b_4\delta)/(8\delta^2 + 2\delta(3b + 8) + \\ +(2 - b)(4 - b)), \end{gathered} \tag{18.13c}$$

$$\begin{gathered} -32\delta^3 c'^2 + 4\delta^2 c'(c'^2 b - 8bc' - 24c' - 8b) + \\ +4\delta(c'^3 b(b + 2) - c'^2(b^2 - 2b + 24) + 4c'b(5 - b) - 2b^2) + \\ +c'(b - 2)^2(b(1 + c')^2 - 8c') > 0, \end{gathered} \tag{18.13d}$$

where

$$\delta = \frac{b_{-2}}{b_3}, \quad c' = \frac{b_{-2}}{b_2}.$$

The boundary of non-uniqueness is defined by the vanishing of the left hand side in (18.13d). For $b = b_*$ and $c' = c'_* = (1 - 2\delta^{1/3})(1 + 2\delta^{1/3})$ a triple zero $x = \delta^{2/3}$ occurs. In limiting cases the conditions (18.13) are simplified.

If $b_3 \to \infty$ then $\delta \to 0$ and (18.13) yields the inequalities

$$b_2 > b_{-2}, \qquad \frac{b_2}{2} \leq b_1 \leq \frac{(b_2 + b_{-2})^2}{8b_{-2}}, \tag{18.14}$$

that were obtained in [**112**]. From (18.13d) one can get more precise expressions for bifurcation boundaries (18.14):

$$b_2/b_1 \simeq 2 - 4\sqrt{b_2/b_{-2} - 1}\sqrt{b_{-2}/b_3} \tag{18.15}$$

(these correspond to the boundary $b_1 = b_2/2$),

$$\frac{b_2}{b_1} \simeq \frac{1}{8}\left(1 + \frac{b_2}{b_{-2}}\right)^2 + \frac{(b_2 + b_{-2})^3}{8\,(b_2 - b_{-2})\,b_3} \tag{18.16}$$

(these correspond to the boundary $b_1 = (b_2 + b_{-2})^2/(8b_{-2})$). Observe that the boundary (18.15) is more sensitive for small values of the parameter b_{-2}/b_3 than the boundary (18.16).

For $b_{-2} = 0$ we deduce from (18.13) the condition

$$b_2 < 2b_1\left(1 + \sqrt{8b_1/b_3}\right),$$

which was obtained in [**149**].

If $c' \ll 1$ then (18.13) implies the inequalities

$$\delta < 1/8, \qquad \xi_1 c' < b < \xi_2 c', \tag{18.17}$$

$$\xi_{1,2} = (20\delta + 1 - 8\delta^2 \mp (1 - 8\delta)^{3/2})/(4\delta).$$

The conditions (18.17) are applied in the domain b_{-2}, $b_3 \ll b_1$. Hence the plurality of s.s. can take place for arbitrary small values of b_3. For example, for $b_1 = 1200$, $b_2 = 1$, $b_{-2} = 10^{-4}$, $b_3 = 1.25 \cdot 10^{-3}$ there exist three s.s.

$$AZ_1 = 0.5141, \qquad BZ_1 = 0.4854,$$

$$AZ_2 = 0.2811, \qquad BZ_2 = 0.7185,$$

$$AZ_3 = 4.427 \cdot 10^{-2}, \qquad BZ_3 = 0.9555.$$

We have thus seen the use of analytic manipulations on the computer allows us to reduce the stationary kinetic model to one equation, and to investigate the number of solutions. The obtained symbolic expressions make it possible to analyze also the limiting cases.

Moreover, we obtain explicit expressions for the stationary solutions and the bifurcation boundaries. In the general situation we need to apply methods from solvability theory (see Sec. 17).

18.3. REALIZATION OF THE METHOD OF GRÖBNER BASES

The REDUCE program which realize the modified method of Gröbner basis described in 17.2 is contained in Application F_5.

For given polynomial the procedure LM finds the leading monomial with respect to the used ordering (in our case the lexicographical one). The names of the variables (their number is NVBL) and the chosen ordering are defined by an array

$$\mathrm{AN}(\mathrm{AN}(1) >_T \mathrm{AN}(2) >_T \ldots >_T \mathrm{AN}(\mathrm{NVBL})).$$

The output of the procedure is: The leading coefficient (LCLP(O)), the leading product of powers (LCLP(1)), its exponents (array DEG0), and the leading monomial LM=LCLP(0)LCP(1). The calculations are carried out by iterated application of the functions REDUCE DEG and LCOF.

The procedure ORM picks out the leading one or two monomials (in the sense of given lexicographical ordering), whose exponents are given in the arrays D1, D2.

The procedure NF realizes the algorithm described in Sec. 17.2 for finding the normal form of a polynomial GNF modulo polynomials given as values of the operator FNF(i), ($i = 1, \ldots, N$). If $N = 0$ then GNF is returned. Otherwise the leading monomial of the polynomial GNF is found, and it is checked if it is smaller than the smallest of the leading monomials of the polynomials FNF(i). If the reply is 'yes' then program stops, otherwise an attempt is made to find a polynomial FNF(L) whose leading monomial is a factor of LM(GNF). If such a polynomial has been found then the reduction

$$\mathrm{GNF} \underset{FNF(L)}{\rightarrow}$$

is realized, otherwise we go to the next monomial (in order) of the polynomial GNF. Before addressing the procedure NF it is necessary to give the following global variables: The array D1 containing the exponents of the minimal among the leading monomials of the polynomials FNF(i); the operator DEG1(JR, K) whose value is the degree of the variable AN(K) in the leading monomial of the polynomial FNF(JR); the operator LMF(L) whose values are the leading monomials of the polynomials FNF. The calculation of these variables has been broken out of the procedure NF in order to avoid superfluous calculations during the basic algorithm. If the value of the global variable NORM is equal to 1 then the output will be in normalized form with the leading coefficient 1. After execution of NF one has found the leading monomial, the leading coefficient and the leading product of powers of AN(j) among the global variables, and the values of the operator DEG NF(K) are the exponents of powers AN(K) in LPPNF for the calculated normal form HNF=NF(N, GNF).

The procedure SP calculates the S-polynomial for a given pair of polynomials (F1, F2) with the leading products LPP1, LPP2 and the leading coefficients LC1, LC2 and

$$\text{LCM(LPP1, LPP2)=LCM.}$$

The procedure REDALL realizes algorithm 'Reduce everything' (see Sec. 17.2). To the sets R, P, G of the algorithm there correspond the operators RB, PB, GB. Here the structure of data is the following: $RB(j,\ 1)$ is j-th polynomial of the set R while

$$RB(j,\ 2),\ RB(j,\ 3),\ RB(j,\ 4)$$

are its leading monomial, leading coefficient and leading product of powers respectively; $RB(j,\ 4+K)$ is the exponents of the variables AN(K) in the leading monomial ($k = 1,\ \dots$, NVBL). The operator BB corresponds to the set B of the algorithm; the values BB(L, 1), BB(L, 2) are the numbers of polynomials of the L-th pair (f_1, f_2) in the set B; the values $BB(L,\ K+2)$, $K = 1,\ \dots$, NVBL, are the exponents of the variables AN(K) in

$$\text{LCM(LPP}(f_1),\ \text{LPP}(f_2)).$$

First an element of the set R is chosen (in our program we choose the last element), and then the set $G \cup P$ and the sets $G0$, $P0$ are formed. The values of the operators $G0$, $P0$ are the numbers of such polynomials from the set G, P which can be reduced modulo the polynomial H =NF($R, G \cup P$), which is being added to the basis. Then the operation

$$R := R \cup G0 \cup P0$$

is realized: From the set B all pairs (f_1, f_2) in which f_1 or f_2 belongs to $G0$ are removed, and from the set G the elements of the set $G0$ are removed. Using the values accumulated in the operator PER, the values $BB(J,\ 1), BB(J,\ 2)$ are replaced by new ones, which are obtained by a renumbering of the elements of the set G. Further the set $P0$ is removed from the set P, and after this procedure the operation $P := P \cup H$ is executed.

The procedure NEWBAS realizes the algorithm 'New basis'. First the operation $G := G \cup P$ is performed, and then this set is enlarged with all pairs (g, p) such that

$$g \in G,\ p \in P,\ g \neq p$$

(first the polynomials g from 'old' set G are taken and the pairs (g,p) are formed, then the pairs (p_i,p_j), $j > i$, are formed). Further, all polynomials $h \in G$ are reduced modulo $G \setminus h$ with the aim of obtaining a reduced basis.

The procedures CRIT1 and CRIT2 realize the algorithms 'Criterion 1' and 'Criterion 2', which serve to avoid superfluous calculations of normal form.

The procedure PRITT(II) provides a detailed printout of the results of the work of the algorithm. The elements of the sets R (if II=1), P, G are printed.

First initial input data are entered into basic algorithm. They must either be written in the file KP.INP, or contain the sentence IN with a reference to the file with the initial data.

The initial data contain:

the number of variables in polynomials of the initial system $\longrightarrow$	NVBL:=
the names of these variables (the other names are interpreted by the program as symbolic parameters) $\longrightarrow$	AN(1):= . . . . . AN(NVBL):=
the indicator of normalization (is executed if set equal to 1) $\longrightarrow$	NORM:=
the indication of printing (if equal to 1 a detailed printout of intermediate results is executed) $\longrightarrow$	IPRI:=
the polynomials of the initial system $\longrightarrow$	RB(1,1):= . . . . . RB(NR,1):=

After this the global variables are initialized, and the procedures REDALL, NEWBAS are addressed. If the set B is not empty then a pair from B is found, and Criterion 1 and 2 are checked. If CRIT1=0 and CRIT2=0 then the S-polynomial of this pair is calculated. This S-polynomial is then in normal form, and if it is not equal to 0 then the set $G0$ is formed. From set K the polynomials occurring in $G0$ are deleted. The calculated normal form is included in the set B, and from set B all pairs containing elements from $G0$ are deleted. Finally,the pairs in B are renumbered, and the procedures REDALL, NEWBAS are addressed over again.

We now consider some examples. (They are all taken from [**27**]. We obtained the same results as in [**27**], except for some misprints in [**27**].) The model corresponding to the nonlinear shock reaction mechanism

$$\begin{array}{ll} 1) & A_2 + 2Z \rightleftarrows 2AZ, \\ 2) & B + AZ \rightleftarrows Z + AB, \end{array} \tag{18.18}$$

has the form

$$\begin{aligned} f_1 &= b_1 Z^2 - b_{-1} AZ^2 - WR = 0, \\ f_2 &= b_2 AZ - b_{-2} Z - 2WR = 0, \\ f_3 &= Z + AZ - 1 = 0, \end{aligned} \tag{18.19}$$

where WR is the stationary reaction rate. Let us now produce its kinetic polynomial. We fix the ordering

$$AZ >_T Z >_T WR$$

and use the described program. The work of the program starts with the algorithm REDALL. It chooses the polynomial f_3 and calculates its normal form. Since the sets G and P are empty so far, the result of the reduction is f_3 itself. The algorithm inserts f_3 in the set P: $P(1) = f_3$ and in the set $R := R(1) = f_1$, $R(2) = f_2$. The following step is a calculation of the normal form of the polynomial f_2 modulo f_3. This result is:

$$f_4 = \frac{Z\,b_{-2} + (Z-1)b_2 + 2\,WR}{b_{-2} + b_2},$$

which is inserted in the set P: $P(2) = f_4$. Then the polynomial $R(1) = f_1$ is reduced modulo f_3 and f_4. The result here is:

$$f_5 = \frac{WR^2(4b_{-1} - 4b_1) + WR(4b_{-1}b_{-2} + b_{-2}^2 + 2b_{-2}b_2 + 4b_1b_2 + b_2^2) + b_{-1}b_{-2}^2 - b_1b_2^2}{4b_{-1} - 4b_1}.$$

This too is inserted in the set P: $P(3) = f_5$. Since the set R has now been exhausted the algorithm REDALL stops. Instead the procedure NEWBAS starts working. First it forms the set G: $G(i) = P(i)$, $i = 1,2,3$, and the set of pairs B: $(1,2)$, $(1,3)$, $(2,3)$. Then the normal forms of polynomials $G(i)$ modulo the other polynomials are calculated. The calculation of the normal form $G(1) = f_3$ modulo $G(2)$, $G(3)$ gives the polynomial

$$f_6 = \frac{(b_{-2} + b_2)AZ - b_{-2} - 2\,WR}{b_{-2} + b_2},$$

which becomes the value of $G(1)$. The polynomials $G(2)$ and $G(3)$ remain unchanged by the algorithm NF.

At this step the calculations of the procedure NEWBAS are finished, and the main algorithm begins to work. For the sequence

$$G(i) \quad (2,3), \quad (1,3), \quad (1,2)$$

of pairs of indices of the polynomials G_i the algorithm checks Criterion 1 and 2. In our case all basis polynomials satisfy that the LCM of the leading products is equal to their product, and hence Criterion 2 comes into play and prevents superfluous calculations of normal forms of S-polynomials, which would here be equal to zero. Therefore the algorithm finishes its work and the desired reduced Gröbner basis consists of the polynomials

$$G(1) = f_6, \quad G(2) = f_4, \quad G(3) = f_5.$$

The polynomial $G(3)$ is the kinetic polynomial obtained in [**154**] for the mechanism (18.18). Its solutions are the possible values of the stationary reaction rate (one of them lacking physical sense). Knowing them it is a simple task to find AZ and Z from equations $GB(1) = 0$ and $GB(2) = 0$, which are linear in Z and AZ

In the case of system (18.19) the main algorithm never comes to work, and the basis is calculated already in the procedures REDALL and NEWBAS. The same happens for many other 'simple' systems.

Application F_6 contains the initial data and the results of Gröbner basis calculation are contained for the model corresponding to the reaction mechanism of hydrogen para–ortho conversion.

$$\begin{aligned} &1) \quad H_2^{(p)} + 2Z \rightleftarrows 2ZH, \\ &2) \quad 2ZH \rightleftarrows 2Z + H_2^{(0)}, \end{aligned} \tag{18.20}$$

where Z and ZH are intermediates, $H_2^{(p)}$ and $H_2^{(0)}$ are para and ortho forms of hydrogen. The basis $GB(1)$, $GB(2)$, $GB(3)$ was calculated for the ordering

$$ZH >_T Z >_T WR.$$

The polynomial $GB(3)$ is the kinetic polynomial for the reaction (18.20).

The initial data and the results of a calculation for the model (18.2) with $b_{-1} = b_{-2} = b_{-3} = 0$ are contained in Application F_7. (This corresponds to an irreversible adsorption mechanism in the reaction (18.1)). The kinetic polynomial for this case is given by

$$GB(4) = \frac{4\,WR^2\,b_1 - WR\,b_2^2}{4\,b_1}. \tag{18.21}$$

On the other hand, substituting the values $b_{-1} = b_{-2} = b_{-3} = 0$ for the coefficients of the polynomial (18.4) (see Application F_1) we get the expression

$$b_3^2\left(WR\left(b_2^2 - 4\,b_1\,WR\right)\right)^2 = 0, \tag{18.22}$$

which indicates that the roots of (18.22) $WR_1 = 0$ and $WR_2 = b_2^2/(4b_1)$ are double. The meaning of this fact is easy to understand by considering the other polynomials of the basis. The equation $GB(3) = 0$ is quadratic in BZ, and therefore to each of the root WR_1, WR_2 there correspond two roots of the equation $GB(3) = 0$. If we keep in mind that the equations $GB(1) = 0$ and $GB(2) = 0$ are linear in Z and AZ, then we find that to each value WR_i there corresponds a family of solutions $BZ_{i(1,2)}$, $AZ_{i(1,2)}$, $Z_{i(1,2)}$. The existence of double roots of (18.22) reflects this fact. From the example it is seen that different methods of elimination of unknowns can lead to different reductions of the initial system and to different representations of its solutions.

Of course, the method of Gröbner bases works also for the systems of linear equations. Application F_8 contains the initial data and results for a model corresponding to the linear single-route reaction mechanism

$$\begin{aligned} &1) \qquad A + Z1 \rightleftarrows Z2, \\ &2) \qquad Z2 \rightleftarrows Z3, \\ &3) \qquad Z3 \rightleftarrows Z1 + B. \end{aligned} \tag{18.23}$$

From the obtained basis $G(1)$, ... , $G(4)$ one finds the solutions in all variables immediately from the independent linear equations $G(1) = 0$, ... , $G(4) = 0$. In the linear case the method of Gröbner bases works in fact like the Gauss method [**27**], which variables being eliminated in every step.

Consider now the more complicated model of a catalytic reactor of an ideal mixture (RIM) in which the reaction proceeds according to mechanism (18.1)+(18.4). It has the form

$$N_{A_2}^0 - N_{A_2} - \alpha\,(W1 + N_{A_2} \cdot R) = 0,$$

$$N_B^0 - N_B - \alpha\,(W2 + W4 + N_B \cdot R) = 0, \tag{18.24}$$
$$2W1 - W3 - W4 = 0,$$
$$W2 - W3 = 0,$$
$$W1 = N_{A_2} z^2 - r_{-1} x^2, \qquad W2 = r_2 N_B z - r_{-2} y,$$
$$W3 = r_3 xy, \quad W4 = r_4 N_B x, \quad z = 1 - x - y, \quad R = W3 - W1 - W2,$$

where N_{A_2}, N_B are the molar parts of the gaseous substances A_2, B; x, y are the surface concentrations of adsorbates, and

$$\alpha,\ r_{-1},\ r_{-2},\ r_3,\ r_4,\ N_{A_2}^0,\ N_B^0$$

are parameters of the model.

Application F_9 contains the results of a Gröbner basis calculation for the model (18.24) with parameter values

$$N_{A_2}^0 = 0.5,\ N_B^0 = 0.25,\ \alpha = r_{-1} = r_{-2} = r_3 = r_4 = 1.$$

Notice that for the chosen ordering the Gröbner basis contains a linear expression, relating the concentrations A_2 and B in the gas. In general case it has the form [**108**]:

$$N_{A_2} = \frac{N_B + p - 2}{p}, \qquad p = \frac{2 - N_B^0}{1 - N_{A_2}^0}. \tag{18.25}$$

The degree of the polynomial $GB(4)$ with respect to N_{A_2} is equal to 10 and the variables x, y are expressed through N_{A_2} in the form of polynomials of degree 9. Of the 10 roots of equation $GB(4) = 0$ only one (N_{A_2} =0.47779941) has physical meaning. The corresponding values are $N_B = 0.17229794$, $x = 0.34460068$, $y = 0.074444287$. We wish to point out that we could not to solve this example on PC AT with the package GROEBNER including in the system REDUCE 3.3.

The program allows for models of fairly high dimension. In Application F_{10} we give the initial data and computation results for a system of 6 algebraic equation, which arises when quasichemical approximation is used to describe the elementary processes on the surface of a catalyst [**161**]. The initial system is reduced to a polynomial $G(8)$ of degree 8 with respect to unknown PAO. The other unknowns are expressed through PAO in the form of polynomials of degree 7. With the parameter values indicated in the sentence LET, the equation $G(6) = 0$ has the following real solutions: PAO_1=0.10909942, PAO_2=0.9733262692, PAO_5=0.12452337, PAO_6=1.2328531, PAO_7=2.1624142, PAO_8=5.2100331. There is a unique PAO solution with a physical sense, and the corresponding values of the other variables are POO=0.28516341, PAA=0.0054376081, PAB=0.064601414, PBO=0.40514981, PBB=0.1151244.

We have also considered some examples from [**11**], which uses results from multidimensional complex analysis. Unfortunately we found some mistakes and misprints in the general examples.

Our experience in applying the method of Gröbner bases has showed that there can arise numerical difficulties while working with polynomials with symbolic coefficients. In the process of calculations there occur very unwieldy expressions, which often contain nontrivial symbolic common factors. Moreover, the algorithm does not always discern the situations in which a normal form of a S-polynomial vanishes. In this case hard calculations, needing large computer resources, are carried out, only to finally end up with a vanishing S-polynomial. Finally, the initial version of the algorithm is not well adapted for real arithmetic. Transformation of floating point data to exact rational form (especially in situations with sharply distinct parameter) leads to unwieldy integer coefficients which can occupy several lines of display. As a result the calculations are slowed down and an emergency stop can occur due to exhaustion of available memory.

A modification of the algorithm allows one to circumvent these difficulties. The idea is the following: Each time a new polynomial is formed in the process of the algorithm (e.g. when S-polynomials and normal forms are being computed) it is replaced by its own 'dummy'. The 'dummy' is a polynomial with indefinite coefficients in which all monomials from the basis polynomials are present. For example, the 'dummy' of the polynomial

$$\frac{b^3x^2y - (a-b)x^2yz + cd}{f}$$

is given by

$$A(1)x^2y + A(2)x^2yz + A(3).$$

The expressions for these indefinite coefficients are then accumulated as the algorithm goes along. In the procedure NF there is a check that next monomial to be reduced really is contained in the polynomial. To this end the value for the coefficient of this monomial is computed for some values of the parameters. In case these values do not belong to an exceptional set (the union of a finite number of hypersurfaces in the parameter space of the initial system) then the coefficient being non-zero for these parameter values means that it also does not vanish in the generic situation.

The key moment of the modification consist in the following: For the checking we take not any real values for the coefficients (symbolic or numerical) but only integer or rational values. The best way to obtain these is to use a generator of random numbers. Having calculated the normal form (or the S-polynomial) we then check that its leading coefficient is not trivial by means of the same substitution. If it is equal to zero then next coefficient is considered etc. In this way the algorithms NF and SP produce either polynomial with nontrivial leading coefficient or the zero polynomial. The substitution can be done so that repeated calculations with the same 'dummy' coefficient are avoided.

The modified algorithm gives as output a basis of polynomials with indefinite coefficients and recursion formulas for them. By checking the consecutive calculations against these formulas we can get the expressions for the true basis polynomials (with symbolic or numerical coefficients).

Thus, checking (once) the calculations with (preferably small) integer or rational numbers we get expressions with symbolic coefficients or, using real arithmetic, expressions with real coefficients. In order to make sure that the test point, in which all checks are conducted, does not belong to the exceptional set, we can perform the calculations with other values of parameters.

Let us consider some examples. The basis for the model (18.19) calculated with the modified algorithm has the form (for ordering $AZ >_T Z >_T WR$)

$$G(1) = (AZ \cdot AM(3,3) + WR \cdot AM(3,2) + AM(3,1))/AM(3,3),$$

$$G(2) = (Z \cdot AM(1,3) + WR \cdot AM(1,2) + AM(1,1))/AM(1,3),$$

$$G(3) = (WR^2 \cdot AM(2,3) + WR \cdot AM(2,2) + AM(2,1))/AM(3,3).$$

The expressions for the coefficients $AM(i,j)$ are accumulated in the operator MAK:

$$\mathrm{MAK}(1,1) = b_2,$$

$$\mathrm{MAK}(1,2) = -2,$$

$$\mathrm{MAK}(1,3) = -b_2 - b_{-2},$$

$$\mathrm{MAK}(2,1) = (-b_{-1} \cdot AM(1,3)^2 - 2b_{-1} \cdot AM(1,1) \cdot AM(1,3) + +(b_1 - b_{-1}) \cdot AM(1,1)^2)/AM(1,3)^2,$$

$$\mathrm{MAK}(2,2) = (-AM(1,3)^2 - 2 \cdot b_{-1} \cdot AM(1,2) \cdot AM(1,3) + +(2b_1 - 2b_{-1}) \cdot AM(1,1) \cdot AM(1,2))/AM(1,3)^2,$$

$$\mathrm{MAK}(2,3) = ((b_1 - b_{-1}) \cdot AM(1,2)^2)/AM(1,3)^2,$$

$$\mathrm{MAK}(3,1) = (-AM(1,3) - AM(1,1))/AM(1,3),$$

$$\mathrm{MAK}(3,2) = -AM(1,2)/AM(1,3),$$

$$\mathrm{MAK}(3,3) = 1.$$

The substitution of the expressions MAK into the operator AM is produced by the single command

$$\text{for all } i,j \quad \text{let} \quad AM(i,j) = \mathrm{MAK}(i,j) \tag{18.26}$$

When this has been executed one obtains expressions which coincide with those written above for the same example.

Consider now the case of the kinetic model (18.2) for the three-route adsorption reaction mechanism (18.1). The results of the calculations for this example are given in Application F_{11}. Using these data and writing the operators $b_1 := 1$, $b_2 := 8/10$, $b_3 := 1000$, $b_{-1} := 1/10$, $b_{-2} := 1/100$, $b_{-3} := 1/10$, on float, as well as the operator (18.26), we obtain the Gröbner basis for these values of parameters b_i:

$$1.839673647 \cdot WR^3 + 0.5297823406 \cdot WR^2 - 2.594329737 \cdot WR + $$

$$+Z - 1.2325350731 \cdot 10^{-2} = 0 \tag{18.27}$$

$$-0.8 \cdot AZ + 119.210523 \cdot WR^3 + 34.32989567 \cdot WR^2 -$$

$$-8.112566972 \cdot WR + 1.3172726443 \cdot 10^{-3} = 0,$$

$$-64663.93507 \cdot BZ - 9516842.982 \cdot WR^3 - 2740624.87 \cdot WR^2 +$$

$$+487978.5602 \cdot WR^4 - 63760.45435 = 0,$$

$$28.11179393 \cdot WR^4 - 6.828539245 \cdot WR^3 - 0.1142106 \cdot WR^2 +$$

$$+7.6913707646 \cdot 10^{-2} \cdot WR - 1.249410287 \cdot 10^{-5} = 0.$$

The solutions of the equations (18.27) have the form $\{WR, Z, AZ, BZ\}$:

1)	0.00016248271,	0.012746870,	$3.4524894 \cdot 10^{8}$,	0.9872531
2)	0.15386307,	0.39225384,	$5.3209078 \cdot 10^{-5}$,	0.60769295
3)	0.18498457,	0.46246234,	0.53746478,	$7.2888333 \cdot 10^{-5}$
4)	−0.096103560,	−0.24025909,	1.240741,	$-1.5032708 \cdot 10^{-5}$

The solutions 1)–3) have a physical sense and correspond to three s.s. of the model.

The considered examples show that the modified algorithm works well both with symbolic and with real data.

In the case of symbolic data it is necessary in process of substitution to watch out for any common symbolic factors and to delete them (for this we can use the command ON GCP and the function FACTORIZE). In the case of unwieldy expressions the calculations of the coefficients $AM(i,j)$ can be carried out step by step in separate session of the system REDUCE.

On the other hand, our experience from working with the method of Gröbner bases has shown that in spite of its generality it is probably not the best elimination method for systems of algebraic equations with symbolic parameters. The best way is to make use of the specific character of system, which is exactly what we shall do in the case considered below.

19. The construction of the resultant

19.1. SYSTEMS OF TYPE (8.2.)

In Sec. 8 (Corollary 8.1) we considered the following system of equations

$$f_j(\mathbf{z}) = z_j^{k_j} + Q_j(\mathbf{z}) = 0, \qquad j = 1, \dots, n, \tag{19.1}$$

where $\mathbf{z} = (z_1, \dots, z_n) \in \mathbf{C}^n$ and $Q_j(\mathbf{z})$ are polynomials whose degrees are strictly less than k_j, $j = 1, \dots, n$.

If $R(\mathbf{z})$ is an arbitrary polynomial of degree K, then the formula (8.9) for finding the power sum S_R of the polynomial R is valid:

$$S_R = \sum_{l=1}^{L} R(\mathbf{z}^{(l)}) =$$

$$= \mathfrak{M}\left[R\Delta\frac{z_1\cdots z_n}{z_1^{k_1}\cdots z_n^{k_n}}\sum_{||\alpha||=0}^{K}(-1)^{||\alpha||}\left(\frac{Q_1}{z_1^{k_1}}\right)^{\alpha_1}\cdots\left(\frac{Q_n}{z_n^{k_n}}\right)^{\alpha_n}\right], \tag{19.2}$$

where $\alpha = (\alpha_1, \dots, \alpha_n)$ multi-index of length $||\alpha|| = \alpha_1 + \dots + \alpha_n$, Δ is the Jacobian of system (19.1), $\mathfrak{M}$ is the linear functional assigning to a Laurent polynomial its constant term, and $\mathbf{z}^{(l)}$ are the roots of system (19.1), $l = 1, 2, \dots, L = k_1 \cdots k_n$.

Consecutively choosing for $R(\mathbf{z})$ in (19.2) the polynomials $z_1, z_1^2, \dots, z_1^L$, we obtain the power sums of the first coordinates of the roots $\mathbf{z}^{(l)}$. Then, by the Newton formulas we can construct a one-variable polynomial in z_1, whose roots are the first coordinates of the roots $\mathbf{z}^{(l)}$ (see Sec. 8). This polynomial is the resultant of system (19.1) in the first coordinate.

Consider now the more complicated system of equations (see Theorem 8.2)

$$f_j(\mathbf{z}) = z_j^{k_j} + \sum_{k=1}^{j-1} z_k \varphi_{jk}(\mathbf{z}) + Q_j(\mathbf{z}) = 0, \quad j = 1, \ldots, n, \tag{19.3}$$

where φ_{jk} are homogeneous polynomials of degree $k_j - 1$ and Q_j are the same as in system (19.1).

The formula (19.2) for the calculation of power sums remains the same but of Q_j have to be replaced by the polynomials $R_j = f_j - z_j^{k_j}$, and the set of indeces (over which summation in (19.2) is performed) is substantially extended. Indeed, if the degree of polynomial R in z_j is equal to m_j and the total degree of R is K, then the summation in (19.2) is performed over all indeces $\boldsymbol{\alpha}$ from the parallelepiped

$$\sigma = \{\boldsymbol{\alpha} = (\alpha_1, \ldots, \alpha_n) : 0 \le \alpha_1 \le K, 0 \le \alpha_2 \le k_1(||\mathbf{m}|| + 1) - m_1 - 2, \ldots, 0 \le \alpha_j \le k_1 \cdots k_{j-1}(||\mathbf{m}|| + 1) - k_2 \cdots k_{j-1}(m_1 + 1) - k_3 \cdots k_{j-1}(m_2 + 1) - \cdots - k_{j-1}(m_{j-2} + 1) - (m_{j-1} + 1), j = 3, \ldots, n\},$$

$$||\mathbf{m}|| = m_1 + \cdots + m_n$$

(see Theorem 8.2).

The calculations of power sums will be substantially simplified by considering (19.1) and (19.3) as systems with parameters. For example, in system (19.1) we isolate the last equation and view the remaining equations as equations of $(n-1)$ variables, with z_n as parameter. As the polynomial R we take the polynomial f_n in the last equation in (19.1) with respect to the variables $z_1, \ldots, z_{n-1}$. Applying formula (19.2) for the polynomials

$$R,\ R^2,\ \ldots,\ R^{k_1 \cdots k_{n-1}}$$

we get the expressions for $S_1(z_n), \ldots, S_{k_1 \cdots k_{n-1}}(z_n)$. Using the Newton formulas (2.1) we then evaluate the resultant. More precisely, these expressions are considered as power sums of the roots of a polynomial in w, with z_n as parameter :

$$\Omega(w) = w^{k_1 \cdots k_{n-1}} + G_1 w^{k_1 \cdots k_{n-1} - 1} + \ldots + G_{k_1 \cdots k_{n-1}}.$$

For finding the coefficients $G_j = G_j(z_n)$ we can use the Newton formulas

$$jG_j = -S_j - S_{j-1}G_1 - \ldots - S_1 G_{j-1},\ j = 1, \ldots, k_1 \cdots k_{n-1}. \tag{19.4}$$

Then desired resultant will then be equal to $G_{k_1 \cdots k_{n-1}}$.

Now we describe the contents of a **MAPLE** procedure which calculates the resultant of system (19.1) using formula (19.2) (see Appl. F_{12} and [**42, 43, 44, 45, 120**]). At the beginning of the procedure the dimension n of the system is introduced from the input file `inp1` . The program then creates four arrays z, k, f and q for storage of the variables $z[1], z[2], \ldots, z[n]$, the degrees of the system $k[1], k[2], \ldots, k[n]$, the equations of the system (19.1)

$$f[1] = 0,\ f[2] = 0,\ \ldots,\ f[n] = 0,$$

and the polynomials $q[1], q[2], \ldots, q[n]$ respectively.

The values of the elements in array k are introduced from the input file `inp2`. Then we can write the equations of system (19.1) in the general form

$$f[j] = z[j]^{k[j]} +$$
$$+ \sum_{k_1=0}^{k[j]-1} \sum_{k_2=0}^{k[j]-1-k_1} \cdots \sum_{k_n=0}^{k[j]-1-k_1-\cdots-k_{n-1}} c(j, k_1, k_2, \ldots, k_n) z[1]^{k_1} z[2]^{k_2} \cdots z[n]^{k_n},$$
$$j = 1, \ldots, n.$$

Since the number of variables $z[j]$ and also the number of indeces of the coefficients $c(j, k_1, \ldots, k_n)$, depends on n, the operator giving $f[j], j = 1, \ldots, n$, is generated in a work file **forout1** beforehand. Note that in the course of the entire program the elements of the array z as well as the variables $c(j, k_1, \ldots, k_n)$ will be free, i.e.their values will be their names.

The elements of the array q are calculated by the formula

$$q[j] = f[j] - z[j]^{k[j]}, \quad j = 1, \ldots, n.$$

With the help of the standard **MAPLE** procedures **det** and **jacobian** we compute the Jacobian $jkob$ of the system.

After evaluating the degree

$$m = k[1] \times k[2] \times \cdots \times k[n]$$

of the desired resultant we produce an array p of dimension m for storage of the degrees of a certain polynomial for which the power sums are calculated by formula (19.2). The polynomial $p[1]$ is introduced from the input file **inp3** and its degree kp is evaluated by the standard **MAPLE** procedure **degree**.

Since the power sum $sp[k]$ of the roots of a system with respect to a general polynomial

$$p[k] = \sum_{k_1=0}^{kp\times k} \sum_{k_2=0}^{kp\times k-k_1} \cdots \sum_{k_n=0}^{kp\times k-k_1-\cdots-k_n} al(k, k_1, k_2, \ldots, k_{n-1}) z[1]^{k_1} z[2]^{k_2} \cdots z[n]^{k_n},$$
$$k = 1, \ldots, m,$$

is given by

$$sp[k] = \sum_{k_1=0}^{kp\times k} \sum_{k_2=0}^{kp\times k-k_1} \cdots \sum_{k_n=0}^{kp\times k-k_1-\cdots-k_n} al(k, k_1, k_2, \ldots, k_{n-1}) s(k_1, k_2, \ldots, k_n),$$
$$k = 1, \ldots, m,$$

where $s(k_1, k_2, \ldots, k_n)$ is the power sum of the monomial

$$z[1]^{k_1} z[2]^{k_2} \cdots z[n]^{k_n},$$

it follows that in order to calculate $sp[k]$ it is sufficient to evaluate the power sums of the corresponding monomials and the coefficients of these monomials in the polynomial $p[k]$. It is clear that $k_1, k_2, \ldots, k_n$ cannot take values outside the interval $[0, m \times kp]$. In order to avoid repeated calculations of the power sums $s(k_1, k_2, \ldots, k_n)$,, which can occur during the work with different elements of the array p, we propose the following procedure: All power sums $s(k_1, k_2, \ldots, k_n)$, with $0 \leq k_i \leq m \times kp$, $i = 1, \ldots, n$, are assigned the value no. After the calculation of the first $s(k_1, k_2, \ldots, k_n)$, for some $k_1, k_2, \ldots, k_n$, its value becomes different from *no*. Thus, if it is later on required to evaluate $s(k_1, k_2, \ldots, k_n)$, then to avoid any repeated calculations it is sufficient to compare the value $s(k_1, k_2, \ldots, k_n)$ with *no*. The calculation of the power sum is executed if and only if its value is equal to *no*.

The calculation of $s(k_1, k_2, \ldots, k_n)$ will be performed by the following algorithm: We find the power sums of the monomials

$$z[1]^{k_1} z[2]^{k_2} \cdots z[n]^{k_n},$$

with $0 \le k_i \le k[i]-1$, $i = 1, \ldots, n$, by formula (19.2). For computing power sums with some $k_i \ge k[i]$, we first reduce the degree of the monomials by writing the system (19.1) in the form

$$-z[j]^{k[j]} =$$
$$\sum_{k_1=0}^{k[j]-1} \sum_{k_2=0}^{k[j]-1-k_1} \cdots \sum_{k_n=0}^{k[j]-1-k_1-\cdots-k_{n-1}} c(j, k_1, k_2, \ldots, k_n) z[1]^{k_1} z[2]^{k_2} \cdots z[n]^{k_n}, \quad (19.5)$$
$$j = 1, \ldots, n.$$

Now, if $k_1 \ge k[1]$ in the monomial

$$z[1]^{k_1} z[2]^{k_2} \cdots z[n]^{k_n},$$

then we write

$$z[1]^{k_1} = z[1]^{k[1]} z[1]^{k_1 - k[1]}$$

and replace $z[1]^{k[1]}$ by the right hand side of formula (19.5). Then given monomial is represented as linear combination of monomials with smaller degrees. For obtained monomials we repeat the same procedure by all variables using again formula (19.5). In this way the power sums of any monomials is represented as a linear combination of $s(k_1, k_2, \ldots, k_n)$, for which $0 \le k_i \le k[i]-1$, $i = 1, \ldots, n$.

To realize the above we create a procedure *pc* with an arbitrary number of parameters $args[i]$ whose valuer will be the coefficient in the polynomial $args[1]$ of the monomial

$$z[1]^{args[2]} z[2]^{args[3]} \cdots z[n]^{args[n+1]}.$$

For the calculation of the coefficients in a polynomial we use the **MAPLE** procedure `coeff`. It should be noted that for the procedure `coeff` to work properly the terms in the polynomial must be ordered beforehand by the **MAPLE** procedure `collect`.

The operator that assigns the initial value *no* to the power sums $s(k_1, k_2, \ldots, k_n)$ with

$$0 \le k_1 \le m \times kp, \; 0 \le k_2 \le m \times kp, \; \ldots, \; 0 \le k_n \le m \times kp,$$

is generated beforehand in the work file `forout2`. In that work file the operators evaluating $s(k_1, k_2, \ldots, k_n)$ for $0 \le k_i \le k[i]-1$, $i = 1, \ldots, n$, are also generated.

The specific character of the **MAPLE** operators `seq` and `sum` requires that the parameters and elements of a sequence and a sum free variables. To this end it is necessary to set free the needed variables by using the **MAPLE** procedure `evaln`.

For the computation of $s(k_1, k_2, \ldots, k_n)$ in the case of arbitrary k_i, $i = 1, \ldots, n$, we generate in the work file `forout3` a recursive function $ss(k_1, k_2, \ldots, k_n)$, whose value is equal to $s(k_1, k_2, \dots, k_n)$ if $k_i < k[i]$ for all $i = 1, \dots, n$ or $s(k_1, k_2, \dots, k_n) \neq no$ or,

$$-\sum_{a_1=0}^{k[j]-1}\sum_{a_2=0}^{k[j]-1-a_1}\cdots\sum_{a_n=0}^{k[j]-1-a_1-\cdots-a_{n-1}} c(j, a_1, a_2, \dots, a_n)\times$$

$$\times ss(a_1 + k_1, \dots, a_{j-1} + k_{j-1}, a_j + k_j - k[j], a_{j+1} + k_{j+1}, \dots, a_n + k_n),$$

if some $k_j \geq k[j]$, $j = 1, \dots, n$.

For storage of the power sums with respect to the polynomials $p[i]$, $i = 1, \dots, m$, and also of the coefficients of the resultant, two arrays sp and g of dimension m are created.

The operators that evaluate the elements of the array sp is generated in the work file `forout4`. The calculation of the coefficients $g[i]$, $(i = 1, \dots, m)$ of the resultant is produced by Newton formulas (19.4). The auxiliary arrays *timesp* and *timeg* are destined for storage of the time of calculation of corresponding values. To this end it is used the **MAPLE** procedure `time`. The results of the calculations are dropped in the file `forg`.

The **MAPLE** program for determining the resultant of the more complicated system (19.3) (see Appl.F_{13}) differs from the previous one in three principal positions.

In the first place, in the work file `forout1` the operator giving the equations of system (19.3) is generated by the following formula

$$f[j] = z[j]^{k[j]} + \sum_{k_1=0}^{k[j]-1}\sum_{k_2=0}^{k[j]-1-k_1}\cdots\sum_{k_n=0}^{k[j]-1-k_1-\cdots-k_{n-1}} c(j, k_1, \dots, k_n) z[1]^{k_1} \cdots z[n]^{k_n} +$$

$$\sum_{l=1}^{j-1} z[l] \sum_{k_1=0}^{k[j]-1}\sum_{k_2=0}^{k[j]-1-k_1}\cdots\sum_{k_{n-1}=0}^{k[j]-1-k_1-\cdots-k_{n-2}} a(l, j, k_1, \dots, k_{n-1}, k[j]-1-k_1-\cdots-k_{n-1})\times$$

$$\times z[1]^{k_1} \cdots z[n-1]^{k_{n-1}} z[n]^{k[j]-1-k_1-\cdots-k_{n-1}}, \quad j = 1, \dots, n.$$

Secondly, the operator generated in work file `forout2` and intended for calculation of $s(k_1, k_2, \dots, k_n)$ for $0 \leq k_i \leq k[i] - 1$, $i = 1, \dots, n$, produces a summation by every index k_i from 0 up to some value $gr[i]$, which is given by the formula

$$gr[1] = k_1 + k_2 + \cdots + k_n,$$

$$gr[i+1] = (gr[i] + 1)k[i] - (k_i + 1) - 1, \quad i = 1, \dots, n-1,$$

and is stored in the array gr of dimension n.

Thirdly, the function $ss(k_1, k_2 \dots, k_n)$ generated in work file `forout3` is calculated by the formula

$$ss(k_1, k_2 \dots, k_n) = -\sum_{a_1=0}^{k[j]-1}\sum_{a_2=0}^{k[j]-1-a_1}\cdots\sum_{a_n=0}^{k[j]-1-a_1-\cdots-a_{n-1}} c(j, a_1, a_2, \dots, a_n)\times$$

$$\times ss(a_1 + k_1, \dots, a_{j-1} + k_{j-1}, a_j + k_j - k[j], a_{j+1} + k_{j+1}, \dots, a_n + k_n)-$$

$$-\sum_{i=1}^{j-1}\sum_{a_1=0}^{k[j]-1}\sum_{a_2=0}^{k[j]-1-a_1}\cdots\sum_{a_{n-1}=0}^{k[j]-1-a_1-\cdots-a_{n-2}} a(j,i,a_1,\dots,a_{n-1},k[j]-1-a_1-\cdots-a_{n-1})\times$$

$$\times ss(a_1+k_1,\dots,a_{i-1}+k_{i-1},a_i+k_i+1,a_{i+1}+k_{i+1},\dots,a_{j-1}+k_{j-1},a_j+k_j-k[j],$$

$$a_{j+1}+k_{j+1},\dots,a_{n-1}+k_{n-1},k[j]-1-a_1-\cdots-a_{n-1}+k_n),$$

if some $k_j \geq k[j], \quad j=1,\dots,n-1$ or

$$ss(k_1,k_2\dots,k_n) = -\sum_{a_1=0}^{k[n]-1}\sum_{a_2=0}^{k[n]-1-a_1}\cdots\sum_{a_n=0}^{k[n]-1-a_1-\cdots-a_{n-1}} c(n,a_1,a_2,\dots,a_n)\times$$

$$\times ss(a_1+k_1,\dots,a_{n-1}+k_{n-1},a_n+k_n-k[n])-$$

$$-\sum_{i=1}^{n-1}\sum_{a_1=0}^{k[n]-1}\sum_{a_2=0}^{k[n]-1-a_1}\cdots\sum_{a_{n-1}=0}^{k[n]-1-a_1-\cdots-a_{n-2}} a(n,i,a_1,\dots,a_{n-1},k[n]-1-a_1-\cdots-a_{n-1})\times$$

$$ss(a_1+k_1,\dots,a_{i-1}+k_{i-1},a_i+k_i+1,a_{i+1}+k_{i+1},\dots,a_{n-1}+k_{n-1},k_n-1-a_1-\cdots-a_{n-1}),$$

if $k_n \geq k[n]$.

As we saw before the calculations of power sums, and hence of the resultant of system (19.1), can be substantially simplified by considering the system as a system with parameter. Let us describe the differences in the MAPLE program for this method (see Appl. F_{14}) compared with the first program. The method consists in considering the first $n-1$ equations of the system as equations with respect to the variables $z[1],,\dots,z[n-1]$, while the variable $z[n]$ is considered as a parameter. The last equation of the system, which also depends on $z[1],\dots,z[n-1]$ and contains the parameter $z[n]$, is selected as the polynomial with respect to which we calculate the power sums. Therefore, after input of the value n from the file `inp1` and creation of the arrays z, k, f and q, the value n is decreased by 1. The operator generated in the work file `forout1` sets (n+1) equations with respect to the variables $z[1],\dots,z[n]$ by formula (19.5). In contrast to program F_{12}, where a polynomial is introduced from the input file `inp3`, here it is calculated by the formula

$$p[1] = f[n+1] - z[n+1]^{k[n+1]}.$$

The evaluations of power sums and also of the coefficients of the resultant of the intermediate system are carried out as the correspondent calculations of the program F_{12}. It is known the resultant of system (19.1) will be the constant term of the resultant of the intermediate system, and in the final step it is necessary to replace the coefficients of the auxiliary system and the polynomial $p[1]$ by the coefficients of the basic system. To this end a work file **change** is generated containing the operators

$$c(i,k_1,k_2,\dots,k_n) = \sum_{j=0}^{k[i]-1-k_1-\dots-k_n} c(i,k_1,\dots,k_n,j)z[n+1]^j$$

for all $i = 1,\dots,n+1, k_1 = 0,\dots,k[i]-1, k_2 = 0,\cdots,k[i]-1-k_1,\dots,k_n = 0,\dots,k[i]-1-k_1-\dots-k_{n-1}$. In case $i = n+1$ and $k_1+k_2+\dots+k_n = 0$ the obtained value of $c(i,k_1,\dots,k_n)$ should be increased by

$$z[n+1]^{k[n+1]}.$$

By an analogous method we can substantially simplify the calculation of the resultant of system (19.3), considering it as a system with parameter (see Appl. F_{15}). The differences described above between the programs for solving the same system by the two different methods take place also in this case. The replacement of the coefficients of the auxiliary system and the polynomial $p[1]$ by the coefficients of the basic system is now provided by the following formulas

$$c(i,k_1,k_2,\dots,k_n) = \sum_{j=0}^{k[i]-1-k_1-\dots-k_n} c(i,k_1,\dots,k_n,j)z[n+1]^j+$$

$$+ \sum_{\substack{1\le j\le i-1 \\ k_j\ge 0 \\ k[i]-k_1-\dots-k_n\ge 0}} a(i,j,k_1,\dots,k_n,k[i]-k_1-\dots-k_n)z[n+1]^{k[i]-k_1-\dots-k_n},$$

$$i = 1,\dots,n+1; k_1 = 0,\dots,k[i]-1; k_2 = 0,\cdots,k[i]-1-k_1,\dots;$$
$$k_n = 0,\dots,k[i]-1-k_1-\dots-k_{n-1};$$
$$a(i,j,k_1,\dots,k_{n-1},k[i]-1-k_1-\dots-k_{n-1}) =$$
$$= a(i,j,k_1,\dots,k_{n-1},k[i]-1-k_1-\dots-k_{n-1},0),$$
$$n>1, i=2,\dots,n+1, j=1,\dots,i-1, k_1=0,\dots,k[i]-1;$$
$$k_2 = 0,\cdots,k[i]-1-k_1,\dots; k_n = 0,\dots,k[i]-1-k_1-\dots-k_{n-1};$$
$$a(2,1,k[i]-1) = a(2,1,k[i]-1,0), n=1.$$

In the case $i = n+1$ and $k_1+k_2+\dots+k_n = 0$ the obtained value $c(i,k_1,\dots,k_n)$ should again be increased by

$$z[n+1]^{k[n+1]}.$$

An analysis of concrete examples shows that in general form the obtaining resultant is unwieldy expression, but its coefficients at small degrees (e.g. the constant term) is usially completely manageable. For example, we considered the system (19.1) of degrees 3,3,2 and the resultant in this case has a size of more than 4 Mbt.

To conclude this section we present some examples.

Consider the system

$$f_1 = z_1^3 + A_1z_1^2 + B_1z_2z_1 + C_1z_2^2 + D_1z_1 + E_1z_2 + F_1 = 0,$$
$$f_2 = z_2^3 + A_2z_1^2 + B_2z_2z_1 + C_2z_2^2 + D_2z_1 + E_2z_2 + F_2 = 0.$$

Setting $R = z_1$ we find that the resultant in the variable z_1 has the form

$$RES = G_0 z_1^9 + G_1 z_1^8 + \dots + G_8 z_1 + G_9.$$

The coefficients G_0, G_1 and G_9 of the resultant are equal to

$$\begin{aligned}
G_0 &= -1,\\
G_1 &= 3A_1,\\
G_9 &= C_2^2C_1F_1^2 - 2C_2C_1^2F_2F_1 - C_2C_1E_2E_1F_1 + C_2C_1E_1^2F_2 -\\
&\quad C_2E_1F_1^2 + C_1^3F_2^2 + C_1^2E_2^2F_1 - C_1^2E_2E_1F_2 -\\
&\quad 2C_1E_2F_1^2 + 3C_1E_1F_2F_1 + E_2E_1^2F_1 - E_1^3F_2 + F_1^3.
\end{aligned}$$

This entire example can be found in Appl. F_{16}.

Consider now the system

$$\begin{aligned}
f_1 &= z_1^2 + A_1z_1 + B_1z_2 + C_1 = 0,\\
f_2 &= z_2^2 + A_2z_1 + B_2z_2 + C_2 = 0,\\
f_3 &= Az_1z_2 + Bz_1 + Cz_2 + D = 0,
\end{aligned}$$

whose coefficients are polynomials in z_3 of arbitrary degree.

We take f_3 as a polynomial R. If

$$\Omega(w) = w^4 + G_1w^3 + G_2w^2 + G_3w + G_4$$

than

$$\begin{aligned}
G_1 &= -3A_2B_1A - B_2A_1A + 2B_2C + 2A_1B - 4D,\\
G_2 &= 3A_2^2B_1^2A^2 - A_2B_2A_1B_1A^2 - A_2B_2B_1AC - A_2A_1^3A^2 + 2A_2A_1^2AC -\\
&\quad A_2A_1B_1AB + 3A_2A_1C_1A^2 - A_2A_1C^2 + 9A_2B_1AD - 3A_2B_1BC -\\
&\quad 4A_2C_1AC - B_2^3B_1A^2 - B_2^2A_1AC + 2B_2^2B_1AB + B_2^2C_1A^2 +\\
&\quad B_2^2C^2 + 3B_2C_2B_1A^2 - B_2A_1^2AB + 3B_2A_1AD + 3B_2A_1BC -\\
&\quad B_2B_1B^2 - 2B_2C_1AB - 6B_2CD + C_2A_1^2A^2 - 2C_2A_1AC -\\
&\quad 4C_2B_1AB - 2C_2C_1A^2 + 2C_2C^2 + A_1^2B^2 - 6A_1BD + 2C_1B^2 + 6D^2.
\end{aligned}$$

The coefficients G_3 and G_4 are also easily written but they are more complicated.

Now replacing the coefficients A, B, C, D in G_4 by the polynomials in z_3, we get the desired resultant in the variable z_3.

More general examples are considered in Appl. F_{17} and F_{18}.

Finally, in Application F_{26} we consider the following example

$$\begin{aligned}
&x^4 + axy^2 + bz^2 + 1 = 0,\\
&y^4 + dyz^2 + cx^2 + 1 = 0,\\
&z^4 + ex^3 + fxz^2 + 1 = 0.
\end{aligned}$$

The resultant of this system in the variable x is about 500 Kbt, and therefore we provide it in arj–form.

19.2. KINETIC POLYNOMIALS, RESULTANTS AND GRÖBNER BASES

In Sec. 14 we considered the construction of the kinetic polynomial for single route catalytic reaction mechanism. It is given by the resultant of the nonlinear algebraic system

$$\begin{aligned} b_1\mathbf{z}^{\alpha^1} - b_{-1}\mathbf{z}^{\beta^1} &= \nu_1 w, \\ &\cdots \\ b_n\mathbf{z}^{\alpha^n} - b_{-n}\mathbf{z}^{\beta^n} &= \nu_n w, \\ z_1 + \ldots + z_n &= 1 \end{aligned} \tag{19.6}$$

in the variable w (stationary reaction rate). Here

$$\boldsymbol{\alpha}^j = (\alpha_1^j, \ldots, \alpha_n^j), \quad \boldsymbol{\beta}^j = (\beta_1^j, \ldots, \beta_n^j)$$

are multi-indeces, $j = 1, \ldots, n$, the integers α_j^i, β_i^j are stoichiometric coefficients of the i-th intermediate in the j-th reaction of the single-route mechanism

$$\sum_{i=1}^{n} \alpha_i^j Z_i \rightleftarrows \sum_{i=1}^{n} \beta_i^j Z_i, \qquad j = 1, \ldots, n, \tag{19.7}$$

$$\mathbf{z}^{\alpha^j} = z_1^{\alpha_1^j} \ldots z_n^{\alpha_n^j}, \quad z_1, \ldots, z_n$$

are concentrations of intermediates, and $\boldsymbol{\nu} = (\nu_1, \ldots, \nu_n)$ is the vector of stoichiometric numbers having the property

$$\sum \nu_i(\boldsymbol{\alpha}^j - \boldsymbol{\beta}^j) = 0.$$

The condition of conservation of active centers implies the relation

$$||\boldsymbol{\alpha}^j|| = \alpha_1^j + \ldots + \alpha_n^j = ||\boldsymbol{\beta}^j|| = r_j.$$

The assumption of single-route mechanism implies that the rank of the matrix

$$\boldsymbol{\Gamma} = \left((\alpha_k^j - \beta_k^j)\right)_{k,j=1}^{n}$$

is equal to $n-1$. Then we have $\nu_j = \Delta_j/t$, where Δ_j are the cofactor of the element $\alpha_n^j - \beta_n^j$ of the matrix $\boldsymbol{\Gamma}$, and t is the greatest common divisor. The theory, its physical and chemical sense and applications are explained in [**110**]. We mention only that it gives a thermodynamically correct description of a complex reaction. This follows from the fact that its constant term B_0 always contains the factor

$$\mathrm{CYC} = b_1^{\nu_1} \ldots b_n^{\nu_n} - b_{-1}^{\nu_1} \ldots b_{-n}^{\nu_n}$$

which in [**150**] is called the cycle characteristic.

In this section, on the basis of the results of Sec. 14, we describe a procedure for calculating the coefficients of the kinetic polynomial, which is carried out by means of computer algebra. As in Sec. 14 we assume that the system (19.6) has no roots for $w = 0$ (for almost all values $b_{\pm i}$), if $\nu_1 \neq 0, \ldots, \nu_n \neq 0$, that the system (19.6) contains no boundary values (so that $z_i \neq 0$ for all i).

The kinetic polynomial has the form

$$B_0 + B_1 w + \ldots + B_N w^N. \tag{19.8}$$

All its roots correspond to solutions of the system (19.6). The coefficients B_i can be calculated by the recursion formulas (see Lemma 14.5 and also [**36, 110**])

$$B_k = \frac{1}{k!}\sum_{j=1}^{n} \frac{B_{k-j}d_j}{(j-1)!}, \qquad 1 \le k \le N. \tag{19.9}$$

The values d_k are logarithmic derivatives of the resultant of system (19.6) in variable w:

$$d_k = \left.\frac{d^k \ln R(w)}{dw^k}\right|_{w=0},$$

where

$$R(w) = \prod_{j=1}^{M_1}\left(b_1 \mathbf{z}_{(j)}^{\alpha^1}(w) - b_{-1}\mathbf{z}_{(j)}^{\beta^1}(w) - \nu_1 w\right),$$

and $\mathbf{z}_{(j)}(w)$ are the roots of the subsystem obtained from (19.6) for fixed w by deleting the first equation, M_1 is the number of solutions of this system counted with multiplicities (see Sec. 14). In [**41, 110, 111**] formulas expressing d_k through solutions of subsystems for $w = 0$ (balanced subsystems) were obtained. But for computer realizations it is more convenient to use the expressions which were obtained with the transformation formula for the Grothendieck residue, i.e. the expressions of the form [**36, 98**] (see Theorem 14.6)

$$d_s = \sum_{l=1}^{n} \sum_{\substack{\|\alpha\|=s,\\ \alpha_1=\ldots=\alpha_{l-1}=0,\\ \alpha_l>0}} (-1)^{n+l-1}\frac{s!}{\boldsymbol{\alpha}!}\boldsymbol{\nu}^{\alpha} \sum_{j_1,\ldots,j_{s-1}=1} C_{j_1\ldots j_{s-1}}\mathfrak{M}[Q(\mathbf{z})], \tag{19.10}$$

where

$$Q(\mathbf{z}) = \frac{(z_1+\ldots+z_n)^{r_l\alpha_l+\ldots+r_n\alpha_n-1}\det \mathbf{A} a_{j_1 l}\ldots a_{j_{s-1}n} J_l}{z_{j_1}^{L+1}\ldots z_{j_{s-1}}^{L+1} z_1^L \ldots z_n^L},$$

$$\boldsymbol{\alpha} = (\alpha_1, \ldots, \alpha_l, \ldots, \alpha_n), \quad \boldsymbol{\alpha}! = \alpha_1!\ldots\alpha_n!, \quad \boldsymbol{\nu}^{\alpha} = \nu_1^{\alpha_1}\ldots\nu_n^{\alpha_n},$$

$$C_{j_1\ldots j_{s-1}} = \beta_1!\ldots\beta_n!$$

if the sequence $j_1, \ldots, j_{s-1}$ contains β_1 ones, β_2 twos etc.

In the formula (19.10) J_l is the Jacobian of the l-th subsystem obtained from (19.6) by deletion of the l-th equation. The key moment in the application of formula (19.10) is the calculation of the matrix

$$\mathbf{A} = (a_{jk}(\mathbf{z}))_{j,k=1}^{n},$$

where the homogeneous coordinates $a_{jk}(\mathbf{z})$, $j, k = 1, \ldots, n$, satisfy the linear relations

$$z_j^{L+1} = \sum_{k=1}^{n} a_{jk}(\mathbf{z})w_k(\mathbf{z}), \qquad j = 1, \ldots, n, \tag{19.11}$$

with the degree L being chosen not to exceed of the value $r_1 + \ldots + r_n - n$, and

$$w_k(\mathbf{z}) = b_k\mathbf{z}^{\alpha^k} - b_{-k}\mathbf{z}^{\beta^k}.$$

Such a sequnce of polynomials a_{jk} exists by Hilbert Nullstellensatz [**142**]. The operator $\mathfrak{M}$ in formula (19.10) assigns to the Laurent polynomial $Q(\mathbf{z})$ its constant term, i.e. it finds in the numerator of $Q(\mathbf{z})$ the coefficients at the monomial standing in denominator of $Q(\mathbf{z})$.

The algorithm for calculating the kinetic polynomial consists of the following steps: 1) The calculation of the matrix $\mathbf{A}$, 2) the calculation of d_s by the formulas (19.10), 3) the calculation B_k from (19.9).

The general method of determination of the matrix $\mathbf{A}$ is the method of indefinite coefficients, which reduces to some system (as a rule, sub-definite) of linear equations in the coefficients of the polynomials $a_{jk}(\mathbf{z})$. It was used in the examples in Sec. 14 (see also [**36, 102**]), where explicit expressions of $a_{jk}(\mathbf{z})$ for systems of special form are contained.

But a more rapid and convenient method for generating the $a_{jk}(\mathbf{z})$ is the method of a Gröbner basis. In [**27**] it is shown that a linear representation of the polynomials of a Gröbner basis can be found through the polynomials f_j of the initial basis F (and conversely).

Suppose such a representation

$$g_i = \sum_{j=1}^{l} f_j X_{ji}, \qquad i = 1, \ldots, m, \tag{19.12}$$

has been found. Now we can solve the following problem (see [**27**]): Find $h_1, \ldots, h_m$ such that

$$f = h_1 g_1 + \ldots + h_m g_m, \tag{19.13}$$

by the given Gröbner basis $G = \{g_1, \ldots, g_m\}$ and some polynomial $f \in \mathrm{Ideal}(F)$.

In view of (19.12) we then get

$$f = h_1 \left(\sum_{j=1}^{l} f_j X_{j1} \right) + \ldots + h_m \left(\sum_{j=1}^{l} f_j X_{jm} \right) = \sum_{j=1}^{l} \left(\sum_{i=1}^{m} h_i X_{ij} \right) f_j. \tag{19.14}$$

Thus we obtained a solution to the problem of finding the representation (19.11). For this we should take $f = z_j^{L+1}$ in (19.12) and the initial basis

$$F = \{w_1(\mathbf{z}), \ldots, w_n(\mathbf{z})\}, \ G = GB(F).$$

Then the elements $a_{jk}(\mathbf{z})$ of the matrix $\mathbf{A}$ can be found as the coefficients of f_j in the representation (19.14). The algorithm for calculating the matrix A in (19.11) has the form:

1) The calculation of a Gröbner basis for the system

$$w_1(\mathbf{z}) = 0, \ \ldots, \ w_n(\mathbf{z}) = 0.$$

In addition to the computation of the coefficients X_{ji} in the representation (19.12) we must include an algorithm which write memorizes the factors of the polynomials $w_k(\mathbf{z})$ which are used when reducing these polynomials to normal forms which constitute the basis G in the reduction process [**27**].

2) The reduction of the monomials z_j^{L+1}, $j = 1, \ldots, n$, to zero modulo

$$G = GB(w_1(\mathbf{z}), \ \ldots, \ w_n(\mathbf{z})).$$

In this reduction process we must sum the occurring factors of the polynomials g_i. The accumulated sums gives the values of the coefficients h_i in the representation (19.13).

3) The calculation of $a_{jk}(\mathbf{z})$ by the formula

$$a_{jk}(\mathbf{z}) = \sum_{i=1}^{m} h_i^{(k)} X_{ji}. \tag{19.15}$$

In this way the Gröbner basis allows us to find the representation (19.11).

19.3. A PROGRAM FOR CALCULATING THE KINETIC POLYNOMIAL

The program consists of two parts: 1) The calculation of the representation (19.11), 2) the calculation of the coefficients of the kinetic polynomial by formulas (19.9), (19.10).

The first part is contained in Application F_5. Let us comment on the addition to the algorithm which is needed to calculate the representation (19.11) (see Application F_{19}). The sets R, P, G (see Sec. 17.2, 18.3) we associate along with the operators RB, GB, PB the new operators RF, GF, PF. The values of the latter are the coefficients in the representation of the polynomials from the sets R, G, P in terms of the polynomials of the initial basis F, for example

$$R(j) = \mathrm{RF}(j,1) \cdot F(1) + \ldots + \mathrm{RF}(j, \mathrm{NFO}) \cdot F(\mathrm{NFO}),$$

$j = 1, \ldots,$ NR. For every change of the sets R, G, P the sets RF, GF, PF are changed accordingly. When entering the algorithm we have $\mathrm{RF}(i,j) = \delta_{ij}$ (δ_{ij} is the Kronecker symbol). Into the procedure NF, which realizes the normal form algorithm, we introduce, in addition, the two operators NF and AFNF. The values of HR(i) are the coefficients in the representation of the reduced polynomial in terms of polynomials of the initial basis. The values of AFNF(i,j) are the coefficients of the i-th polynomial modulo which the reduction is being performed. In the process of calculating the normal form the values of the operator HR are changed in the text of the program by means of the sentence HR(JNO):=HR(JNO)−BU∗AFNF(L,JNO). The operator HR is used for the same purpose in the procedure SP. After the calculation of the Gröbner basis there is a print-out of the polynomials entering in this basis, and also of the values of the operator GF (i.e. the coefficients of the Gröbner basis through polynomials of the initial basis).

The reduction of z_j^{L+1} modulo the Gröbner basis is carried out by the procedure NF1. It differs from procedure NF by the presence of the operator HR1, whose values are the coefficients h_i (see (19.13)), which get accumulated during the reduction process . This is realized by the operator

$$\mathrm{HR1}(L) := \mathrm{HR1}(L) + \mathrm{BU}.$$

After execution of procedure NF1 the values of the operator HR1 are printed. If the result of the reduction is equal to zero, then the elements of the matrix $\mathbf{A}$ are calculated according to formula (19.15). If this is not the case (which may be due to a mistake in the input data, or because the algorithm has been applied to a system non satisfying the conditions under which formula (19.10) was obtained) then a diagnostic report gets printed out. The calculated matrix A is put into the file 'kp.out' in a form which allows it to be used as initial data for the second part of the algorithm.

We now briefly describe the second part of the program, which is contained in Application F_{19}. The procedure GENAL generates the indices α in formula (19.10) for given integer values S, L, N (here $S \geq 1$ is the index of the coefficients of the kinetic polynomial to be computed, $L = 1, \ldots, n$ is the index l in (19.10), and N is the dimension of the problem). The indices α have the properties:

$$\sum_{i=1}^{n} \alpha_i = S, \qquad \alpha_1, \ldots, \alpha_{l-1} = 0, \qquad \alpha_l > 0.$$

They are the values of the operator

$$\text{ALF}(j,i) \quad (j = 1, \ldots, \text{NAL}, \ i = 1, \ldots, N).$$

Having calculated the values ALF(j,i) (they are built up in the form of a tree) the program goes on to compute the values of the operator KAL(j,i) (i.e. the values of the column indices $l, \ldots, n$ of the quantities

$$a_{j_1 l}, \ldots, a_{j_s-1 n}$$

in formula (19.10)). From the deduction of (19.10), contained in [**98**], it follows that these indices are uniquely correspond to collections of indices α according to the following rule:

$$(\alpha_1, \ldots, \alpha_n) \longrightarrow (\underbrace{l, \ldots, l}_{\alpha_{l-1} \text{ times}}, \underbrace{(l+1), \ldots, (l+1)}_{\alpha_{l+1} \text{ times}}, \underbrace{\ldots, n, \ldots, n}_{\alpha_n \text{ times}}).$$

The procedure GENJ1S (N, S) generates the collections of indices $j_1, \ldots, j_{s-1}$ (the operator JI) and the corresponding coefficients

$$C_{j_1 \ldots j_{s-1}}$$

in (19.10) (the operators CJ1JS).

The procedure CPM(nmul) is an implementation of the operator

$$\mathfrak{M}[Q(\mathbf{z})]$$

in formula (19.10) i.e. it calculates the coefficients of the monomial

$$z_{j_1}^{L+1} \ldots z_{j_{s-1}}^{L+1} z_1^L \ldots z_n^L$$

in the numerator of the expression $Q(\mathbf{z})$. We have considered two algorithms for calculating $\mathfrak{M}[Q(\mathbf{z})]$. The first one resembles a chain process. It starts with an initiation: In the first factor of the numerator of $Q(\mathbf{z})$ those monomials which divide the denominator of $Q(\mathbf{z})$ get memorized. Then the second factor of the numerator of $Q(\mathbf{z})$ is considered. Here it is those monomials, whose products with previously memorized monomials divide the denominator, that get singled out. This is iterated up to the (nmul-1)-th factor. In this way 'competing' chains are constructed. At the last step it is checked, for all chains and for all monomials of the last (nmul-th) factor of the numerator of $Q(\mathbf{z})$, that the product of powers of z_j from the chain with the monomial gives the denominator of $Q(\mathbf{z})$. As a result one obtains a family of chains, for which the corresponding sum of products of coefficients is computed by means of the procedure CPM. However, this method is close to the calculation 'by hand' of (19.10), which turned out not to be effective for realization in the system REDUCE. This is due to large dimensions of cofactors and of required memory volume for storage of intermediate results.

Therefore a simpler and more effective algorithm has been suggested, which essentially is based on the possibilities of the command LET. It consists in the following: Step by step one picks out those terms in which the degree of z_j is equal to the degree of Z_j in the denominator of $Q(\mathbf{z})$ (these degrees are written in the operator 'degmon'). At each step it is essential to generate the sentences

$$\text{LET } z_j ** \text{degmon}(j) + 1 = 0$$

which guarantees the elimination of all terms in which the degree of z_j is greater than degmon(j). This significantly shortens the expressions. To generate the commands LET and the corresponding sentences CLEAR the external files 'cpm.out' and 'cpm1.out' are used.

The main program begins with the input of initial data, written in file 'kp.inp', and collecting the matrix **A** from file 'kp.inp'. Further, the determinant of **A** (DETA) and the Jacobians of the subsystems (JAL(L)) are computed, and by formula (19.10) the values d are formed. (The number of coefficients SMAX to be calculated is given in the input data). When performing the calculations we notice that in the output of the algorithm CPM there will usually be a power of the power of expressions

$$\text{CYC} = b_1^{\nu_1} \ldots b_n^{\nu_n} - b_{-1}^{\nu_1} \ldots b_{-n}^{\nu_n}$$

occurring as a common factor of the numerator and the denominator. Instead of using the expensive command ON GCD, we apply the LET command generated in within the program to make suitable substitutions which automatically delete the factor CYC during the calculation. When the coefficients of the kinetic polynomial have been computed, an inverse substitution is performed and the resulting output is given in terms of the parameters b_i, b_{-i}.

Let us consider the examples. The calculation of the matrix **A** for the model corresponding to the parallel three-stage adsorption mechanism (18.1) is carried out by the program, in the case of ordering Z $>_T$ AZ $>_T$ BZ, as follows. The main algorithm forms the set R: $R(i) = f_i$, where

$$\begin{aligned} f_1 &= -\text{BR1}*\text{AZ}^2 + \text{Z}^2 * \text{B1}, \\ f_2 &= \text{Z} * \text{B2} - \text{BZ} * \text{BR2}, \\ f_3 &= \text{BZ} * \text{B3} * \text{AZ} - \text{Z}^2 * \text{BR3}, \end{aligned}$$

and addresses the procedure REDALL, which first inserts f_3 in the set $P: P(1) = f_3$. After an unsuccessful attempt to reduce the polynomial f_2 modulo $P(1)$ the algorithm discovers that the polynomial $P(1)$ can be reduced modulo f_2, and the polynomial $P(1)$ is inserted in the set P0 (see Sec. 17.2, 18.3). Therefore the new sets R, P have the form:

$$R(1) = f_1, \quad R(2) = f_3, \quad P(1) = f_2.$$

Reduction of $R(2)$ modulo $P(1)$ now leads to the polynomial

$$f_4 = \frac{BZ * B3 * (B2)^2 * AZ - BZ^2 * BR3 * (BR2)^2}{(B2)^2},$$

which is inserted in the set P: $P(2) = f_4$. The remaining polynomial $R(1) = f_2$ is then reduced modulo $P(1) = f_2$ and $P(2) = f_4$, and as a result we obtain the polynomial

$$f_5 = \frac{-BR1 * (B2)^2 * AZ^2 + BZ^2 * (BR2)^2 * B1}{(B2)^2},$$

which is inserted in the set P: $P(3) = f_5$. The algorithm REDALL thereby completed. The algorithm NEWBAS then starts working and forms the set G : $(G(1) = f_2,\ G(2) = f_4,\ G(3) = f_5)$, as well as the set of pairs B : (1,2), (1,3), (2,3). The main algorithm chooses the pair (2,3), with the corresponding S-polynomial

$$f_6 = \frac{-BZ^2 * BR3 * (BR2)^2 * BR1 * AZ + BZ^3 * (BR2)^2 * B3 * B1}{BR1 * (B2)^2}.$$

The normal form of f_6 modulo G is

$$f_7 = \frac{BZ^3 * (BR2)^2 * B3^2 * (B2)^2 * B1 - BZ^3 * (BR3)^2 * (BR2)^4 * BR1}{BR1 * B3 * (B2)^4}.$$

Since the set R is empty, the procedure REDALL does not come into play and the procedure NEWBAS forms the set G ($G := G \cup P$), including the polynomials

$$G(1) = f_2,\ G(2) = f_4,\ G(3) = f_5,\ G(4) = f_7,$$

and the set of pairs B:

$$(1,2),\ (1,3),\ (1,4),\ (2,4),\ (3,4).$$

The main algorithm now chooses the pair (2,4) whose S-polynomial is given by

$$f_8 = -\frac{BZ^4 * BR3 * (BR2)^2}{(B2)^2}.$$

But since the normal form of f_8 modulo G is equal to zero the basis remains the same. The remaining pairs, chosen by the algorithm in the order (3,4), (1,4), (1,2), (1,3) do not 'pass' by Criterion 2. At this place the calculation of the Gröbner basis therefore terminates. Next the monomials Z^3, AZ^2, BZ^3 are reduced to zero modulo G, which allows us to compute the coefficients $HR(i)$ in the linear representation (19.13) of them in terms of the Gröbner basis polynomials, and by formula (19.15) the elements $AA(i,j)$ of the matrix A are calculated (see Application F_{20}). It should be noted that the entries in the matrix A are not uniquely defined. When using the Gröbner basis method the form of A depends on the chosen ordering. For example, if we use the ordering

$$BZ >_T Z >_T AZ$$

we get the following elements in the matrix A:

AA(1,1) = (B2^5*B3^2*Z)/(B1*B2^2*B3^2*BR2^3–BR1*BR2^5*BR3^2),
AA(1,2) = (AZ^2*B2^4*B3^2*BR1+AZ*B2^3*B3*BR1*BR2*BR3*Z–
–B1*B2^4*B3^2*Z^2–B1*B2^3*B3^2*BR2*BZ*Z–
–B1*B2^2*B3^2*BR2^2*BZ^2 +B2^2*BR1*BR2^2*BR3^2*Z^2+
+B2*BR1*BR2^3*BR3^2*BZ*Z +BR1*BR2^4*BR3^2*BZ^2)/
/(B1*B2^2*B3^2*BR2^3–BR1*BR2^5*BR3^2),
AA(1,3) = (AZ*B2^4*B3*BR1+B2^3*BR1*BR2*BR3*Z)/
/(B1*B2^2*B3^2*BR2^2–BR1*BR2^4*BR3),
AA(2,1) = (B2^2*B3^2*Z)/(B1*B2^2*B3^2–BR1*BR2^2*BR3^2),
AA(2,2) = (AZ^2*B2*B3^2*BR1+AZ*B3*BR1*BR2*BR3*Z)/
/(B1*B2^2*B3^2–BR1*BR2^2*BR3^2),
AA(2,3) = (AZ*B2*B3^2*BR1*BR2+BR1*BR2^2*BR3*Z)/
/(B1*B2^2*B3^2–BR1*BR2^2*BR3^2),
AA(3,1) = (–AZ*B1*B2^2*B3^2+AZ*BR1*BR2^2*BR3^2+
B1*B2*B3*BR2*BR3*Z)/(B1*B2^2*B3^2*BR1–BR1^2*BR2^2*BR3^2),
AA(3,2) = (AZ^2*B1*B3*BR1*BR2*BR3+AZ*B1^2*B2*B3^2*Z)/
/(–BR1^2*BR2^2*BR3^2+B1*B2^2*B3^2*BR1),
AA(3,3) = (AZ*B1*BR1*BR2^2*BR3+B1^2*B2*B3*BR2*Z)/
/(B1*B2^2*B3^2*BR1–BR1^2*BR2^2*BR3^2).

The second part of the program calculates the determinant of the matrix A, the Jacobians of the subsystems, the quantities d_I, and coefficients of the kinetic polynomial. For the given example only the first two coefficients KINPOL(1), KINPOL(2) of the kinetic polynomial are computed.

Application F_{21} contains the input data and the computation of the kinetic polynomial for the nonlinear shock mechanism (18.19), see Example 14.3. The resulting expressions are identical to the formulas for the coefficients of the polynomials f_5, which were obtained in 18.3 by means of Gröbner bases to the full model (18.19). In this example it was given that $SMAX = 3$. But f_5 is a second degree polynomial. Hence the program gives the output KINPOL(3)=0.

An example of the reaction mechanism (18.21) is contained in Application F_{22}. A special feature of this example is the fact that the quantity

$$\mathrm{CYC} = b_1 b_2 - b_{-1} b_{-2}$$

is contained in free form in the second degree, and it is a factor of the first coefficient.

This is reflected in the expressions KINPOL(1), KINPOL(2) and the identical coefficients of the polynomial $GB(3)$ from Application F_6. In the general case the algorithm automatically detects the appearance of powers of CYC in the coefficients of the polynomial. The results obtained for three-stage linear mechanism (18.24) (Application F_{23}) correspond to the formulas for the general case of a n-stage mechanism (see Example 14.2). For instance, the inverse matrix

$$\mathbf{A}^{-1} = \begin{pmatrix} b_1 & -b_{-1} & 0 \\ 0 & b_2 & -b_{-2} \\ -b_{-3} & 0 & b_3 \end{pmatrix}$$

corresponds to the matrix

$$\mathbf{A} = \frac{1}{\mathrm{CYC}} \begin{pmatrix} b_2 b_3 & b_{-1} b_3 & b_{-1} b_{-2} \\ b_{-2} b_{-3} & b_1 b_3 & b_{-2} b_1 \\ b_{-3} b_2 & b_{-1} b_{-3} & b_1 b_2 \end{pmatrix}$$

obtained by algorithm.

In the situation of a nonlinear mechanism without stages of interaction between different substances the algorithm gets the expressions for the elements of A on another way, suggested in [36]. As an example of this Application F_{24} contains the mechanism

$$\begin{array}{ll} 1)\ 2Z1 \rightleftarrows 2Z2, & 2)\ Z2 \rightleftarrows Z3, \\ 3)\ 2Z3 \rightleftarrows 2Z4, & 4)\ Z4 \rightleftarrows Z1. \end{array} \qquad (19.16)$$

In the formulas given in [36] it is assumed that the number $L+1$ in (19.11) is equal to the product of the powers of the polynomials (19.6), which is equal to 4 in the case (19.16). But the algorithm chooses the value $L = 2+1+2+1-4 = 2$, so that $L+1 = 3$, and hence expressions for a_{ij} different from ones in [36] obtained.

Consider now the case of a mechanism without interactive stages and for which one stage is linear while the other stages all are the same order α (see Application F_{25}):

$$\begin{array}{l} 1)\ 2Z1 \rightleftarrows 2Z2, \\ 2)\ 2Z2 \rightleftarrows 2Z3, \\ 3)\ Z3 \rightleftarrows Z1. \end{array}$$

Here the algorithm gives the degree $L+1 = 2+2+1-3+1 = 3$ which is greater than in the general consideration from [36], where $L+1 = \alpha = 2$. The result (i.e. the kinetic polynomial) is of course still the same. So, with respect to the first coefficient, we have

$$\frac{K(1)}{K(0)} = \frac{\alpha^{n-2}\left(\alpha^2(b_1 \ldots b_{n-1} b_n^{\alpha-1} + b_{-1} \ldots b_{-n}^{\alpha-1}) + b_2 \ldots b_{n-1}(b_n + b_{-n})^{\alpha}\right)}{b_1 \ldots b_{n-1} b_n^{\alpha} - b_{-1} \ldots b_{-(n-1)} b_{-n}^{\alpha}}.$$

The expression for KINPOL(1) with $\alpha = 2$, $n = 3$ (see F_{25}), is of the form

$$\mathrm{KINPOL}(1) = \frac{2(4(b_1 b_2 b_3 + b_{-1} b_{-2} b_{-3}) + \ldots + b_2 b_3^2 + b_2 b_{-3}^2 + 2 b_2 b_3 b_{-3}}{b_1 b_2 b_3^2 - b_{-1} b_{-2} b_{-3}^2}.$$

Notice also that the constant terms of the polynomials from Applications F_{24} and F_{25} contain the square of the cycle characteristic CYC. This follows from a general property (see Sec. 14): The power of CYC is equal to the GCD of the cofactors of the matrix

$$\mathbf{\Gamma} = ((\alpha_k^j - \beta_k^j))_{k,j=1}^n$$

which indeed is equal 2 in this case.

We have thus completed the computer realization of our algorithm for calculating the coefficients of the kinetic polynomial for a nonlinear single-route reaction mechanism. A distinctive feature is the 'hybrid' character of the program: In the first step of the algorithm it is the method of Gröbner bases that is being used, whereas in the second step the calculations are conducted by formulas obtained from multidimensional complex analysis (see Ch. 1–3). This permitted us to realize a rather effective program in the system REDUCE.

The specific character of the kinetic equations (19.6) plays an essential role here. But we have still not made full use of it. This is reflected in the way our intermediate calculations (such as the logarithmic derivatives of the resultant $D(i)$) are often much more unwieldy than our final results. The next problem to tackle should therefore to obtain explicit expressions for the coefficients of the kinetic polynomial by means of formulas of the types (19.9) and (19.10). Graph theory would seem to be the adequate language for structuring these expressions and making them more visible. The expressions for the coefficients should be connected with the type of reaction graph. The theory of linear reaction mechanisms [**150**], for which computer realization have be created [**71, 152**], gives a hint as to how this problem could be solved.

In this chapter we have showed how computer algebra may be used to realize algorithms for elimination of unknowns. We described various algorithms and programs for investigating systems of nonlinear equations of general and special (chemical kinetics) forms. In their present state these programs provide solutions to the following problems: Determination of solutions to systems of algebraic equations, calculations of bifurcation conditions for these solutions and the bifurcation boundaries. It is essential that these programs allow one to work both with symbolic and numerical data.

The developed methods and programs proved to be quite effective (even in the case of limited computer capacity) for rather complicated problems of chemical kinetics. An interesting feature of our algorithm is the synthesis of traditional computer algebra methods (Gröbner bases) and new results from several complex variables. Problems that were hard to solve by traditional methods turned out to be effectively solvable by our 'hybrid' algorithms.

The approach developed on the above examples of kinetic equations can quite easily be adapted to other systems, since the key moments have already been carried out (the calculations of linear representations, and the calculations by formulas of type (19.10)).

Naturally, the subject still contains a lot of interesting and more difficult problems (for example, the problems of solvability theory described in this chapter), but we are convinced that there will be rapid progress. The general character of the problems, their high complexity and the appealing solution processes that are used, ought to guarantee a continued interest in this field.

LIST OF APPLICATIONS

Application F_1**.** Simple `Maple` procedure for calculating the resultant of two polynomials with an example of determination of the kinetic polynomial for a three-stage parallel adsorption mechanism.

Application F_2**.** Standard `Maple` procedure for calculating the resultant of two polynomials with an example of determination of the kinetic polynomial for a three-stage parallel adsorption mechanism.

Application F_3**.** `Maple` procedure for calculating the resultant of two polynomials with an example of determination of the kinetic polynomial for a three-stage parallel adsorption mechanism by rate of reaction.

Application F_4**.** Steady-state multiplicity and bifurcation boundary for a model corresponding the two-route mechanism of CO oxidation at T=450K on the plane PO2-PCO: discriminant coefficients and roots.

Application F_5**.** Kinetic polynomial computation: Gröbner basis. Part 1.

Application F_6**.** Elimination of variables for a model corresponding the mechanism of a catalytic Hydrogen para-ortho conversion reaction.

Application F_7**.** Elimination of variables for a model corresponding the irreversible three-stage adsorption mechanism of a catalytic reaction.

Application F_8**.** Elimination of variables for a model corresponding the linear mechanism of a catalytic reaction.

Application F_9**.** Elimination of variables for a catalytic reactor of ideal mixture.

Application F_{10}**.** Elimination of variables for a model corresponding the quasi-chemical approximation for description of elementary catalytic processes.

Application F_{11}**.** Example of application of the modified Gröbner basis algorithm for a three-stage adsorption mechanism.

Application F_{12}**.** `Maple` procedure for determining the coefficients of the resultant for system (19.1).

Application F_{13}**.** `Maple` procedure for determining the coefficients of the resultant for system (19.3).

Application F_{14}**.** `Maple` procedure for determining the coefficients of the resultant for system (19.1) with parameter.

Application F_{15}**.** `Maple` procedure for determining the coefficients of the resultant for system (19.3) with parameter.

Application F_{16}**.** Example of system (19.1) and determination of its resultant.

Application F_{17}**.** Example of system (19.1) with parameter and determination of its resultant.

Application F_{18}**.** Example of system (19.3) with parameter and determination of its resultant.

Application F_{19}**.** Kinetic polynomial computation: calculation of coefficients. Part 2.

Application F_{20}**.** Calculation of the kinetic polynomial coefficients for the three-stage adsorption mechanism of a catalytic reaction.

Application F_{21}**.** Kinetic polynomial computation: nonlinear shock two-stage mechanism.

Application F_{22}. Kinetic polynomial computation: mechanism of Hydrogen para–ortho conversion.

Application F_{23}. Kinetic polynomial computation: linear three–stage mechanism.

Application F_{24}. Kinetic polynomial computation: 4–stage mechanism without interaction between different intermediates.

Application F_{25}. Kinetic polynomial computation: mechanism with a single linear stage and equal orders of the other stages.

Application F_{26}. Example of system (19.1) of three equations with parameter and determination of its resultant (arj–file).

Bibliography

[1] Adams, W.W., Loustaunau, P., *An introduction to Gröbner bases*, Grad. Studies in Math., AMS, Providence, RI, (1994).

[2] Aizenberg, L.A., On a formula of the generalized multidimensional logarithmic residue and the solution of system of nonlinear equations, *Dokl. Akad. Nauk* **234**(3) (1977), 505–508; English. Transl. in *Sov. Math. Dokl.* **18** (1977), 691–695.

[3] Aizenberg, L.A., Bolotov, V.A. and Tsikh, A.K., On the solution of systems of nonlinear algebraic equations using the multidimensional logarithmic residue. On the solvability in radicals, *Dokl. Akad. Nauk*, **252**(1) (1980), 11–14; English transl. in *Sov. Math. Dokl.* **21** (1980), 645–648.

[4] Aizenberg, L.A., Bykov, V.I. and Kytmanov, A.M., Determination of all steady states of equations of chemical kinetics with the help of modified method of elimination. 1. Algorithm, *Fizika goreniya i vzryva* **19**(1) (1983), 60–66 (Russian).

[5] Aizenberg, L.A., Bykov, V.I., Kytmanov, A.M. and Yablonskii, G.S., Determination of all steady states of equations of chemical kinetics on the base of multidimensional logarithmic residue, *Doklady Akad. Nauk* **257**(4) (1981), 903–907 (Russian).

[6] Aizenberg, L.A., Bykov, V.I., Kytmanov, A.M. and Yablonskii, G.S., Determination of all steady states of equations of chemical kinetics with the help of modified method of elimination. 2. Applications, *Fizika goreniya i vzryva* **19**(1) (1983), 66–73 (Russian).

[7] Aizenberg, L.A., Bykov, V.I., Kytmanov, A.M. and Yablonskii, G.S., Search of steady states of chemical kinetics equations with the modified method of elimination. 1. Algorithm, *Chem. Eng. Sci.* **38**(9) (1983), 1555–1560.

[8] Aizenberg, L.A., Bykov, V.I., Kytmanov, A.M. and Yablonskii, G.S., Search for all steady states of chemical kinetic equations with the modified method of elimination. 2. Applications, *Chem. Eng. Sci.* **38**(9) (1983), 1561–1567.

[9] Aizenberg, L.A. and Kytmanov, A.M., Multidimensional analogs of Newton formulae for systems of nonlinear algebraic equations and some their applications, *Sibirsk. Mat. Zh.* **22**(2) (1981), 19–30; English transl. in *Sib. Math. J.* **22** (1981), 81–89.

[10] Aizenberg, L.A. and Tsikh, A.K., Application of multidimensional logarithmic residue to systems of nonlinear algebraic equations, *Sibirsk. Mat. Zh.* **20**(4) (1979), 669–707; English transl. in *Sib. Math. J.* **20** (1979), 485–491.

[11] Aizenberg, L.A. and Yuzhakov, A.P., *Integral representations and residues in multidimensional complex analysis,* Transl. Math. Monographs, AMS, Providence, R.I., (1983).

[12] Alekseev, B.B., The construction on computer of bases of polynomial ideal, *Abstracts of Conf. "Systems for analytic transformations in mechanics",* Gorkii State Univ., Gorkii, (1984), 66–67 (Russian).

[13] Arnold, V.I. and Il'yashenko, Yu.S., Ordinary differential equations, *Itogi nauki i tekhniki. Sovr. probl. Mat.,* VINITI, Moscow, **1** (1985), 7–149 (Russian).

[14] Arnold, V.I., Varchenko, A.N. and Gusein-Zade, S.M., *Singularities of differential mappings,* Nauka, Moscow, (1982) (Russian).

[15] Arnon, P.S. and McCallum, S., Cylindrical algebraic decomposition by quantifier elimination, *Lect. Notes in Comput. Sci.* **162** (1983), 215–222.

[16] Balabaev, N.K. and Lunevskaya, L.V., Motion along curve in n-dimensional space, in *Algorithms and programs on FORTRAN,* Work Collect., ONTI NIVT AN SSSR, Puschino, **1** (1978) (Russian).

[17] Balakotiach, V. and Luss, D., Global analysis of multiplicity features of multi-reaction lumped-parameter system, *Chem. Eng. Sci.* **39**(5) (1984), 867–881.

[18] Balk, A.M., On computer realization of method of Newton diagrams, *Zh. Vych. Mat. i Mat. Fiziki* **24**(7) (1984), 972–985 (Russian).

[19] Ben-Or, N. and Cohen, P., The complexity of elementary algebra and geometry, *Proc. 16 ACM Symp. Th. Comput.* (1984), 457–464.

[20] Bernstein, D.N., The number of roots of a system of equations, *Funkt. Analiz i Pril.* **9**(3) (1975), 1–4 (Russian).

[21] Bernstein, D.N., Kushnirenko, A.G. and Khovanskii, A.G., Polyhedrons of Newton, *Uspekhi Math. Nauk* **31**(3) (1976), 201–202 (Russian).

[22] Bolotov, V.A., The correspondence between roots and coefficients of multidimensional systems of algebraic equations, in *Some Questions of Multidimensional Complex Analysis,* Inst. of Physics, Krasnoyarsk, (1980), 225–230 (Russian).

[23] Bordoni, L., Cologrossi, A. and Miola, A., Linear algebraic approach for computing polynomial resultant, *Lect. Notes in Comput. Sci.* **162** (1983), 231–237.

[24] Borisyuk, R.M., Stationary solutions of a system of ordinary differential equations, in *Algorithms and programs on FORTRAN,* Work Collect., ONTI NIVT AN SSSR, Puschino, **6** (1981) (Russian).

[25] Bose, N.K., Problems and progress in multidimensional systems theory, *Proc. IEEE* **65**(6) (1977), 824–840.

[26] Bourbaki, N., *Algebra,* Hermann, Paris, **2** (1961).

[27] Buchberger, B., Gröbner bases. Algorithmic method in theory of polynomial ideals, in *Symbolic and algebraic computation,* Springer–Verlag, Vienna–New York, (1982).

[28] Buchberger, B., Collins, G.E., and Luss, R., (eds) *Symbolic and algebraic computations,* Springer–Verlag, Vienna–New York, (1982).

[29] Bykov, V.I., On simple model of oscillating catalytic reactions, *Zh. Fizich. Khimii* **59**(11) (1985), 2712–2716 (Russian).

[30] Bykov, V.I., *Modeling of critical phenomenons in chemical kinetics,* Nauka, Moscow, (1988) (Russian).

[31] Bykov, V.I., Chumakov, G.A., Elokhin, V.I. and Yablonskii, G.S., Dynamic properties of a heterogeneous catalytic reaction with several steady states, *React. Kinet. Catal. Lett.* **4**(4) (1976), 397–403.

[32] Bykov, V.I., Elokhin, V.I. and Yablonskii, G.S., The simplest catalytic mechanism admitting several steady states of the surface, *React. Kinet. Catal. Lett.* **4**(2) (1978), 191–198.

[33] Bykov, V.I. and Gorban, A.N., Quasi-thermodynamics of reactions without interaction of different substances, *Zh. Fizich. Khimii* **57**(12) (1983), 2942–2948 (Russian).

[34] Bykov, V.I. and Kytmanov, A.M., Estimation of a number of steady states for three-step adsorption mechanisms, 1. Parallel scheme, *Kinetics and catalysis* **25**(5) (1984), 1276–1277, Dep. VINITI No.4772-84 (Russian).

[35] Bykov, V.I. and Kytmanov, A.M., Estimation of a number of steady states for three-step adsorption mechanisms, 2. Sequential scheme, *Kinetics and catalysis* **25**(5) (1984), 1277–1278, Dep. VINITI, No.4773-84 (Russian).

[36] Bykov, V.I. and Kytmanov, A.M., *The algorithm of construction of coefficients of kinetic polynomials for nonlinear single-route mechanism of catalytic reaction,* Preprint 44M of Institute of Physics, Krasnoyarsk, (1987), 26 pp. (Russian).

[37] Bykov, V.I., Kytmanov, A.M. and Lazman, M.Z., The modified elimination methods in computer algebra of polynomials, *Abstracts of Intern. Congress on Comp. Syst. and Appl. Math.,* St. Petersburg State Univ., St. Petersburg, (1993), 106–107.

[38] Bykov, V.I., Kytmanov, A.M., Lazman, M.Z. and Yablonskii, G.S. Quadratic neighborhood of thermodynamic balance and kinetic polynomial, *Proc. of V Conf. "Application of mathematical methods for description and study of physical and chemical equilibrium"*, INKhSO AN SSSR, Novosibirsk, (2) (1985), 99–102 (Russian).

[39] Bykov, V.I., Kytmanov, A.M., Lazman, M.Z. and Yablonskii, G.S., Kinetic polynomial for single-route n-stages mechanism, in *Theoretical problems in Kinetics,* Work Collect., Inst. Khim. Fiz. AN SSSR, Chernogolovka, (1985), 67–74 (Russian).

[40] Bykov, V.I., Kytmanov, A.M., Lazman, M.Z. and Yablonskii, G.S., Kinetic polynomial for nonlinear single–route n-stage catalytic reaction, in *Dynamics of chemical and biological systems,* Work Collect., Nauka, Novosibirsk, (1989), 174–191 (Russian).

[41] Bykov, V.I., Kytmanov, A.M., Lazman, M.Z. and Yablonskii, G.S., Kinetic polynomial for single-route n-stage catalytic reaction, in *Mathematical problems of chemical kinetics,* Work Collect., Nauka, Novosibirsk, (1989), 125–149 (Russian).

[42] Bykov, V.I., Kytmanov, A.M. and Osetrova, T.A., Computer algebra of polynomials and its applications, *Abstracts of Intern. Conf. AMCA-95,* NCC Publisher, Novosibirsk, (1995), 60–61.

[43] Bykov, V.I., Kytmanov, A.M. and Osetrova, T.A., Computer algebra of polynomials and its applications, *Proc. Russian Conf. "Computer logic, algebra and intellect control",* Computer Center of Rus. Acad. Sci., Irkutsk, **3** (1995) (Russian).

[44] Bykov, V.I., Kytmanov, A.M. and Osetrova, T.A., Computer algebra of polynomials. Methods and applications, in *Computing technologies,* Work Collect., Inst. Comp. Technol. of Rus. Acad. Sci., Novosibirsk, **4**(10) 1995, 79–88 (Russian).

[45] Bykov, V.I., Kytmanov, A.M. and Osetrova, T.A., Computer algebra of polynomials. Modified method of elimination of unknowns, *Dokl. Ross. Akad. Sci.* (1996) (to appear).

[46] Bykov, V.I. and Yablonskii, G.S., Stationary kinetic characteristics of shock and adsorption mechanisms, *Kinetics and Catalysis* **18**(5) (1977), 1305–1310 (Russian).

[47] Bykov, V.I. and Yablonskii, G.S., Simplest model of catalytic oscillator, *React. Kinet. Catal. Lett.* **16**(4) (1981), 872–875.

[48] Bykov, V.I., Yablonskii, G.S. and Elokhin, V.I., Stationary rate of catalytic reaction with three-step adsorption mechanism, *Kinetics and Catalysis* **20**(4) (1979), 1029–1032 (Russian).

[49] Bykov, V.I., Yablonskii, G.S. and Elokhin, V.I., Phase portrait of simple catalytic schemes permitting multiplicity of steady states, *Kinetics and Catalysis* **20**(4) (1979), 1033–1038 (Russian).

[50] Bykov, V.I., Yablonskii, G.S. and Kim, V.F., On one simple model of kinetic autooscillations in catalytic reaction of oxidation CO, *Dokl. Akad. Nauk* **242**(3) (1978), 637–640 (Russian).

[51] Bykov, V.I., Yablonskii, G.S. and Kuznetsova, I.V. Simple catalytic mechanism admitting a multiplicity of catalyst steady states, *React. Kinet. Catal. Lett.* **10**(4) (1979), 307–310.

[52] Bykov, V.I., Zarkhin, Yu.G. and Yablonskii, G.S., The domain of multiplicity of steady states in oxidation reaction CO on Pt, *Teor. i Eksp. Khimiya* **16**(4) (1980), 487–491 (Russian).

[53] Cattani, E., Dickenstein, A. and Sturmfels, B., *Computing multidimensional residues,* Preprint of Cornell University, (1993), 24pp.

[54] Char, B.W., Geddes, K.O. etc., *Maple 5. Language. Reference. Manual,* Springer–Verlag, Berlin–Heidelberg–New York, (1991).

[55] Char, B.W., Geddes K.O. etc., *Maple 5. Library. Reference. Manual,* Springer–Verlag, Berlin–Heidelberg–New York, (1991).

[56] Chen, W.K., On direct trees and directed k-trees of digraph and their generation, *J. SIAM Appl. Math.* **14**(3) (1966), 530–560.

[57] Chistov, A.L., The algorithm of polynomial complexity for decomposition of polynomials and determination of components of variety in subexponential time, in *Theory of complexity of calculations,* Work Collect., Proc. of LOMI AN SSSR, Leningrad, **137** (1984), 124–188 (Russian).

[58] Chubarov, M.A., Dolgov, G.A. Kiseleva, L.V. etc., Structure of automatized system for research of stability linearized models, in *Package of applied programs. Analytic transformations,* Nauka, Moscow, (1988), 36–62 (Russian).

[59] Clarke, B.L., Stability of complex reaction networks, *Adv. Chem. Phys.* **43** (1980), 7–215.

[60] Clarke, B.L., Quality dynamics and stability of chemical reaction systems, in *Chemical Appl. of Topology and Graph Theory,* Work Collect., Mir, Moscow, (1987), 367–406 (Russian).

[61] Cohen, P.A., *A simple proof of Tarski's theorem on elementary algebra,* Mimeographed manuscript, Stanford University, CA, (1967).

[62] Davenport, J.H., *On the integration of algebraic functions,* Springer–Verlag, Berlin–Heidelberg–New York, (1981).

[63] Collins, G.E., Quantifier elimination for real closed fields by cylindrical algebraic decomposition, *Lect. Notes in Comput. Sci.* **33** (1975), 134–183.

[64] Collins, G.E., The calculation of multivariate polynomial resultants, *J. ACM* **18**(4) (1978), 515–532.

[65] Collins, G.E., Factorization in cylindrical algebraic decomposition, *Lect. Notes in Comput. Sci.* **162** (1983), 212–213.

[66] Courant, R., *Geometrische funktionentheorie,* Springer–Verlag, Berlin–Göttingen–New York, (1964).

[67] Davenport, J., Siret, Y. and Tournier, E., *Calcul formul. Systemes et algorithmes de manipulations algebriques,* Massou, Paris–New York–Barselone–Milan–Mexico–San Paulo, (1987).

[68] Dubovitskii, V.A., *Modified Sturm algorithm in a problem of determination of a number of real roots of a polynomial satisfying polynomial inequalities,* Preprint of Institute of Chemical Physics, Chernogolovka, (1990), 21 pp. (Russian).

[69] Edelson, V., Interpretation program of chemical reactions for modeling of complex kinetics, in *Artificial intellect: application in chemistry,* Work Collect., Nauka, Moscow, (1988), 143–150 (Russian).

[70] Evstigneev, V.A., Yablonskii, G.S. and Bykov, V.I., General form of stationary kinetic equation of complex catalytic reaction (many-route linear mechanism), *Dokl. Akad. Nauk* **245**(4) (1979), 871–874 (Russian).

[71] Evstigneev, V.A., Yablonskii, G.S. and Semenov, A.L., Use of methods of graphs theory for analysis on computer of kinetics of complex reactions, *Abstracts of Conference "Use of computers in spectroscopy of molecules and chemical researches",* Novosibirsk, (1983), 104–105 (Russian).

[72] Fedotov, V.Kh., Kol'tsov, N.I., Alekseev, B.V. and Kiperman, S.L., Self-oscillations in three-step catalytic reactions, *React. Kinet. Catal. Lett.* **23**(3–4) (1983), 301–306 (Russian).

[73] Field, R.J. and Burger, M.,(eds.) *Oscillations and traveling waves in chemical systems,* Wiley, New York, (1985).

[74] From, N.J., Computer assisted derivation of steady state rate equations, *Methods in Enzymology,* Paris, A, **63** (1982), 84–103.

[75] Gantmakher, F.R., *Matrices theory,* Nauka, Moscow, (1967) (Russian).

[76] Gerdt, V.P., Tarasov, O.V. and Shirokov, D.V., Analytic calculations on computer with applications to physics and mathematics, *Uspekhi Fiz. Nauk* **22**(1) (1980), 113–147 (Russian).

[77] Goldstein, V.M., Kononenko, L.I., Lazman, M.Z., Sobolev, V.A. and Yablonskii, G.S., The quality analysis of dynamic properties of catalytic isothermal reactor of ideal mixture, in *Mathematical problems of chemical kinetics,* Work Collect., Nauka, Novosibirsk, (1989), 176–204 (Russian).

[78] Gorban, A.N., *Equilibrium encircling. Equations of chemical kinetics and their thermodynamic analysis,* Nauka, Novosibirsk, (1984) (Russian).

[79] Gorban, A.N., Bykov, V.I. and Yablonskii, G.S., *Sketches on chemical relaxation,* Nauka, Novosibirsk, (1986) (Russian).

[80] Griffiths, F. and Harris, J., *Principles of algebraic geometry,* Wiley, New York, (1978).

[81] Grigoryev, D.Yu., Complexity of decision of theory of the first order of real closed fields, in *Theory of complexity of calculations,* Work Collect., Proc. of LOMI AN SSSR, Leningrad, **174** (1988), 53–100 (Russian).

[82] Grigoryev, V.Yu., Decomposition of polynomials over finite field and resolution of systems of algebraic equations, in *Theory of complexity of calculations,* Work Collect., Proc. of LOMI AN SSSR, Leningrad, **137** (1984), 20–79 (Russian).

[83] Grosheva, N.V. and Efimov, G.B., On systems of analytic transformations on computer, in *Package of applied programs. Analytic transformations,* Work Collect., Nauka, Moscow, (1988), 15–29 (Russian).

[84] Hearn, A.C., (ed.) *Reduce user's manual. Version 3.2,* Rand Corporation, Santa Monica, (1985).

[85] Herve, M., *Several complex variables. Local theory,* Oxford Univ. Press, Bombay, (1963).

[86] Hirai, K. and Sawai, N., A general criterion for jump resonance of nonlinear control systems, *IEEE Trans. Automat. Contr.* **23**(5) (1978), 896–901.

[87] Hoenders, B.J. and Slump, C.H., On the calculation of the exact number of zeroes of a set of equations, *Computing,* **30**(2) (1983), 137–147.

[88] Hörmander, L., *The analysis of linear partial differential operators. II. Differential operators with constant coefficients,* Springer–Verlag, Berlin–Heidelberg–New York–Tokyo, (1983).

[89] Imbihl, R. and Ertl, G, Oscillatory kinetics in heterogeneous catalysis, *Chem. Rev.* **95**(3) (1995), 697–733.

[90] Ivanova, A.N., Furman, G.A., Bykov, V.I. and Yablonskii, G.S., Catalytic mechanisms with auto-oscillations of reaction rate, *Dokl. Akad. Nauk* **242**(4) (1978), 872–875 (Russian).

[91] Ivanova, A.I. and Tarnopolskii, S.L., On one approach to solving of a series of quality problems for kinetic systems and its realization on computer (bifurcational conditions, auto-oscillations), *Kinetics and Catalysis* **20**(6) (1979), 1541–1548 (Russian).

[92] Jury, E.I., *Inners and stability of dynamic systems,* Wiley, New York–London–Sydney–Toronto, (1974).

[93] King, S.L. and Altman, J., A schematic method of deriving the rate laws for enzyme-catalyzed reactions, *J. Phys. Chem.* **60**(10) (1956), 1375–1381.

[94] Krein, M.G. and Naimark, M.A., *Method of symmetric and Hermitian forms in theory of localization of roots of algebraic equations,* KhGU, Kharkov, (1936) (Russian).

[95] Kubicek, M. and Marek, M., *Computation methods in bifurcation theory and dissipative structures,* Springer–Verlag, New York–Berlin–Tokyo, (1983).

[96] Kuchaev, V.L. and Temkin, M.I., Study of mechanism of reaction of H with O on Pt with help of derived ion-ion emission, *Kinetics and Catalysis,* **13**(4) (1972), 1024–1031 (Russian).

[97] Kurosh, A.G., *Course of higher algebra,* Nauka, Moscow, (1971) (Russian).

[98] Kytmanov, A.M., A transformation formula for Grothendieck residue and some its application, *Sibirsk. Mat. Zh.* **29**(3) (1988), 198–202; English transl. in *Sib. Math. J.* **29** (1988), 573–577.

[99] Kytmanov, A.M., On a system of algebraic equations encountered in chemical kinetics, *Selecta Math. Sov.* **8**(1) (1989), 1–9.

[100] Kytmanov, A.M., The logarithmic derivative for resultants of systems of algebraic equations, *Sibirsk. Mat. Zh.* **31**(6) (1990), 96–103; English transl. in *Sib. Math. J.* **31** (1990), 956–962.

[101] Kytmanov, A.M., On the number of real roots of systems of equations, *Izv. Vyssh. Uchebn. Zaved., Mat.* (6) (1991,) 20–24; English trasl. in *Sov. Math.* **35** (1991), 19–22.

[102] Kytmanov, A.M. and Bykov, V.I., *On one modification of method of construction of resultants of a system of nonlinear algebraic equations,* Preprint 45M of Institute of Physics, Krasnoyarsk, (1988), 44 pp. (Russian).

[103] Kytmanov, A.M. and Mkrtchyan, M.A., On uniqueness of restoration of a system of nonlinear algebraic equations by simple roots, *Sibirsk. Mat. Zh.* **24**(6) (1983), 204–206 (Russian).

[104] Lazard, D., Systems of algebraic equations, *Lect. Notes in Comput. Sci.* **79** (1979), 88–94.

[105] Lazard, D., Resolution des systems d'equation algebriques, *Theor. Comp. Sci.* **15** (1981), 77–110.

[106] Lazard, D., Gröbner bases, Gaussian elimination and resolution of systems of algebraic equations, *Lect. Notes in Comput. Sci.* **162** (1983), 146–156.

[107] Lazard, D., Systems of algebraic equations (algorithms and complexity), in *Computational algebraic geometry and commutative algebra*, Cambridge Univ. Press, London, (1993).

[108] Lazman, M.Z., *Research of nonlinear models of kinetics of heterogeneous catalytic reactions*, Thesis of candidate in science, Institution of Catalysis, Novosibirsk, (1986) (Russian).

[109] Lazman, M.Z., Spivak, S.I. and Yablonskii, G.S. Kinetic polynomial and a problem of determination of relations between kinetic constants under solving reverse problem, *Khim. Fizika* **1**(4) (1985), 479–483 (Russian).

[110] Lazman, M.Z. and Yablonskii, G.S., Kinetic polynomial: a new concept of chemical kinetics, in *Pattern and Dynamics in Reactive Media*, Work Collect., The IMA Volumes in Mathematics and its Applications, Springer–Verlag, New York, (1991), 117–150.

[111] Lazman, M.Z., Yablonskii, G.S. and Bykov, V.I., Stationary kinetic equation. Nonlinear single-route mechanism, *Khim. Fizika* **2**(2) (1983), 239–248 (Russian).

[112] Lazman, M.Z., Yablonskii, G.S. and Bykov, V.I., Stationary kinetic equation. Adsorption mechanism of catalytic reaction, *Khim. Fizika* **2**(3) (1983), 413–419 (Russian).

[113] Lazman, M.Z., Yablonskii, G.S., Vinogradova, G.M. and Romanov, L.N., Application of kinetic polynomial for description of stationary dependencies of reaction rate, *Khim. Fizika* **4**(5) (1985), 691–699 (Russian).

[114] Lyberatos, G., Kuszta, B. and Bailey, J.G., Steady-state multiplicity and bifurcation analysis via the Newton polyhedron approach, *Chem. Eng. Sci.* **39**(6) (1984), 947–960.

[115] Macaulay, F.S., *Algebraic theory of modular systems,* Cambridge, (1916).

[116] Mikhalev, A.V. and Pankratyev, E.V., The problem of algebraic simplification in commutative and differential algebra, *Abstracts of Conf. "Systems for analytic transformation in mechanics"*, Gorkii State Univ., Gorkii, (1984), 131–133 (Russian).

[117] Möller, H.M. and Buchberger, B., The construction of multivariate polynomials with preassiqued zeroes, *Lect. Notes in Comput. Sci.* **162** (1983), 24–31.

[118] Moses, J., Solution of systems of polynomial equations by elimination, *Commun. of ACM* **9**(8) (1966), 634–637.

[119] Mysovskikh, I.P., *Interpolational cubic formulas,* Nauka, Moscow, (1981) (Russian).

[120] Osetrova, T.A., `Maple` procedure for determination of resultants of systems of nonlinear algebraic equations, in *Complex Analysis and Differential Equations,* Work Collect., Krasnoyarsk State Univ., Krasnoyarsk, (1996), 123–131 (Russian).

[121] Ostrovskii, G.M., Zyskin, A.G. and Snagovskii, Yu.S., A system for automation of kinetic computations, *Comput. Chem.* **14**(2) (1987), 85–96.

[122] *Package of applied programs. Analytic transformations,* Work Collect., Nauka, Moscow, (1988) (Russian).

[123] Pavell, R., MACSYMA: Capabilities and application to problems in engineering and science, *Lect. Notes on Computer,* **203** (1986), 19–32.

[124] Pedersen, P., Roy, M.-F. and Szpirglas, A., Counting real zeros in the multivariate case, *Computational Algebraic Geometry,* Proc. MEGA–92, (1992), 203–223.

[125] Polya, G. and Szegö, G., *Aufgaben und lehrsätze aus der analysis,* Springer–Verlag, Berlin–Göttingen, (1964).

[126] Reznikov, I.G., Stelletskii, V.I. and Topunov, V.L., Some standard refal-modules, *Abstracts of Conference by Methods of Translation,* Novosibirsk, (1981), 174–176 (Russian).

[127] Shabat, B.V., *Introduction a l'analyse complexe,* Mir, Moscow,**1** (1990).

[128] Schwarts, J.T., Probabilistic algorithms for verification of polynomial identities, *Lect. Notes in Comput. Sci.* **72** (1979), 200–215.

[129] Seidenberg, A., A new decision method for elementary algebra, *Ann. Math.* **60** (1954), 365–374.

[130] Sheintuch, M. and Luss, D., Dynamic features of two ordinary differential equations with widely separated time scales, *Chem. Eng. Sci.* **40**(9) (1985), 1653–1664.

[131] Silvestri, A.J. and Zahner, J.C., Algebraic manipulations of the rate equations of certain catalytic reactions with digital computers, *Chem. Eng. Sci.* **22**(3) (1967), 465–467.

[132] Tarkhanov, N.N., On the calculation of the Poincaré index, *Izv. Vyssh. Ucheb. Zaved., Mat.* (9) (1984), 47–50; English transl. in *Sov. Math.* **28** (1984), 63–66.

[133] Tarski, A., *A decision method for elementary algebra and geometry,* Manuscript, Berkeley, CA, (1951).

[134] Titchmarsh, E.C., *The theory of functions,* Oxford Univ. Press, Oxford, (1939).

[135] Tsikh, A.K., Bezout theorem in theory of function space. On solution of systems of algebraic equations, in *Some Questions of Multidimensional Complex Analysis,* Work Collect., Inst. of Physics, Krasnoyarsk, (1980), 185–196 (Russian).

[136] Tsikh, A.K., *Multidimensional residues and their applications,* Transl. Math. Monographs, AMS, Providence, R.I. (1992).

[137] Tsikh, A.K. and Yuzhakov, A.P., Properties of the complete sums of residues with respect to a polynomial mapping and their applications, *Sibirsk. Mat. Zh.* **25**(4) (1984), 208–213; English Transl. in *Sib. Math. J.* **25** (1984), 677–682.

[138] Tsiliginuis, C.A., Steady-state bifurcation and multiplicity conditions for lumped systems, *Chem. Eng. Sci.* **43**(1) (1988), 134–136.

[139] Vainberg, M.M. and Trenogin, V.A., *The theory of brunching solutions of nonlinear equations,* Nauka, Moscow, (1969); English transl. in Nordhoff, (1974).

[140] Vol'pert, A.I. and Ivanova, A.N., Mathematical models in chemical kinetics, in *Mathematical modeling of nonlinear differential equations of mathematical physics,* Work Collect., Nauka, Moscow, 1987, 57–102 (Russian).

[141] Vorob'ev, N.P. and Grigoryev, D.Yu., The decision of systems of polynomial inequalities over real closed field in subexponential time, in *Theory of complexity of calculations,* Proc. of LOMI AN SSSR, Leningrad, **174** (1988), 1–36 (Russian).

[142] Van der Waerden, B.L. *Algebra,* Springer–Verlag, Berlin–Heidelberg–New-York, (1971).

[143] Winkler, F., Buchberger, B., Lichtenberger, F. and Polletschk, H., An algorithm for constructing canonical bases of polynomial ideals, *ACM Trans. Math. Software* **11**(1) (1985), 66–78.

[144] Wolfram, S., *Mathematica: a system for doing mathematics by computer*, Addison–Wesley, Reading, MA, (1991).

[145] Yablonskii, G.S. and Bykov, V.I., Simplified form of writing of kinetic equation of n-stage single-route catalytic reaction, *Dokl. Akad. Nauk* **233**(4) (1977), 642–645 (Russian).

[146] Yablonskii, G.S. and Bykov, V.I., Simplified form of writing of kinetic equation of complex catalytic reaction with single-route mechanism, *Kinetics and Catalysis* **18**(6) (1977), 1561–1567 (Russian).

[147] Yablonskii, G.S. and Bykov, V.I., Structureable kinetic equations of complex catalytic reactions, *Dokl. Akad. Nauk* **238**(3) (1978), 615–618 (Russian).

[148] Yablonskii, G.S. and Bykov, V.I., Analysis of structure of kinetic equation of complex catalytic reaction (linear single-route mechanism), *Teor. i Experiment. Khimiya* **15**(1) (1979), 41–45 (Russian).

[149] Yablonskii, G.S., Bykov, V.I. and Elokhin, V.I., *The kinetics of model reactions of heterogeneous catalysis,* Nauka, Novosibirsk, 1984 (Russian).

[150] Yablonskii, G.S., Bykov, V.I. and Gorban, A.N., *Kinetic models of catalytic reactions,* Nauka, Novosibirsk, (1983) (Russian).

[151] Yablonskii, G.S., Bykov, V.I., Gorban, A.N. and Elokhin, V.I., *Kinetic models of catalytic reactions,* Elsevier, Amsterdam–New York, (1991).

[152] Yablonskii, G.S., Evstigneev, V.A. and Bykov, V.I., Graphs in chemical kinetics, in *Applications of graph theory in chemistry,* Work Collect., Nauka, Novosibirsk, (1988), 70–143 (Russian).

[153] Yablonskii, G.S., Evstigneev, V.A., Noskov, A.S. and Bykov, V.I., Analysis of general form of stationary kinetic equation of complex catalytic reaction, *Kinetics and Catalysis* **22**(3) (1981), 738–743 (Russian).

[154] Yablonskii, G.S., Lazman, M.Z. and Bykov, V.I., Stoichiometric number, molecularity and multiplicity, *React. Kinet. Catal. Lett.* **20**(1–2) (1982), 73–77.

[155] Yablonskii, G.S., Lazman, M.Z. and Bykov, V.I., Kinetic polynomial, molecularity and multiplicity, *Dokl. Akad. Nauk* **269**(1) (1983), 166–168 (Russian).

[156] Yablonskii, G.S., Lazman, M.Z. and Bykov, V.I., Kinetic polynomial, its properties and applications, *Abstracts of Conf. "Thermodynamics of irreversible processes and its applications"*, Chernovtsy, (42) (1984), 311–312 (Russian).

[157] Yuzhakov, A.P., On the computation of sum of values of a polynomial in the roots of a system of algebraic equations, in *Some Questions of Multidimensional Complex Analysis*, Work Collect, Inst. of Physics, Krasnoyarsk, (1980), 197–214 (Russian).

[158] Yuzhakov, A.P., On the computation of the complete sum of residues relatively to a polynomial mapping in $\mathbf{C}^n$, *Dokl. Akad. Nauk* **275**(4) (1984), 817–820; English transl. in *Sov. Math. Dokl.* **29** (1984), 321–324.

[159] Zarkhin, Yu.G. and Kovalenko, V.N., *The determination of solutions of a system of two algebraic equations*, Preprint of NIVT AN SSSR, Puschino, (1978), 13 pp. (Russian).

[160] Zeldovich, Ya.B. and Zysin, Yu.A., To theory of thermal strength. 2. Accounting of heat transfer in process of reaction. *Zh. Tekhnich. Fiziki* **11**(6) (1941), 501–508 (Russian).

[161] Zhdanov, V.P., *Elementary physical and chemical processes on surface*, Nauka, Novosibirsk, (1988) (Russian).

Index

Other *Mathematics and Its Applications* titles of interest:

Yu.L. Dalecky and S.V. Fomin: *Measures and Differential Equations in Infinite-Dimensional Space.* 1991, 338 pp. ISBN 0-7923-1517-0

W. Mlak: *Hilbert Space and Operator Theory.* 1991, 296 pp. ISBN 0-7923-1042-X

N.Ja. Vilenkin and A.U. Klimyk: *Representation of Lie Groups and Special Functions. Volume 1: Simplest Lie Groups, Special Functions, and Integral Transforms.* 1991, 608 pp. ISBN 0-7923-1466-2

N.Ja. Vilenkin and A.U. Klimyk: *Representation of Lie Groups and Special Functions. Volume 2: Class I Representations, Special Functions, and Integral Transforms.* 1992, 630 pp. ISBN 0-7923-1492-1

N.Ja. Vilenkin and A.U. Klimyk: *Representation of Lie Groups and Special Functions. Volume 3: Classical and Quantum Groups and Special Functions.* 1992, 650 pp ISBN 0-7923-1493-X
(Set ISBN for Vols. 1, 2 and 3: 0-7923-1494-8)

K. Gopalsamy: *Stability and Oscillations in Delay Differential Equations of Population Dynamics.* 1992, 502 pp. ISBN 0-7923-1594-4

N.M. Korobov: *Exponential Sums and their Applications.* 1992, 210 pp. ISBN 0-7923-1647-9

Chuang-Gan Hu and Chung-Chun Yang: *Vector-Valued Functions and their Applications.* 1991, 172 pp. ISBN 0-7923-1605-3

Z. Szmydt and B. Ziemian: *The Mellin Transformation and Fuchsian Type Partial Differential Equations.* 1992, 224 pp. ISBN 0-7923-1683-5

L.I. Ronkin: *Functions of Completely Regular Growth.* 1992, 394 pp. ISBN 0-7923-1677-0

R. Delanghe, F. Sommen and V. Soucek: *Clifford Algebra and Spinor-valued Functions. A Function Theory of the Dirac Operator.* 1992, 486 pp. ISBN 0-7923-0229-X

A. Tempelman: *Ergodic Theorems for Group Actions.* 1992, 400 pp. ISBN 0-7923-1717-3

D. Bainov and P. Simenov: *Integral Inequalities and Applications.* 1992, 426 pp. ISBN 0-7923-1714-9

I. Imai: *Applied Hyperfunction Theory.* 1992, 460 pp. ISBN 0-7923-1507-3

Yu.I. Neimark and P.S. Landa: *Stochastic and Chaotic Oscillations.* 1992, 502 pp. ISBN 0-7923-1530-8

H.M. Srivastava and R.G. Buschman: *Theory and Applications of Convolution Integral Equations.* 1992, 240 pp. ISBN 0-7923-1891-9

A. van der Burgh and J. Simonis (eds.): *Topics in Engineering Mathematics.* 1992, 266 pp. ISBN 0-7923-2005-3

Other *Mathematics and Its Applications* titles of interest:

F. Neuman: *Global Properties of Linear Ordinary Differential Equations.* 1992, 320 pp. ISBN 0-7923-1269-4

A. Dvurečenskij: *Gleason's Theorem and its Applications.* 1992, 334 pp. ISBN 0-7923-1990-7

D.S. Mitrinović, J.E. Pečarić and A.M. Fink: *Classical and New Inequalities in Analysis.* 1992, 740 pp. ISBN 0-7923-2064-6

H.M. Hapaev: *Averaging in Stability Theory.* 1992, 280 pp. ISBN 0-7923-1581-2

S. Gindinkin and L.R. Volevich: *The Method of Newton's Polyhedron in the Theory of PDE's.* 1992, 276 pp. ISBN 0-7923-2037-9

Yu.A. Mitropolsky, A.M. Samoilenko and D.I. Martinyuk: *Systems of Evolution Equations with Periodic and Quasiperiodic Coefficients.* 1992, 280 pp. ISBN 0-7923-2054-9

I.T. Kiguradze and T.A. Chanturia: *Asymptotic Properties of Solutions of Nonautonomous Ordinary Differential Equations.* 1992, 332 pp. ISBN 0-7923-2059-X

V.L. Kocic and G. Ladas: *Global Behavior of Nonlinear Difference Equations of Higher Order with Applications.* 1993, 228 pp. ISBN 0-7923-2286-X

S. Levendorskii: *Degenerate Elliptic Equations.* 1993, 445 pp. ISBN 0-7923-2305-X

D. Mitrinović and J.D. Kečkić: *The Cauchy Method of Residues, Volume 2.* Theory and Applications. 1993, 202 pp. ISBN 0-7923-2311-8

R.P. Agarwal and P.J.Y Wong: *Error Inequalities in Polynomial Interpolation and Their Applications.* 1993, 376 pp. ISBN 0-7923-2337-8

A.G. Butkovskiy and L.M. Pustyl'nikov (eds.): *Characteristics of Distributed-Parameter Systems.* 1993, 386 pp. ISBN 0-7923-2499-4

B. Sternin and V. Shatalov: *Differential Equations on Complex Manifolds.* 1994, 504 pp. ISBN 0-7923-2710-1

S.B. Yakubovich and Y.F. Luchko: *The Hypergeometric Approach to Integral Transforms and Convolutions.* 1994, 324 pp. ISBN 0-7923-2856-6

C. Gu, X. Ding and C.-C. Yang: *Partial Differential Equations in China.* 1994, 181 pp. ISBN 0-7923-2857-4

V.G. Kravchenko and G.S. Litvinchuk: *Introduction to the Theory of Singular Integral Operators with Shift.* 1994, 288 pp. ISBN 0-7923-2864-7

A. Cuyt (ed.): *Nonlinear Numerical Methods and Rational Approximation II.* 1994, 446 pp. ISBN 0-7923-2967-8

G. Gaeta: *Nonlinear Symmetries and Nonlinear Equations.* 1994, 258 pp. ISBN 0-7923-3048-X

Other *Mathematics and Its Applications* titles of interest:

V.A. Vassiliev: *Ramified Integrals, Singularities and Lacunas.* 1995, 289 pp.
ISBN 0-7923-3193-1

N.Ja. Vilenkin and A.U. Klimyk: *Representation of Lie Groups and Special Functions.* Recent Advances. 1995, 497 pp. ISBN 0-7923-3210-5

Yu. A. Mitropolsky and A.K. Lopatin: *Nonlinear Mechanics, Groups and Symmetry.* 1995, 388 pp. ISBN 0-7923-3339-X

R.P. Agarwal and P.Y.H. Pang: *Opial Inequalities with Applications in Differential and Difference Equations.* 1995, 393 pp. ISBN 0-7923-3365-9

A.G. Kusraev and S.S. Kutateladze: *Subdifferentials: Theory and Applications.* 1995, 408 pp. ISBN 0-7923-3389-6

M. Cheng, D.-G. Deng, S. Gong and C.-C. Yang (eds.): *Harmonic Analysis in China.* 1995, 318 pp. ISBN 0-7923-3566-X

M.S. Livšic, N. Kravitsky, A.S. Markus and V. Vinnikov: *Theory of Commuting Nonselfadjoint Operators.* 1995, 314 pp. ISBN 0-7923-3588-0

A.I. Stepanets: *Classification and Approximation of Periodic Functions.* 1995, 360 pp.
ISBN 0-7923-3603-8

C.-G. Ambrozie and F.-H. Vasilescu: *Banach Space Complexes.* 1995, 205 pp.
ISBN 0-7923-3630-5

E. Pap: *Null-Additive Set Functions.* 1995, 312 pp. ISBN 0-7923-3658-5

C.J. Colbourn and E.S. Mahmoodian (eds.): *Combinatorics Advances.* 1995, 338 pp.
ISBN 0-7923-3574-0

V.G. Danilov, V.P. Maslov and K.A. Volosov: *Mathematical Modelling of Heat and Mass Transfer Processes.* 1995, 330 pp. ISBN 0-7923-3789-1

A. Laurinčikas: *Limit Theorems for the Riemann Zeta-Function.* 1996, 312 pp.
ISBN 0-7923-3824-3

A. Kuzhel: *Characteristic Functions and Models of Nonself-Adjoint Operators.* 1996, 283 pp. ISBN 0-7923-3879-0

G.A. Leonov, I.M. Burkin and A.I. Shepeljavyi: *Frequency Methods in Oscillation Theory.* 1996, 415 pp. ISBN 0-7923-3896-0

B. Li, S. Wang, S. Yan and C.-C. Yang (eds.): *Functional Analysis in China.* 1996, 390 pp.
ISBN 0-7923-3880-4

P.S. Landa: *Nonlinear Oscillations and Waves in Dynamical Systems.* 1996, 554 pp.
ISBN 0-7923-3931-2

A.J. Jerri: *Linear Difference Equations with Discrete Transform Methods.* 1996, 462 pp.
ISBN 0-7923-3940-1

Other *Mathematics and Its Applications* titles of interest:

I. Novikov and E. Semenov: *Haar Series and Linear Operators*. 1997, 234 pp. ISBN 0-7923-4006-X

L. Zhizhiashvili: *Trigonometric Fourier Series and Their Conjugates*. 1996, 312 pp. ISBN 0-7923-4088-4

R.G. Buschman: *Integral Transformation, Operational Calculus, and Generalized Functions*. 1996, 246 pp. ISBN 0-7923-4183-X

V. Lakshmikantham, S. Sivasundaram and B. Kaymakcalan: *Dynamic Systems on Measure Chains*. 1996, 296 pp. ISBN 0-7923-4116-3

D. Guo, V. Lakshmikantham and X. Liu: *Nonlinear Integral Equations in Abstract Spaces*. 1996, 350 pp. ISBN 0-7923-4144-9

Y. Roitberg: *Elliptic Boundary Value Problems in the Spaces of Distributions*. 1996, 427 pp. ISBN 0-7923-4303-4

Y. Komatu: *Distortion Theorems in Relation to Linear Integral Operators*. 1996, 313 pp. ISBN 0-7923-4304-2

A.G. Chentsov: *Asymptotic Attainability*. 1997, 336 pp. ISBN 0-7923-4302-6

S.T. Zavalishchin and A.N. Sesekin: *Dynamic Impulse Systems*. Theory and Applications. 1997, 268 pp. ISBN 0-7923-4394-8

U. Elias: *Oscillation Theory of Two-Term Differential Equations*. 1997, 226 pp. ISBN 0-7923-4447-2

D. O'Regan: *Existence Theory for Nonlinear Ordinary Differential Equations*. 1997, 204 pp. ISBN 0-7923-4511-8

Yu. Mitropolskii, G. Khoma and M. Gromyak: *Asymptotic Methods for Investigating Quasiwave Equations of Hyperbolic Type*. 1997, 418 pp. ISBN 0-7923-4529-0

R.P. Agarwal and P.J.Y. Wong: *Advanced Topics in Difference Equations*. 1997, 518 pp. ISBN 0-7923-4521-5

N.N. Tarkhanov: *The Analysis of Solutions of Elliptic Equations*. 1997, 406 pp. ISBN 0-7923-4531-2

B. Riečan and T. Neubrunn: *Integral, Measure, and Ordering*. 1997, 376 pp. ISBN 0-7923-4566-5

N.L. Gol'dman: *Inverse Stefan Problems*. 1997, 258 pp. ISBN 0-7923-4588-6

S. Singh, B. Watson and P. Srivastava: *Fixed Point Theory and Best Approximation: The KKM-map Principle*. 1997, 230 pp. ISBN 0-7923-4758-7

A. Pankov: G-*Convergence and Homogenization of Nonlinear Partial Differential Operators*. 1997, 263 pp. ISBN 0-7923-4720-X

Other *Mathematics and Its Applications* titles of interest:

S. Hu and N.S. Papageorgiou: *Handbook of Multivalued Analysis*. Volume I: Theory. 1997, 980 pp. ISBN 0-7923-4682-3 (Set of 2 volumes: 0-7923-4683-1)

L.A. Sakhnovich: *Interpolation Theory and Its Applications*. 1997, 216 pp. ISBN 0-7923-4830-0

G.V. Milovanović: *Recent Progress in Inequalities*. 1998, 531 pp. ISBN 0-7923-4845-1

V.V. Filippov: *Basic Topological Structures of Ordinary Differential Equations*. 1998, 530 pp. ISBN 0-7293-4951-2

S. Gong: *Convex and Starlike Mappings in Several Complex Variables*. 1998, 208 pp. ISBN 0-7923-4964-4

A.B. Kharazishvili: *Applications of Point Set Theory in Real Analysis*. 1998, 244 pp. ISBN 0-7923-4979-2

R.P. Agarwal: *Focal Boundary Value Problems for Differential and Difference Equations*. 1998, 300 pp. ISBN 0-7923-4978-4

D. Przeworska-Rolewicz: *Logarithms and Antilogarithms*. An Algebraic Analysis Approach. 1998, 358 pp. ISBN 0-7923-4974-1

Yu. M. Berezansky and A.A. Kalyuzhnyi: *Harmonic Analysis in Hypercomplex Systems*. 1998, 493 pp. ISBN 0-7923-5029-4

V. Lakshmikantham and A.S. Vatsala: *Generalized Quasilinearization for Nonlinear Problems*. 1998, 286 pp. ISBN 0-7923-5038-3

V. Barbu: *Partial Differential Equations and Boundary Value Problems*. 1998, 292 pp. ISBN 0-7923-5056-1

J. P. Boyd: *Weakly Nonlocal Solitary Waves and Beyond-All-Orders Asymptotics*. Generalized Solitons and Hyperasymptotic Perturbation Theory. 1998, 610 pp. ISBN 0-7923-5072-3

D. O'Regan and M. Meehan: *Existence Theory for Nonlinear Integral and Integrodifferential Equations*. 1998, 228 pp. ISBN 0-7923-5089-8

A.J. Jerri: *The Gibbs Phenomenon in Fourier Analysis, Splines and Wavelet Approximations*. 1998, 364 pp. ISBN 0-7923-5109-6

C. Constantinescu, W. Filter and K. Weber, in collaboration with A. Sontag: *Advanced Integration Theory*. 1998, 872 pp. ISBN 0-7923-5234-3

V. Bykov, A. Kytmanov and M. Lazman, with M. Passare (ed.): *Elimination Methods in Polynomial Computer Algebra*. 1998, 252 pp. ISBN 0-7923-5240-8

W.-H. Steeb: *Hilbert Spaces, Wavelets, Generalised Functions and Modern Quantum Mechanics*. 1998, 234 pp. ISBN 0-7923-5231-9

Other *Mathematics and Its Applications* titles of interest:

X. Xu: *Introduction to Vertex Operator Superalgebras and Their Modules*. 1998, 356 pp.
ISBN 0-7923-5242-4

E.E. Rosinger: *Parametric Lie Group Actions on Global Generalised Solutions of Nonlinear PDFs* and a Solution to Hilbert's Fifth Problem. 1998, 234 pp. ISBN 0-7923-5232-7